Popular Geopolitics

This book brings together scholars from across a variety of academic disciplines to assess the current state of the subfield of popular geopolitics. It provides an archaeology of the field, maps the flows of various frameworks of analysis into (and out of) popular geopolitics, and charts a course forward for the discipline. It explores the real-world implications of popular culture, with a particular focus on the evolving interdisciplinary nature of popular geopolitics alongside interrelated disciplines including media, cultural, and gender studies.

Robert A. Saunders is Professor in History, Politics, and Geography at Farmingdale State College, a campus of the State University of New York (SUNY). His research explores the impact of popular culture and mass media on geopolitics, nationalism, and religious identity. His scholarship has appeared in *Progress in Human Geography*, *Europe-Asia Studies*, *Slavic Review*, *Nations and Nationalism*, and *Geopolitics*, among other journals. He is the author of four books, the most recent being *Popular Geopolitics and Nation Branding in the Post-Soviet Realm* (2017). He is also curator of the 'Popular Culture and IR' blog at E-International Relations.

Vlad Strukov is Associate Professor in Film and Digital Culture at the University of Leeds, specialising in world cinemas, visual culture, digital media, intermediality, and cultural theory. He explores theories of empire and nationhood, global journalism and grassroots media, and consumption and celebrity by considering the Russian Federation and the Russian-speaking world as his case study. He is the founding and principal editor of the journal *Studies in Russian, Eurasian and Central European New Media* (www.digitalicons.org). He is the author of *Contemporary Russian Cinema: Symbols of a New Era* (2016), and other publications on film.

Routledge Geopolitics Series

Series Editors

Klaus Dodds, Professor of Geopolitics at the Department of Geography, Royal Holloway University of London, Egham, Surrey, UK. k.dodds@rhul.ac.uk

Reece Jones, Professor of Geography at the Department of Geography, University of Hawai'i at Manoa, Hawai'i, USA. reecej@hawaii.edu

Geopolitics is a thriving area of intellectual enquiry. The *Routledge Geopolitics Series* invites scholars to publish their original and innovative research in geopolitics and related fields. We invite proposals that are theoretically informed and empirically rich without prescribing research designs, methods and/or theories. Geopolitics is a diverse field making its presence felt throughout the arts and humanities, social sciences and physical and environmental sciences. Formal, practical and popular geopolitical studies are welcome as are research in areas informed by borders and bordering, elemental geo-politics, feminism, identity, law, race, resources, territory and terrain, materiality and objects. The series is also global in geographical scope and interested in proposals that focus on past, present and future geopolitical imaginations, practices and representations.

As the series is aimed at upper-level undergraduates, graduate students and faculty, we welcome edited book proposals as well as monographs and textbooks which speak to geopolitics and its relationship to wider human geography, politics and international relations, anthropology, sociology, and the interdisciplinary fields of social sciences, arts and humanities.

Popular Geopolitics
Plotting an Evolving Interdiscipline
Edited by Robert A. Saunders and Vlad Strukov

For more information about this series, please visit: https://www.routledge.com/Routledge-Geopolitics-Series/book-series/RFGS

Popular Geopolitics

Plotting an Evolving Interdiscipline

Edited by Robert A. Saunders and Vlad Strukov

Routledge
Taylor & Francis Group
LONDON AND NEW YORK

First published 2018
by Routledge

2 Park Square, Milton Park, Abingdon, Oxfordshire OX14 4RN
52 Vanderbilt Avenue, New York, NY 10017

Routledge is an imprint of the Taylor & Francis Group, an informa business

First issued in paperback 2020

British Library Cataloguing-in-Publication Data
A catalogue record for this book is available from the British Library

Library of Congress Cataloging-in-Publication Data
A catalog record has been requested for this book

ISBN: 978-0-8153-8403-8 (hbk)
ISBN: 978-0-367-59162-5 (pbk)

Typeset in Times New Roman
by Deanta Global Publishing Services, Chennai, India

To our families for their patience and support.

Contents

Figures

Tables

Foreword

An odd couple? Popular culture and geopolitics

Iver B. Neumann

Without categories that order cognition, humans cannot grasp the world, and so will find it hard to act upon it. Categories, that is, come between humans and the world which they make. French philosopher Jacques Derrida (1967) once tried to make this point by stating that there is nothing outside of language, or more accurately that there is no hypertext (*hors-texte*) that may explain the text itself. What he meant was not that things do not exist before they are categorised. To take a natural phenomenon as an example, that would imply that what we now know as the galaxy next to our own, the Andromeda Galaxy, was not there before we spotted it, and that is a bit of a stretch. Derrida rather had in mind that things come to exist as discursive phenomena only once they enter language. Tenth-century Persian maps of the sky marked the phenomenon in question as a little cloud. Early modern Europeans categorised it as a nebula. In the twentieth century, it became a galaxy. Those who have been exposed to the categories that present-day scientists use to order 'the sky', will now immediately see a galaxy when they look at Andromeda. The entity is the same as the one ancient astronomers knew as 'a small cloud'. Those who do not possess the relevant knowledge may stare right at Andromeda and see nothing - or perhaps they will see a little speck. Derrida's point is that this kind of categorisation always happens in and through language. There is nothing outside of language that may help us sharpen our discursive categories. If we are interested in researching what physical and social worlds look like, then, one place to begin is by looking at which categories make up those world, how those categories combine, and with what effects (for example, how those categories conspire to give certain groups better life chances than others). Perhaps there is no better example of the power of discursive construction than the establishment of the modern state, a theme that is certainly implicit throughout this volume.

Some worlds in question are international (that is, revolving around state-to-state relations), or even global (where the worlds in question consist of all kinds of agents and relations between them). How do we get at the categories that make up these worlds? Historically, there were two major answers to this question. One was to look at travelogues, which purported to inform their intended readers about what adventurers had seen, heard, smelled, and otherwise experienced as they ventured to places where their readers themselves had not been. One example of this. is how a number of ninth to eleventh-century Arab travellers wrote about what we now know as Russia and Northern Europe (Ibn Fadlan 2012). The other major way to study the categories that people used to order the world would be to go to cosmologies, which were almost always religious in kind. Such cosmologies would order the world according to the principle of sacred geography, where there is usually a timeless centre such as Mecca or Jerusalem, around which ever more distant peripheries are ordered. The Arab travelogues just mentioned are of this kind: the basic category used to make sense of what is seen is that this is the end of the world, the 'seventh climate' (where the first climate is of course the one that these travellers themselves hail from).

This volume presents a different answer to the question of where to look in order to get a handle on the emergent production of global worlds, namely in the realm of popular culture. The basic preconditions for this answer are to do with technology and information, and particularly with information technology. The advent of newspapers was important. For example, when, in 1876, Ottomans massacred revolting Bulgarians, a British newspaper-reading public caused an uproar, and demanded that their government do something about it. In other words, newspaper reporting led not only to a strengthening of geographical categorisations, as Bulgaria was 'put on the map', as we so aptly say, but also had the knock-on effect of creating public pressure on the British Government to intervene. Newspaper reporting had a geopolitical effect. As it happened, there was no intervention, but when the Ottoman Empire turned to Britain for help against the Russian Empire in the following year, the British Government referred to the public uproar about Ottoman action in Bulgaria as one reason why it did not want to help. Scholars who were interested in the production of global social worlds had received a new place to look, namely in news coverage. Motion pictures and radio soon presented themselves as yet other sources of data for such an undertaking.

Newspapers and radios notwithstanding, up until the 1960s, most people's knowledge of the world was severely circumscribed. Although literacy was high in certain parts of the globe, people did not read all that much, and little of what they read was about distant worlds. With the coming of mass education, mass tourism and mass communication, all that began to change. The world came seeping in to more and more everyday experience. The opening chapters below chart scholarly responses to this. The Birmingham School of Cultural Studies opened in 1964. The late 1980s and early 1990s saw the first rumblings of a popular geopolitics, where 'popular' simply denotes a willingness to look at how categories and social worlds are being produced, not only in and by established genres and agents such as diplomats, artists, and preachers, but also in, say, TV soap operas, Hollywood films, and restaurants.

Popular geopolitics came into being as a response to how developments in information technologies and increases in global interaction density changed way global social worlds were being produced. Popular geopolitics were also as a politically driven project, for it asked exactly *whose* global worlds should be singled out for scholarly attention. Should it be the social worlds of decision-makers only, or should it also be the social worlds of mass audiences, of minorities and of the underprivileged? Developments over the last three decades have only intensified the factors that gave birth to popular geopolitics. There are more social media, more users of social media in more locations and more human time spent on social media than ever before. By the same token, the particularisation of audiences has only continued. There are more language-specific social media than ever before, more specialised blogs, more specialised discussion groups. The scholarly challenges increase accordingly. We need to understand the new production modalities involved—how, for example, do social worlds that emerge on Twitter depend on the unseen code that is a part of each tweet for its existence, and how does the way such code works privilege what groups? We need to understand the new content that emerges on these platforms. Given that the gaming market has been larger than the film market for years already, the reflections in this volume on the content of video games is an excellent example. And we need to understand the effects of all this not only on the emergent global social worlds of different groups, but also on geopolitical policy outcomes.

There are, I think, two additional reasons why the study of popular geopolitics has taken on a new urgency over the last decade or so. The first one is to do with our increased knowledge of human cognition, and the second one concerns how the global political landscape is changing. First, cognition. Social worlds are being produced in language: they are inter-subjective. The subjects that do the construction, however, trust their brains to produce the input that goes into the intersubjective production. As social scientists, how the brain works is not and should not be our *métier*, but it is certainly one of the preconditions for how the social works, and therefore also for what we do. It is, therefore, of immediate interest to us that contemporary neuroscience has things to say about what the brain does when it needs information that is not available from direct or first-order experience. It reaches for knowledge that is available from other sources than one's own lived life, that is, from so-called synthetic experience. For most people, most second-order knowledge will probably come from popular culture. As summed up by Paul Musgrave and J. Furman Daniel:

> The notion of synthetic experiences rests on the idea that comprehending an argument or a narrative requires, even for a moment, accepting it as true. This argument in turn has roots in 'dual-process model' of cognition, perhaps best known from Daniel Kahneman's (2011) *Thinking Fast and Slow*. Kahneman describes rapid, automatic processing of stimuli as 'System 1' and more difficult, conscious and explicit reasoning processes as 'System 2'. [...] The implication of dual-process models for how people comprehend fiction is important. Dual-process models "call into serious question the notion that information is ever greeted with default disbelief" (emphasis added) and instead "point to people's inherent credulity—their tendency to allow any information, reliable or unreliable, to gain entry into their store of knowledge and to influence their beliefs about the world".
>
> (Musgrave and Daniel 2018, forthcoming, 9)

However slowly we pride ourselves on thinking, there is no denying that 'fast thinking' is the mode that dominates our lives. Even those of us that are highly critical of, say, newspaper coverage of international affairs, because we know so much of it is inaccurate and misleading, tend to accept coverage of other issue areas than the ones we specialise as bona fide. In other words, we will easily find ourselves in the throes of fast thinking once we move outside of our areas of expertise.

The political implications of this 'inherent credulity', and what seems to be a hard-wired human, drive to create meaning out of whatever information is at hand are stark. The *locus classicus* of what happens when political leaders wilfully set aside facts in order to dominate the production of social space is Walter Benjamin's ([1936] 1968) essay 'The Work of Art in the Era of Mechanical Reproduction', which demonstrates how Nazi Germany drew on what we may call popular culture—the use of icons, the staging of rallies, and so on—in creating totalising control or *Gleichschaltung* of society. This volume gives contemporary examples, such as the use made of the colour saffron by the Hindutva movement that is now ruling India. We find parallels in how Recep Tayyip Erdoğan draws on a badly defined 'Neo-Ottomanism' to root out a multiplicity of aesthetics in Turkey, in Vladimir Putin's laws against 'homosexual propaganda' and general social conservatism cast Russia as a paragon of 'traditional values', and in the efficacy to which U.S. president Donald J. Trump puts his extensive experience from reality TV shows in discrediting everybody and everything that may come across as challenging his view of the world. The scopic regimes inside which such *Gleichschaltung* happens are different, and the degree in which these attempts at re-homogenising a fast-globalising world are successful differ, but the trend is the same everywhere: popular culture is an ever-more contested field, and the political importance of these contestations is on the rise. In addition to general reasons that have to do with social and scientific developments, we have here a tangible political reason why the study of popular geopolitics has taken on new urgency at the present juncture, and why this volume is a timely contribution to this growing interdiscipline.

Acknowledgements

The editors thank our contributors for their work, cooperation, and patience, especially Chris Homewood, Kyle Grayson, and Federica Caso whose early enthusiasm for the project was invaluable. We would also like to thank Paul Cooke and Klaus Dodds for their comments on the first draft of the Introduction. The contributors express their gratitude to the various entities that granted permission to reproduce the visuals included in the volume, particularly Activision, Charis Tsevis, Ranka Grgic Posavec, Alexander Wells, and Malath Abbas. The editors are indebted to the financial support of the Leeds Russian Centre of the School of Languages, Cultures and Societies of the University of Leeds which sponsored Robert Saunders' research fellowship during which this project was originally conceived. Lastly, we wish to thank our friends and colleagues who generously sent materials pertinent to our project, including Pedro Hernandez for his technical assistance, as well as the personnel at Routledge for their support at all stages of the publication process.

Introduction

Theorising the realm of popular geopolitics

Robert A. Saunders and Vlad Strukov

In this volume, we understand popular geopolitics to be an overlay of discursive fields of popular culture, International Relations (IR), and geopolitics. Closely connected to the disciplines of cultural studies, gender studies, and popular culture studies, popular geopolitics is a form of post-language, or an occurrence that exceeds structuralist categories and yet makes creative use of them. At the same time, popular geopolitics invents and projects its own powers, symbols, and iterations about individual and collective entities, which are also somewhat greater than connoted by the familiar categories of 'consumer', 'audience', and/or 'citizen'. Although the relationships between artefacts of popular culture and items of political meaning in the arena of interstate relations is a long-standing one, the history of their academic investigation is quite short. The complexity of the issue is in that popular geopolitics simultaneously defines and produces what it studies. The most significant features of this dualistic relationship include: (a) the use of popular culture to construct and promote a specific worldview; (b) the dissolution of 'real politics' in favour of hyper-mediated, impression-based politics on the world stage; and (c) a disciplinary approach to the study of contemporary and historical phenomena. In this volume we propose to consider these relationships from an interdisciplinary standpoint, or a conceptual framework which allows scholars to explore phenomena as fluctuations, cross-overs, and hybridisations, not simply essentialities or identities. Plotting the interdiscipline implies tracing its origins, theorising its evolution, and putting forward new conceptual frameworks for the future. This is where the alternative meaning of the verb 'to plot' becomes useful, because—in a mildly conspiratorial fashion—it emphasises the speculative nature of research on contemporary phenomena. To confirm, it is a type of research that is always evolving and adjusting its methodological and theoretical apparatus. Popular geopolitics, as we see it, is perennially capable of breaking away—categorically—from the orthodoxies of existing disciplines, and thus able to lay the foundations for an entirely new range of (inter)disciplines in the twenty-first century.

As the first text to theorise popular geopolitics as an interdiscipline, our efforts are focussed on examining key feedback loops among interrelated, though often artificially distanced disciplines such as cultural studies, gender studies, media studies, film studies, and others. This is accomplished through a pioneering mixture of theoretical and methodological analyses, as well as ground-breaking, primary source-based case studies and interviews. Given the interdisciplinary focus of the collection and its diverse approaches to scholarly inquiry, we interrogate and assess anew key concepts such as power, agency, and identity by adopting a critical orientation towards the poststructuralist paradigm. This task is undertaken with the aim of advancing research associated with popular culture by linking such work to 'real-world' outcomes in world politics, and vice versa: claiming that feedback loops exist not only between geopolitical entities such as nation-states, but also via various actors at the sub-state level. We view popular geopolitics as a process as much as an outcome, and, as an *interdisciplinary* encounter, we contend that it provides productive models for *disciplinary* conversations in geography, media studies, and IR.[1] In this endeavour, we are particularly inspired by new forms of publishing outputs such as Roland Bleiker's recent edited volume *Visual Global Politics* (2018), which collects essays from more than 50 thinkers from a wide variety of academic backgrounds, to address the everyday reality of the visual age by examining the images and visual artefacts that shape international events and our understanding(s) of them.

By bringing together our own group of scholars from multiple disciplines (geography, IR, film studies, political theory, etc.) and various geographic areas (UK, U.S., continental Europe, Australia, etc.) to assess the importance of maintaining a cross-disciplinary focus in popular geopolitics, we hope that this text will serve as a preliminary framework for new avenues of inquiry in the 'popular culture-world politics (PCWP) continuum' (Grayson, Davies, and Philpott 2009). In this introduction to the volume we start by conceptualising popular culture for the twenty-first century (Section 1) and we reevaluate the scope and direction of the field of popular geopolitics vis-à-vis established disciplines (Section 2). In the subsequent sections, we focus on the relationship between popular geopolitics and the state (Section 3), and on the issue of global citizenship as it has been articulated in popular culture (Section 4). It in these two sections that we theorise popular geopolitics as an interdiscipline and determine our pioneering contribution to the field. We conclude the introduction by assessing conceptually the volume's structure and scope (Section 5).

1 Why popular culture? Why now?

Popular culture is a language.[2] It displays universal principles and local variations. Like Mandarin, Swahili, or Faroese, it has syntax, morphology, phonetics, and phonology, as well as a grammar, dialects, stylistics, and semiotics. Like any language, it has a history that informs who uses it and why. But most importantly, popular culture is a medium of communication: it conveys information, structures thinking, brings people together and forces them apart. A James Bond film would thus be equivalent to a paragraph, a season of *The Bridge* mimics an extended speech, while the gameplay of *S.T.A.L.K.E.R.* is comparable to a multi-party, polyvalent conversation. In our framework, popular culture is a type of iteration that makes use of different items of meaning—images, practices, arrangements, responses, and so on—available to us across a range of communication channels, platforms, and traditions. It is specific insofar as it relates to large social groups, not a single class of people; yet popular culture consists of its own subsections and overlapping clusters. Ultimately, we are concerned with how popular culture articulates political meanings, and essentially, geopolitical spaces. To paraphrase, we are interested in how popular culture constructs and reveals spatial and political fields of meaning.

Geopolitics, on the other hand, is more like a religion. It has its doctrines, denominations, rituals, and liturgies; its institutions and structures of power; its holy places and distant gods; its believers and clerics; and—of course—its schismatics, heretics, and critics (and we should not forget its victims). In its orthodox form, geopolitics exists as a field of power projection, commanded by policy-makers and think tanks. In its critical form, geopolitics is a way of seeing the world, a roadmap for comprehending how space, place, and people interact in the imagined realm(s) of international affairs. Geopolitics utilises its own language to account for phenomena, advance agendas, and to impact communities, while simultaneously providing the means to consider events and phenomena in terms of how they articulate interstate and other forms of transnational power. In this regard, the release of a new James Bond film will be interpreted as an affirmation of British soft power and a reflection on the country's geopolitical concerns. Conversely, *The Bridge* serves as a state-funded exercise self-criticism, laying bare a slew of problems that afflict the mature Nordic welfare state. Equally, *S.T.A.L.K.E.R.* reveals the experience of living in a post-apocalyptical world where areas of exclusion signify intolerances of the present-day society and manifest power in relation to habitable spaces.

In the realm of orthodox IR, the role reserved for the study of popular culture has long been marginal at best, and treated as a folly at worst. While a number of journals – especially *Politics*, *Geopolitics*, and *Millennium* – have increasingly 'made space' for empirical and theoretical analyses of popular culture artefacts and essays on the importance of popular culture in/as political culture, many of the field's premier publications remain impenetrable fortresses when it comes to such academic fare. (In this regard the development of the field of popular geopolitics vis-à-vis IR is reminiscent of how cultural studies had evolved in opposition to traditional humanities.) Consequently, research into the international political connotations of popular culture has often found fecund ground in other disciplines, from geography to cultural studies to sociology. This state of affairs does a disservice to academe. In sequestering the study of popular culture, we argue that IR fails to achieve its most basic goal, i.e. the study of relationships between (contemporary) political entities which include not only institutions of the state but institutions of popular culture, too. Our wish here is not to establish a new orthodoxy by operationalising popular culture as the main modus for understanding how the state works, but instead to provoke new ways of thinking about the larger realm of what constitutes IR, i.e. the vast continuum of items and actions that bind and divide polities, corporations, individuals and other congeries (including national governments).

The digital revolution that began in 1990s has remade the global political realm, challenging all aspects of foreign affairs and upending long-held notions about the way politics get done. Recognising the spatial (Lefebvre 1991), visual (Rose 2001), aesthetic (Bleiker 2001), and digital (Nicholson 2013) 'turns' in the social sciences and humanities, this volume serves as a nascent attempt at integrating popular culture more deeply into scholarship across multiple arenas. Given the undeniable relevance of Twitter, Facebook, television series, films, and videogames on how we see the world and the way it works demands that such research be brought into the IR mainstream. Therefore, in this volume we challenge the very notion of the political, its entities, institutions and other forms of being and affecting. We argue that the political, and by extension IR, is closely linked to the *popular* insofar as popular culture artefacts play a similar role to political institutions in informing our perceptions of the world. Equally we maintain that the term popular has acquired new connotations in the twenty-first century. For example, the popularity of a phenomenon is now measured in millions of likes and billions of views, not in circumstantial evidence of reviews, memoires, and oral histories. Also, myriad phenomena now reach a global audience at unprecedented and hitherto unimaginable rates of distribution and (re)transmission, thus raising concerns about political interpretation of speed and acceleration generally (see Connolly 2002).

The transformations caused by the digital turn compel us to seek new theoretical tools to account for the popular geopolitics in the new millennium. We break ground by focusing on three specific areas, hoping that the ideas and methods presented in the volume will advance theoretical considerations in other fields, as well as in related disciplines. First, as a number of scholars have pointed out, the lines between celebrities and politics grows thinner by the day (see, for instance, Street 2004; Marsh, 't Hart, and Tindall 2010; Wood, Corbett, and Flinders 2016), and there is a growing list of individuals who cannot be easily qualified as either one or the other, including Imran Khan, George Galloway, Cicciolina (Ilona Staller), Manny Pacquiao, Silvio Berlusconi, Arnold Schwarzenegger, and Donald J. Trump. As a result, the very structure of *politics*, including foreign affairs, has been yoked to the *popular*, from the United Nations' appointment of the (fictional) comic book superhero Wonder Woman as an Honorary Ambassador for the Empowerment of Women and Girls to the intervention of Angelina Jolie, Bono, Ai Weiwei, Kseniia Sobchak, and other celebrities in geopolitical conflicts from Sudan to Myanmar.

Second, we identify an increasing porousness in popular culture production, distribution, and consumption on the global scale, or to be more accurate, we see a great bleeding and blending of national, regional, and geolinguistic popular cultures into each other. This is the result of new platforms which have expanded access and diversified content (e.g. Instagram, Netflix, Weibo). This is also an outcome of the influence of neoliberal capitalism which employs hybridised and globally oriented cultural production (see Dean 2005). As we have argued elsewhere, this situation often results in recursive feedback loops that flow over, under, and across national borders (Saunders and Strukov 2017). These feedback loops rely on new types of agency—personalised, networked, adaptive, and deterritorialised—that often overlap. Here space, or '–scaping' (the relationship of places to global economic, political, social and cultural factors) (Appadurai 1996) is used as a means to examine and analyse popular geopolitics feedback loops and to reconsider the political as a form of knowledge construction. In doing so, we are careful to keep in mind Connolly's admonition that 'diverse elements' are always infiltrating politics, folding, bending, blending, emulsifying, and dissolving into each other 'forging a qualitative assemblage' that defies classical modes of analysis and explanation (2005, 870).

Third, there is increasing evidence that popular culture is being explicitly recognised as a battlefield for the hearts and minds of citizenries around the world (Weldes 2003; Shapiro 2008; Takacs 2015). While Hollywood once reigned nearly unchallenged in its employment of the popular as a political tool, cultural producers in Beijing, Istanbul, Johannesburg, Lagos, Moscow, Mumbai, and São Paulo (among other sites) are vying for influence, tethering national narratives to universal truths through various media. Perhaps equal in importance to the explicitly geopolitical hueing of transnational popular culture is the increasing diversity of voices within these purportedly 'national' milieus. The dramatic lowering of financial, technological, and spatial barriers to cultural production via cyberspace has only added to complexity of this new world order (Saunders 2010), allowing underrepresented groups (women, LGBTQ, minorities, indigenous peoples, etc.) to stake their own political claims via popular culture (Enloe 1989; Dodds 2010; Shepherd 2013). We argue that this new constellation of political agency requires a new, horizontal global framework for investigation, and this volume is the first step in this direction.

Thus, our approach focusses on popular geopolitics as an interdiscipline, recognising that no single field of knowledge production is capable of applying the necessary set of methodological, theoretical, empirical, and epistemological frameworks for linking the popular to the geopolitical. Here we extend the work of van Munster and Sylvest, who in their *Documenting World Politics*, make the following claim:

> It is increasingly recognised that popular culture and world politics are deeply imbricated. Elements of such culture(s) are more than just pedagogical tools for teaching complex theories and concepts; they can trigger events; they reflect and shape ideas, norms and power relations in global politics; and they constitute sites of contestation, reproduction and innovation of political languages and imaginaries.
>
> (van Munster and Sylvest 2015, 5)

Van Munster and Sylvest focus on non-fiction film as a means to understanding the imbrication between popular culture and world politics, while we examine other forms of cultural production from Hollywood and Bollywood films to videogames and advertisements. Yet, we share these authors' orientation that scholarly engagement, via the contextualisation of images and interrogation of ways of seeing, must be approached from multiple angles, therein demanding that popular geopolitics must be treated (and operate) as an 'interdiscipline' (van Munster and Sylvest 2015, 6). Consequently, we have embarked on this study to conceptualise popular geopolitics as an interdisciplinary meta-phenomenon, hoping that a more rigorous understanding of where the field is going will help it better serve the interests of those polities who are directly impacted by the artefacts we study. In addition to pragmatic benefits, the consideration of popular geopolitics as an interdiscipline has the potential to impact the configuration and constellation of academic disciplines and theories, thus transforming the present-day system of (institutionalised) knowledge.

2 Reconfiguring the direction of the interdiscipline[3]

The rapid changes in what we understand to be the realms of the popular and the political mean that, as an academic discipline, popular geopolitics is evolving at a breakneck speed. However, its evolution—at the scale of common perceptions—is marked by an association—and which is faulty!—with another discipline, that of geopolitics. To be more accurate, popular geopolitics evolved out of critical geopolitics, a fairly narrow niche within the discipline of political geography, but clearly is shaped by the tradition of classical geopolitics. Geopolitics has a rather chequered past, given its historical association with Nazi Germany, most notably in the person of Karl Haushofer, whose student Rudolf Hess went on to become Deputy Führer of the Third Reich until his arrest in Scotland in 1941. Often derided as a form of 'Nazi social science' (see Guzzini 2012), *Geopolitik* was rebranded during the Cold War, with scholars returning to the foundational works of the field, particularly those of American naval strategist Alfred Thayer Mahan and British academic Halford Mackinder. With the advent of global superpowers, nuclear weapons, and the expansion of military technology into outer space, geopolitics became a critical avenue of intellectual inquiry for policy-makers, academics, and journalists. However, in the 1970s, French geographer Yves Lacoste challenged the status quo - built on a rather neat division between state/practical geopolitics and elite/classical geopolitics - to call for a greater level of critical inquiry into how to see the world as it really is (see Klinke 2009), and to recognise that geopolitics is fundamentally about the making of war (Claval 2000). It is with Lacoste and the establishment of the journal of radical French geography, *Hérodote*, that we witness the beginning of critical geopolitics. Following the 'discursive turn' in a variety of disciplines (as well as the spatial, visual, aesthetic, and digital ones referenced above), there was an increasing call to engage the totality of geopolitical thought and action, triggering a comparative flood of new approaches to geopolitics by the mid-1990s, spearheaded by the writings of Gearóid Ó Tuathail (1996) and Simon Dalby (1996).

Out of this flurry of activity emerged a distinct subfield of critical geopolitics which sought to interrogate popular culture as a key site of knowledge production, effectively establishing a third plank of geopolitics to complement those of the state and academia. Joanne Sharp (1993, 1996, 2000) pioneered this new frontier of analysis with her interrogation of the long-running American publication *Reader's Digest*, and how it influenced the U.S. Cold War imagination. She was soon joined by other geographers seeking to link the popular to the political, including Klaus Dodds (2003, 2005), who engaged James Bond films to analyse their role in the narration of post-Imperial Britain, and Jason Dittmer (2005, 2007, 2013), who explored Captain America and how a comic superhero makes the United States' place in the world. Geographers, however, did not hold a monopoly on the rapidly growing field of study. In fact, they were quickly joined by a host of IR scholars employing similar methodologies and approaching their subjects from common theoretical orientations. Among the more prominent studies which overlap with popular geopolitics are: Jutta Weldes' (1999) research on the television and film series *Star Trek* and its impact on the notion of 'benevolent empire' in the (post-)Cold War context; Cynthia Weber's (2005) analysis of filmic mediation of American identity at war; Iver Neumann and Daniel Nexon's (2006) explanations of how the *Harry Potter* books and film franchise explain and reinforce world politics; and more recently, Daniel Drezner's (2011) efforts at using zombies to advance IR theory. This powerful confluence of scholars from different fields turning their critical gaze to 'low cultural products' such as films (Crampton and Power 2005), television series (Buzan 2010), popular fiction (Mukharji 2015), song lyrics (Boulton 2008), comic books (Dunnett 2009), political cartoons (Dodds 2007), animation (Thorogood 2016), tabloid journalism (Debrix 2008), radio broadcasts (Pinkerton and Dodds 2009), videogames (Salter 2011), stand-up comedy (Purcell, Scott Brown, and Gokmen 2010), and social networking sites (Saunders 2014) to explain the way the world works reflects what we argue is the genuinely—even radically—interdisciplinary nature of what has come to be called 'popular geopolitics'. Despite the obvious need to cross disciplinary boundaries, with only a few exceptions, these studies have operated from field-specific launch pads and thus tended to remain tethered to disciplinary concerns.

Our aim in this volume is to loosen the epistemological shackles that have hitherto characterised research in popular geopolitics; we do so via normative grounding in interdisciplinarity rather than working with a disciplinary structure. By moving across such intellectual borders, both real and imagined, our theoretical project means to reflect the very nature of popular culture and world politics transgressing of the boundaries that purportedly keep these fields of knowledge and action separate. By treating our field of enquiry as an interdiscipline, we wish to accentuate our position 'in-between', not emanating from some imaginary centre or periphery. We, of course, make use of Edward Said's (1978) trenchant concept of Orientalism and his intellectual excavation of colonial discourse and 'world-making' rooted in works of literature, art, and other forms of both high and low culture. Like other scholars working in the field of popular geopolitics, we start our inquiries with the recognition of the power of imagination in determining (geo)political realities on the ground. However, unlike many of our peers, we (i.e. the contributors) extend it to political, social, and cultural institutions, and, ultimately, we consider imagination itself as a form of reality and as a discursive field where power is located.

Popular geopolitics unabashedly affirms the ocularcentrism of dominant geographic understandings. By exploring 'ways of seeing' (Cosgrove 2008) that go beyond maps and physical surveys, scholars of popular geopolitics engage a variety of geopolitical phenomena outside the realm of policy and practice, from the aesthetic (Bleiker 2001) and the emotional (Pain 2009) to the humorous (Ridanpää 2009), fantastical (Ruane and James 2008), and cinematic (Shapiro 2008). As with our critique of imagination, we wish to employ ocularcentric strategies to investigate popular culture and its geopolitical concerns, and we also wish to consider the ocular as a space and means of production of the political. To confirm, we approach the apparatus of vision not only as a means of communicating knowledge but as a kind of knowledge, too. The eye, we argue, produces epistemologies which reveal their own spatial politics. That being stated, we are also attentive to Saunders and Holland's call for IR scholars to 'go beyond the visual, incorporating a broader range of sensory processes', including taste, touch, smell, and hearing (2018, 126).

Our innovative interpretation of key concepts requires that we should reexamine our methodological basis. The scholarship of popular geopolitics (and critical IR research in the vein of the works discussed above) employs a fairly wide set of methodologies and foci (see, for instance, Weldes 1999; Grayson, Davies, and Philpott 2009; Dittmer and Gray 2010; Scharf 2013; Caso and Hamilton 2015). The most common modalities identified in these studies are:

- Analysis of institutions and processes involved in production.
- Analysis of the 'geopolitical moments' present in media representations.
- Analysis of the ideological structure of the text and its possible meanings.
- Analysis of the reception of that structure and those meanings and their potential transformation by their audience.

In the case of institutions and processes, the gaze of the scholar focusses on the cultural producer and the environment in which the text is crafted. The importance of movie studios, television networks, publishing houses, videogame developers, and other culture industries is a major part of the interrogation. Methodologies for such projects often involve interviews with cultural producers, data collection related to funding sources, and/or interrogation of supporting apparatuses which enabled the project. Research on the 'geopolitical moments' in a given text (or set of texts) is a very common form of popular geopolitics. Such approaches are typically done in conjunction with research on the ideological structure of popular culture texts as discussion of the former is often difficult to separate from interrogation of the latter. Lastly, popular geopolitics is gravitating towards audience studies, while also looking carefully at how political elites 'use' popular culture for their own ends. This is perhaps the most underdeveloped modality of analysis, yet the one which generates the greatest level of enthusiasm among researchers active in the field.

Our own contribution in this volume seeks to go beyond this schema, identifying through the use of case studies opportunities for new ways of conceptualising flows of ideas via geopolitical popular culture, and how these transmissions feed into and alter a complex system of self-perception and ideas about (geopolitical) Others. In addition to interdisciplinary borrowings and cross-overs, we invite researchers to see popular culture and the political from the interdisciplinary perspective. We advocate this 'inter-', 'in-between' position as a methodological standpoint, not as a set of instrumentalised suppositions. In Part I, by tracing the relationship of popular geopolitics to other interdisciplines such as gender studies and cultural studies, we demonstrate how popular geopolitics provides a new theorisation of the popular and the political. In Part II, we focus on showcasing how reading the world from an interdisciplinary perspective raises new concerns about the origins, history, directions, and scope of popular geopolitics—as a realm of human activity and as a research enquiry—in an increasingly globalised world.

3 Popular culture and the state: a changing dynamic

Popular geopolitics interrogates contemporary myth-making (Saunders 2012), seeking to uncover the cultural forms and reception of those forms that inform quotidian understandings of space, place, and people (Dittmer 2010), and how these phenomena impact international politics, diplomacy, and global affairs. Hitherto, this has produced scholarship that has tended to be markedly state-centric, and focussed on artefacts that, on the surface at least, are quite conservative in nature (masculinist, Eurocentric, pro-imperial, etc.). Researchers operating in the field and at its edges have demonstrated a particular interest in how popular culture 'leaks' (Dodds 2005) into the real-world actions of policymakers, diplomats, and citizenries, from support for foreign policy initiatives to attitudes toward the military to national animosities. Until quite recently, however, the states which were the depository of such leakage have been the U.S., UK, Australia, and other globally oriented Anglophone countries. As Dittmer and Dodds (2008) point out, the blurring of the lines between the popular and the political during the 'Second Cold War' was especially profound, particularly when President Ronald Reagan began using Hollywood as his own personal tool-kit to achieve broad foreign policy goals, drawing on everything from George Lucas's space opera *Star Wars* to John Rambo's aggrieved libertarianism to Rocky Balboa's pugilistic anti-Communism.

While there has been a geographic and geolinguistic preference for what Vucetic (2011) labels the 'Anglosphere', scholarship into the realm of the popular has a quasi-normative bent to it, promoting an understanding of geopolitical orders, codes, and visions (see Dijkink 1996), while interrogating and often critiquing the representation of national images of the 'self' and the 'other'. Hitherto, prominent themes in popular geopolitical studies have included: patriotism and nationalism (Sage 2008; Dittmer and Larsen 2007); imperialism and military conquest (Holland 2012); gender roles and the nation (Emad 2006; Burrows 2010; Shepherd 2013); the exotic or dangerous 'Other' (Semmerling 2006; Al-Rawi 2008; Dunnett 2009; Morrissette 2014); promotion of and/or clashes between ideologies (Muntean and Payne 2009; Ciută and Klinke 2010); and protecting the 'homeland' via surveillance and securitisation (MacDonald, Hughes, and Dodds 2010; Dodds 2011). Reflecting such a scope, this volume is quite attentive to the 'past' character of the field and critically readdresses the current state of the literature; however, we are in fact more concerned with what the current thinkers in the field believe is to come.

As we and other researchers have discussed, the early focus of popular geopolitics on the Anglo-American geopolitical project has led to criticism of the field as needlessly limited in scope, and one which can greatly benefit from expanding its gaze (see Saunders and Strukov 2017; as well as Holland 2012; Mukharji 2015). Specifically, we argue for an expansion in scope beyond the Anglosphere, i.e. to both large (Arabic, Chinese, Hindi, Russian, Spanish, etc.) and medium-sized (Indonesian, German, Japanese Korean, Portuguese, Turkish, etc.) geolinguistic realms, with a focus on how the modes of popular geopolitics work outside of the Anglo-American framework. Moreover, we advocate the popular geopolitics widens its gaze to include 'smaller things', i.e. objects of study that are less-often linked to international scales of power such as the body, the home, flora/fauna, and interpersonal experiences (e.g. laughter, sex, etc.).

As a separate theoretical strand, we propose an innovative consideration of the relationship between popular geopolitics and the Cold War. Contributions to the volume reveal that the origins of the popular geopolitics are in the transformation of the political discourse in aftermath of the dissolution of the USSR. Although the focus of the discipline has been on the Anglophone popular culture, its very emergence, we argue, has to do with the shift from hard power to soft power concerns after the end of the Cold War. Most recently international media and some researchers have voiced concerns about the return to some of the confrontations of the Cold-War era, or even a wholescale 'new Cold War' (see, for instance, Lucas 2008; Ciută and Klinke 2010; Gaufman and Wałasek 2014; Legvold 2014). On the one hand, these fears are based on Russia's aggressive actions in Georgia and Ukraine, as well as more recent forms of asymmetrical actions against Western governments; on the other, they reveal the inability of the West to articulate new foreign policy agendas, prompting a lusty return to previously tested vantages, orientations, and arrangements of contestation.

For more than a decade, the War on Terror provided the West with an ideological scaffolding and a sense of purpose; however, it was shattered after the 2008 financial crisis and new frameworks of military engagement in Iraq and Syria. It is possible to claim that putting Russia at the centre of geopolitical concerns obscures actual challenges existing in the globalised world. At the same time, we wish to emphasise that in actual terms the rhetoric of the new Cold War indeed is an admission of the powers of other states, including the powers of discourses present in their popular culture. In this volume we prepare the ground for the study of popular geopolitics in the Russian and other non-Anglophone contexts. Moreover, by conceptualising our approach as that of the interdiscipline, we aim to direct the attention of future research to areas where new tensions and confrontations emerge. 'Inter' does not mean lingering in no man's land, but rather zooming in on the areas of productions of new meanings and orders. In this regard, the rhetoric of the new Cold War implies an imperative to reconsider the position of the West vis-à-vis other global players such as China, India, and others. Therefore, studies presented in this volume do not aim simply to sample popular geopolitics outside the domain of Anglophone popular culture, but rather to propose a new conceptualisation of popular geopolitics in the global(ised) perspective.

Owing to this 'worlded' perspective (Spivak 1985), our distinct contribution to the existing literature associated with popular culture and politics, including the disciplines of political geography, critical IR, cultural studies, political communication, and media studies, is threefold. First, we provide a normative contribution with a distinct focus on identifying and revisiting explicit and implicit socio-cultural and geopolitical codes which have been presented in the literature thus far. We are careful to attenuate such criticisms, particularly given that the vast majority of scholars have produced research that actually condemns patriarchy, heteronormativity, nationalism, classism, racism, and other forms of prejudice and privilege. Yet, despite the irony, this volume makes the case that there is more to be done, particularly in the selection of artefacts, an embrace of the global perspective, and an expansion of methods and modalities. In effect, we demand that research begin tearing down the walls of the ideological prisons that have long hemmed in what popular geopolitics is ultimately capable of. We argue that—in addition to expanding the non-Anglophone scope of popular geopolitics—by directly reconnecting the field to certain streams of inquiry, specifically those coming from cultural studies and gender studies, popular geopolitics will be better positioned to move forward in terms of its ability to speak to the pressing issues of the day, as well as provide a meaningful and wide understanding of how the world really works.

Second, in a theoretical contribution, we reframe popular geopolitics as an interdiscipline, and in doing so, aim to expand the ways in which scholars in the field (and those whose work overlaps) are freed from preexisting expectations of what popular geopolitics *is* and *does*. Grounded in critical approaches to geopolitics, IR, and humanities such as cultural studies, gender studies, film studies and so on, this book charts the evolution of scholarship aimed at understanding the increasingly dense bindings of popular culture and world politics, and how mediation informs geopolitical understanding(s) of the diplomacy, conflict, and various flows of globalisation. By examining new media and cutting-edge communication technologies as catalysts, facilitators, and inhibitors of praxis and habitus, in conjunction with the persistence of 'old media' forms, we—for the first time—reflect on the ways in which popular culture (as text, discourse, and practice) now flows across international borders and transforms and repurposes the spaces within those borders and the borders themselves.

Our pioneering contribution is in documenting and analysing new tensions and new areas of conflict and collaboration that have yet to be studied. As a pertinent example, we identify an increasing trend towards a form of transnational traditionalism, or even conservatism, which is not particularly state-centric, but instead bubbles up within societies as disparate as Britain, India, Russia, Thailand, Turkey, and the U.S. Whether one talks of the current 'Viking wave' or the consumption of Russian-made conservative television series in China, it is evident that the global media landscape has been transformed and that popular culture has documented and responded to these changes and challenges by depicting an interest in place-based identity, among myriad other trends. These alterations, aberrations, and flows reveal a reversal of a century-long development towards cosmopolitanism. Or, arguably, these phenomena suggest that we should reexamine what it means to be a citizen of the global society, which is possible thanks to popular culture and the work of imagination delinked from institutions, space, and time. Consequently, we present a new conceptualisation of the relationship between the (nation-)state and media, one which—through our engagement with the meta-discourses that touch on the changing fields of identity and post-identity, citizenship (both local and global), and self/non-self—we hope will have ramifications far beyond the scope of popular geopolitics.

Last, we make a conceptual input to the study of the 'popular culture-world politics' continuum by having invited our contributors to reflect on a variety of cultural artefacts and eras, from Yugoslavia's Cold War-era imaginings of 'America' to Bollywood's use of science fiction to comment on India's role in the world to the zombie's undying ability to speak truth to power. Their findings presented in the subsequent chapters enable us to suggest that in the realm of popular culture and political relations there is an attempt to query our modern-day concepts of reality. Equally, the contributions attest that there is an ongoing process of reevaluation and recontextualisation of what we actually mean by culture, including popular culture. From a repository of artefacts to economic activity and socially oriented practice to culture as imaginary environments to culture as a realm of effects and affects to culture as a form of post-ontology facilitated by the emergence of artificial intelligence and networked iterations, our conceptualisation of culture evolves to identify new areas of investigation and to establish new theoretical frameworks.

4 From global citizenship to global structures of knowledge

To respond to these global challenges and accentuate our theoretical contribution, we have structured the volume around two conceptual nodes. Part I provides a critical reflection on the discipline of popular geopolitics, thus demanding that it should be framed theoretically as an interdiscipline, i.e. as a form of knowledge that exists in-between, through relation and as a mode of enhancing citizen participation in the critical construction of global citizenship. Part II supplies a set of new concepts emanating from the analysis of previously overlooked fields of cultural and political production. In the style of the interdiscipline which privileges speculative, innovative research, contributions in Part II engage with fluid phenomena by adopting a bi-focal, 'inter' perspective on global events. The two parts form a singular theoretical narrative which culminates in the Conclusion where we formulate new modes of enquiry, global structures of knowledge, and new modes of ontological experience.

Beginning with an interview of two of the seminal figures in the field (Jo Sharp and Klaus Dodds) conducted by one of its most well-known practitioners

(Jason Dittmer), we present a primer for the reader, focusing on how the field began and where it is going. This is complemented by the subsequent chapter in which Kyle Grayson revisits the claims he made (with Davies and Philpott 2009) about the popular culture-world politics continuum. From here, the chapters by Strukov and Caso relate popular geopolitics to other interdisciplines (cultural studies and gender studies, respectively), making profound arguments for a deepening of exchange between all three fields. Saunders' chapter interrogates what happens when popular culture 'gets real' (i.e. when it triggers geopolitical reactions), thus challenging our assumptions about what constitutes social and political realities and realms of imagination and creativity. Each of these chapters advances a specific theoretical concern such as the issue of hegemony, and demonstrates the growing efficacy of popular geopolitics to investigate key issues in world politics and (popular) culture. The chapters in the second part simultaneously expand the scope of popular geopolitics beyond the traditional focal areas (i.e. U.S./UK national identity and empire) to new realms of inquiry (i.e. Yugoslavia, China, India), *and* employ novel modes of analysis to explore two subject areas that have provoked some of the most fecund explorations of the PCWP continuum (i.e. zombie fiction and war videogames). Whilst these chapters engage popular geopolitics as an interdiscipline in the contemporary setting, they also supply material and conceptualisations for reading popular geopolitics in a historical setting.

5 Volume outline, or possible ways of engagement

There are multiple ways in which the reader can engage with the volume. They can focus on contextual readings first and then consider theoretical conceptualisations. Alternatively, they can approach the materials in the volume from an historical perspective, tracing the permutations of the discipline. The outline of the collection which we provide here is one of many potentialities, and is by no means prescriptive. This is consistent with our emphasis on the concept of interdiscipline which allows for multi-directional, multi-purpose engagements with critical thought. In this regard, the Conclusion to the volume is not a summary of the book's major findings but rather an invitation to consider them in new contexts, thus reflecting on the contemporary condition and signposting new directions of research.

Jason Dittmer's 'The origins and evolution of popular geopolitics: an interview with Jo Sharp and Klaus Dodds' (Chapter 1) is a wide-ranging and collegial interview with Jo Sharp, Deputy Head of the School of Geographical and Earth Sciences at the University of Glasgow and author of *Condensing the Cold War: Reader's Digest and American Identity* (2000), and Klaus Dodds, Professor of Geopolitics at Royal Holloway University of London and author of *The World is Not Enough: The Geographies, Genders and Geopolitics of James Bond* (2017, with Lisa Funnell). Bringing together two of the field's originators, this chapter sketches out the fundaments of popular geopolitics, presenting the reader with an intimate portrait of how the discipline was forged by examining its genetic makeup, with a particular focus on the Falkland Islands War and the end of the

Cold War. This chapter sets the tone for the volume, investigating the curious interplay between popular culture and politics, paying tribute to the theorists who informed the first generation of scholars, from Gramsci to Said to Shapiro.

Following Dittmer is Kyle Grayson's 'Popular geopolitics and popular culture in world politics: pasts, presents, futures' (Chapter 2). Grayson revisits his seminal 2009 essay 'Pop Goes IR? Researching the Popular Culture–World Politics Continuum' (co-authored with Newcastle University colleagues Matt Davies and Simon Philpott) in an effort to assess why movies, television series, comics, and other media are so vitally important to our understanding of global affairs. While Dittmer's interview represents an interpersonal remembering of the origins of the (inter)discipline, Grayson's chapter offers more textually centric reporting of how popular geopolitics came into vogue, and makes some important predictions about where it is headed.

The next two chapters shift the focus away from the internal workings of the discipline, instead looking at how popular geopolitics can learn from (and influence) other interdisciplines. The first of these is Vlad Strukov's 'Towards a new paradigm of resistance: theorizing popular geopolitics as an interdiscipline' (Chapter 3). This argute contribution to the collection carefully discerns relevant parallels between cultural studies and popular geopolitics, noting important overlaps and key differences between these two fields of study. Strukov's analysis presents a normative argument for scholars to take popular geopolitics to the natural next step of its evolution, i.e. a scholarship of resistance. While extant research tends to support the neoliberal world-order wrought by the dissolution of the Soviet Union and the spread of market-style capitalism, he identifies ample opportunities for the audience and cultural producers to speak back to the system, thus channelling various subversive and subaltern 'ways of knowing' associated with Bhabha, Spivak, and Gilroy among others. The chapter critiques popular geopolitics from the perspective of cultural studies whilst pointing out that popular geopolitics should develop its own project of resistance.

Pairing neatly with Strukov's analysis is Federica Caso's probing chapter entitled 'Gender studies and popular geopolitics: indispensable bedfellows for interdisciplinarity' (Chapter 4). Structurally mirroring the previous chapter while centring its gaze more specifically on notions of gendered and sexed representation, Caso's incisive contribution focusses on methodological and theoretical elements of gender studies which would better serve popular geopolitics were they to be adopted. With a shared interest in the relationship between the everyday and global flows of power, she argues for popular geopolitics to take more interest in gender, sexuality, and the body, making the claim that such a shift will enable scholars to better understand the myriad subjectivies that are part of the constellation of what we call IR.

Robert A. Saunders' 'Crossing the boundary: "real-world" geopolitical responses to the popular' (Chapter 5) synthesises his decade-long work on the intersection between the popular and the political on the international stage. Saunders examines approximately one dozen instances of popular culture triggering actual geopolitical outcomes, from 'Hollywood' films that provoked foreign governments into action to YouTube videos that disrupted the geopolitical balance of the Middle East. His analysis focusses on the ways in which states react to controversial depictions in Western media. By splitting popular geopolitics into two separate forms (i.e. phronesis and praxis), he argues that both producers and consumers of geopolitically infused pop culture become part of the foreign policy landscape, thus bearing at least some responsibility for how IR gets done in the contemporary globalised realm. Beginning with a brief reprisal of his ground-breaking work on the role Sacha Baron Cohen's 'Borat' parody played in Kazakhstan's diplomatic and nation-branding efforts and ending with an analysis of two of the most dramatic popular geopolitical interventions by foreign governments in the U.S. (the North Korean hack of Sony and Russia's use of social media networks to influence the outcome of the 2016 presidential election), Saunders provides a holistic overview of how popular culture makes world politics what it is today.

Leading off the second half of the collection, 'The convenient fiction of geopolitics: rethinking "america" in the geopolitical imagination of yugoslav culture' (Chapter 6) by Maša Kolanović expands the field of popular geopolitics by looking at the U.S. from the outside in. Adapting methodologies from the burgeoning field of imagology, or image studies, Kolanović's meticulously researched chapter delineates the changing perceptions of 'America' in Yugoslavia's popular consciousness through depiction of the US in a wide variety of artefacts. In her choice of material, she is able to provide the reader with a novel and refreshing perspective on the national image of the United States in a country that, while neutral in geopolitical terms, was part of the socialist bloc.

Her contribution is followed by Ashvin Devasundaram's 'Beyond brand Bollywood: alternative articulations of geopolitical discourse in new Indian films' (Chapter 7), which examines the world's largest motion picture-producing country (India). Devasundaram's chapter engages the notion of self-orientalism as a popular geopolitical undertaking, employing two recent films – *Court (2015)* and *PK (2014)* – to interrogate the often-conflicting streams of representation inherent in Bollywood films, i.e. domestic and global. Ranging from science fiction to religion to gender relations, this chapter presents an evocative model for future work in the field and argues that greater attention be paid to pop culture production sites beyond Hollywood and London studios.

Shifting from one of the world's fastest growing economies to another, Chris Homewood's '"Directed by Hollywood, edited by China"? Chinese soft power, geo-imaginaries, and neo-Orientalism(s) in recent U.S. blockbusters' (Chapter 8) recognises the increasing economic heft of the Chinese marketplace in international film distribution. This contribution examines a few of the Western movies that have dispensed with hoary stereotypes of China as the 'exotic, underdeveloped east' and Chinese as 'recondite schemers', instead offering up depictions of China as a bastion of (non-Orientalised) technological superiority and civilisational development where Chinese characters are forthright and heroic (especially in relation to 'American' villains). Mixing film studies with the study of geographical imagination, this chapter examines the role of global flows of capital in influencing how foreign cultural producers can be compelled to alter long-standing norms when it comes to popular geopolitical representation.

Co-authored by the Australia-based scholars Roxanne Chaitowitz and Shannon Brincat, '"Warning! Zombies Ahead…": determinate negation, predatory capitalism and globalised place-making' (Chapter 9) delves into the philosophical and material meanings of the undead in an effort to explain how the global economic system actually works, as well as how its insidious tendrils frame our darkest fears (currently manifested in popular culture via the ultimate 'Other', the flesh-eating ghoul). With its gaze on borders, consumption, transactions, and collectives, this piquant chapter reanimates the analysis of zombies as geopolitical stand-ins, arguing for a new understanding of neoliberalism and its discontents.

From the realm of the undead, we move into the virtual world of videogames with Daniel Bos' 'Popular geopolitics and the landscapes of virtual war' (Chapter 10). Bos explores the representative power of war videogames, assessing the increasing importance of immersion in virtual worlds and how such escapism informs real-world views of distant places and peoples. In his treatment of the popular videogame series *Call of Duty*, Bos focusses on landscapes of violence and the connection to militarisation, masculinity, and nationalism, arguing that by moving through these imagined worlds, one experiences something more-than-representation. In their ability to stimulate a variety of emotions and affectual associations, such artefacts have an acute influence on notions of (in)security.

The concluding chapter to the volume provides a theoretical extrapolation of the key ideas presented in the chapters, while also reflecting the contributors' realisation of the role of researchers as producers (as well as consumers) of popular geopolitics. The conclusion includes images and text-based sections that offer critical analyses of specific aspects of popular geopolitics as an interdiscipline. Whilst each of the sections refers to a specific chapter of the volume, together they form a singular narrative aimed at providing yet another tier of theoretical reflection. Here images are not mere illustrations; instead they are used as a means of conducting analysis and conveying ideas in popular geopolitics: it is an attempt to theorise in images, not to speak through images. This way we pay respect to the visual turn in popular geopolitics and recognise the ocularcentrism of the present-day world. We also adopt a critical position vis-à-vis our own role as researchers: what matters is not only how we interpret images, or how we collect them, but how we *image* our own research world (see Mitchell 1986, 2002). Importantly, these images were collected by the chapters' authors and the accompanying curations were authored jointly by the contributors and the editors. This approach reveals our collaborative effort to advance popular geopolitics as an interdiscipline, that is, to think and to work collectively on the margins of discourse. This is indicative of the rapidly changing field of popular geopolitics and the constantly changing academic environment.

Notes

1 However, we hasten to point out that in the field of geography, there is less need to provoke such discussions as feminist and critical geographers are responsible for the inauguration of popular geopolitics as a subfield. Moreover, these scholars continue to push the discipline into increasingly diverse areas, from the everyday to the body to realms of affect.

2 We do not advocate a structuralist reading of (popular) culture; instead we use the language metaphor to indicate the complexity of formalist and post-structuralist readings of discourse.
3 Portions of this section appeared in our co-authored article 'The Popular Geopolitics Feedback Loop: Thinking beyond the "Russia versus the West" Paradigm', *Europe-Asia Studies* 69 (2): 303–324.

References

Al-Rawi, Ahmed. 2008. 'Iraqi Stereotypes in American Culture: The Case of Video Games and Films'. *International Journal of Contemporary Iraqi Studies* 2 (2):225–249.

Bleiker, Roland. 2001. 'The Aesthetic Turn in International Political Theory'. *Millennium: Journal of International Studies* 30 (3):509–533.

Bleiker, Roland. 2018. *Visual Global Politics*. London and New York: Routledge.

Boulton, Andrew 2008. 'The Popular Geopolitical Wor(l)ds of: Post-9/11 Country Music'. *Popular Music and Society* 31 (3):373–387.

Burrows, Karen K. 2010. 'Alias America: 9/11 and the Televisual Quest for a National Identity'. In *Politics and Popular Culture*, edited by Leah A. Murray, 126–135. Newcastle-upon-Tyne: Cambridge Scholars Publishing.

Buzan, Barry. 2010. 'America in Space: The International Relations of *Star Trek* and *Battlestar Galactica*'. *Millennium: Journal of International Studies* 39 (1):175–180

Caso, Federica, and Caitlin Hamilton. 2015. *Popular Culture and World Politics: Theories, Methods, Pedagogies*. Bristol, UK: E-International Relations.

Ciută, Felix, and Ian Klinke. 2010. 'Lost in Conceptualization: Reading the 'New Cold War' with Critical Geopolitics'. *Political Geography* 29:323–332.

Claval, Paul. 2000. '*Hérodote* and the French Left'. In *Geopolitical Traditions: Critical Histories of a Century of Geopolitical Thought*, edited by David Atkinson and Klaus Dodds, 239–267. London and New York: Routledge.

Connolly, William E. 2002. *Neuropolitics: Thinking, Culture, Speed*. Minneapolis: University of Minnesota Press.

Connolly, William E. 2005. 'The Evangelical-Capitalist Resonance Machine'. *Political Theory* 33 (6):869–886.

Cosgrove, Denis E. 2008. *Geography and Vision: Seeing, Imagining and Representing the World*. London: I. B. Tauris.

Crampton, Andrew, and Marcus Power. 2005. 'Reel Geopolitics: Cinemato-graphing Political Space'. *Geopolitics* 10:193–203.

Dalby, Simon, and Gearóid Ó. Tuathail. 1996. 'The Critical Geopolitics Constellation: Problematizing Fusions of Geographical Knowledge and Power'. *Political Geography* 15 (6–7):451–456.

Dean, Jodi. 2005. 'Communicative Capitalism: Circulation and the Foreclose of Politics'. *Cultural Politics* 1 (1):51–74.

Debrix, François. 2008. *Tabloid Terror: War, Culture, and Geopolitics*. London and New York: Routledge.

Dijkink, Gertjan. 1996. *National Identity and Geopolitical Visions: Maps of Pride and Pain*. New York and London: Routledge.

Dittmer, Jason. 2005. 'Captain America's Empire: Reflections on Identity, Popular Culture, and Post-9/11 Geopolitics'. *Annals of the Association of American Geographers* 95 (3):626–643.

Dittmer, Jason. 2007. 'The Tyranny of the Serial: Popular Geopolitics, the Nation, and Comic Book Discourse'. *Antipode* 39:247–268.

Dittmer, Jason. 2010. *Popular Culture, Geopolitics, and Identity*. Lanham, MD: Rowman & Littlefield.

Dittmer, Jason. 2013. *Captain America and the Nationalist Superhero: Metaphors, Narratives, and Geopolitics*. Philadelphia: Temple University Press.

Dittmer, Jason, and Soren Larsen. 2007. '*Captain Canuck*, Audience Response, and the Project of Canadian Nationalism'. *Social & Cultural Geography* 8 (5):735–753.

Dittmer, Jason, and Klaus Dodds. 2008. 'Popular Geopolitics Past and Future: Fandom, Identities and Audiences'. *Geopolitics* 13 (3):437–445.

Dittmer, Jason, and Nicholas Gray. 2010. 'Popular Geopolitics 2.0: Towards Methodologies of the Everyday'. *Geography Compass* 4 (11):1664–1677.

Dodds, Klaus. 2003. 'License to Stereotype: Popular Geopolitics, James Bond and the Spectre of Balkanism'. *Geopolitics* 8 (2):125–156.

Dodds, Klaus. 2005. 'Screening Geopolitics: James Bond and the Early Cold War Films (1962–1967)'. *Geopolitics* 10 (2):266–289.

Dodds, Klaus. 2007. 'Steve Bell's Eye: Cartoons, Geopolitics and the Visualization of the 'War on Terror''. *Security Dialogue* 38:157–177.

Dodds, Klaus. 2010. 'Popular Geopolitics and Cartoons: Representing Power Relations, Repitition and Resistance'. *Critical African Studies* 2 (4):113–131.

Dodds, Klaus. 2011. 'Gender, Geopolitics, and Surveillance in *The Bourne Ultimatum*'. *Geographical Review* 101 (1):88–105.

Drezner, Daniel W. 2011. *Theories of International Politics and Zombies*. Princeton and Oxford: Princeton University Press.

Dunnett, Oliver. 2009. 'Identity and Geopolitics in Hergé's *Adventures of Tintin*'. *Social & Cultural Geography* 10 (5):583–599.

Emad, Mitra C. 2006. 'Reading Wonder Woman's Body: Mythologies of Gender and Nation'. *Journal of Popular Culture* 39 (6):954–984.

Enloe, Cynthia. 1989. *Bananas, Beaches, and Bases: Making Feminist Sense of International Politics*. London: Pandora Press.

Funnell, Lisa, and Klaus Dodds. 2017. *Geographies, Genders and Geopolitics of James Bond*, New York. Palgrave Macmillan.

Gaufman, Elizaveta, and Katarzyna Wałasek. 2014. 'The New Cold War on the Football Field: .ru vs .pl'. *Digital Icons: Studies in Russian, Eurasian and Central European New Media* (12):55–75.

Grayson, Kyle, Matt Davies, and Simon Philpott. 2009. 'Pop Goes IR? Researching the Popular Culture-World Politics Continuum'. *Politics* 29 (3):155–163.

Guzzini, Stefa 2012. *The Return of Geopolitics in Europe?: Social Mechanisms and Foreign Policy Identity Crises*. Cambridge: Cambridge University Press.

Holland, Edward C. 2012. '"To Think and Imagine and See Differently": Popular Geopolitics, Graphic Narrative, and Joe Sacco's "Chechen War, Chechen Women"'. *Geopolitics* 17 (1):105–129.

Klinke, Ian. 2009. 'Five Minutes for Critical Geopolitics: A Slightly Provocative Introduction'. *Exploring Geopolitics*, available at http://www.exploringgeopolitics.org/publication_klinke_ian_five_minutes_for_critical_geopolitics_a_slightly_provocative_introduction/ (last accessed 30 January 2015).

Lefebvre, Henri. 1991. *The Production of Space*. Cambridge: Blackwell.

Legvold, Robert. 2014. 'Managing the New Cold War'. *Foreign Affairs* 93 (4):74–84.

Lucas, Edward. 2008. *The New Cold War: Putin's Russia and the Threat to the West*. New York: Palgrave Macmillan.

MacDonald, Fraser, Rachel Hughes, and Klaus Dodds. 2010. *Observant States: Geopolitics and Visual Culture*. London and New York: I. B. Tauris.

Marsh, David, Paul 't Hart, and Karen Tindall. 2010. 'Celebrity Politics: The Politics of the Late Modernity?' *Political Studies Review* 8 (3):322–340.

Mitchell, W. J. T. 1986. *Iconology: Image, Text, Ideology*. Chicago: University of Chicago Press.

Mitchell, W. J. T. 2002. 'Showing Seeing: A Critique of Visual Culture'. *Journal of Visual Culture* 1 (2):165–181.

Morrissette, Jason. 2014. 'Zombies, International Relations, and the Production of Danger: Critical Security Studies versus the Living Dead'. *Studies in Popular Culture* 36 (2):1–28.

Mukharji, Projit Bihari. 2015. 'Technospatial Imaginaries: Masud Rana and the Vernacularization of Popular Cold War Geopolitics in East Pakistan, 1966–1971'. *History & Technology* 31 (3):324–340.

Muntean, Nick, and Matthew Thomas Payne. 2009. 'Attack of the Living Dead: Recalibrating Terror in Post-September 11 Zombie Film'. In *The War on Terror and American Popular Culture: September 11 and Beyond*, edited by Andrew Schopp and Matthew B. Hill, 239–258. Madison and Teaneck, NJ: Fairleigh Dickinson Press.

Nexon, Daniel H., and Iver B. Neumann. 2006. *Harry Potter and International Relations*. Lanham, MD: Rowman & Littlefield.

Nicholson, Bob. 2013. 'The Digital Turn'. *Media History* 19 (1):59–73.

Ó Tuathail, Gearóid. 1996. *Critical Geopolitics: The Politics of Writing Global Space*. Minneapolis: University of Minnesota Press.

Pain, Rachel 2009. 'Globalized Fear? Towards an Emotional Geopolitics'. *Progress in Human Geography* 33 (4):466–486.

Pinkerton, Alasdair, and Klaus Dodds. 2009. 'Radio Geopolitics: Broadcasting, Listening and the Struggle for Acoustic Spaces'. *Progress in Human Geography* 33 (1):10–27.

Purcell, Darren, Melissa Scott Brown, and Mahmut Gokmen. 2010. 'Achmed the Dead Terrorist and Humor in Popular Geopolitics'. *Geoforum* 75:373–385.

Ridanpää, Juha. 2009. 'Geopolitics of Humour: The Muhammad Cartoon Crisis and the *Kaltio* Comic Strip Episode in Finland'. *Geopolitics* 14 (4):729–749.

Rose, Gillian. 2001. *Visual Methodologies: An Introduction to the Interpretation of Visual Materials*. Thousand Oaks, CA: Sage Publications.

Ruane, Abigail E., and Patrick James. 2008. 'The International Relations of Middle-earth: Learning from *Lord of the Rings*'. *International Studies Perspectives* 9:377–394.

Sage, Daniel. 2008. 'Framing Space: A Popular Geopolitics of American Manifest Destiny in Outer Space'. *Geopolitics* 13:27–53.

Said, Edward. 1978. *Orientalism*. New York: Vintage Books.

Salter, Mark B. 2011. 'The Geographical Imaginations of Video Games: Diplomacy, Civilization, *America's Army* and *Grand Theft Auto IV*'. *Geopolitics* 16 (2):359–388.

Saunders, Robert A. 2010. *Ethnopolitics in Cyberspace: The Internet, Minority Nationalism, and the Web of Identity*. Lanham, MD: Lexington Books.

Saunders, Robert A. 2012. 'Undead Spaces: Fear, Globalisation, and the Popular Geopolitics of Zombiism'. *Geopolitics* 17 (1):80–104.

Saunders, Robert A. 2014. 'Mediating New Europe-Asia: Branding the Post-Socialist World via the Internet'. In *New Media in New Europe-Asia*, edited by Jeremy Morris, Natalya Rulyova and Vlad Strukov, 143–166. London: Routledge.

Saunders, Robert A., and Vlad Strukov. 2017. 'The Popular Geopolitics Feedback Loop: Thinking beyond the "Russia versus the West" Paradigm'. *Europe-Asia Studies* 69 (2):303–324.

Saunders, Robert A., and Jack Holland. 2018. 'The Ritual of Beer Consumption as Discursive Intervention: Effigy, Sensory Politics, and Resistance in Everyday IR'. *Millennium: Journal of International Studies* 46 (2): 119–141.

Scharf, Inge. 2013. 'Staging the Border: National Identity and the Critical Geopolitics of West German Film'. In *Cinema and Popular Geo-politics*, edited by Marcus Power, 182–201. London and New York: Routledge.

Semmerling, Tim Jon. 2006. *'Evil' Arabs in American Popular Film: Orientalist Fear*. Austin: University of Texas Press.

Shapiro, Michael J. 2008. *Cinematic Geopolitics*. Abingdon, UK: Taylor & Francis.

Sharp, Joanne P. 1993. 'Publishing American Identity: Popular Geopolitics, Myth and *The Reader's Digest*'. *Political Geography* 12 (6):491–503.

Sharp, Joanne P. 1996. 'Hegemony, Popular Culture and Geopolitics: *The Reader's Digest* and the Construction of Danger'. *Political Geography* 15 (6–7):557–570.

Sharp, Joanne P. 2000. *Condensing the Cold War: Reader's Digest and American Identity*. Minneapolis: University of Minnesota Press.

Shepherd, Laura J. 2013. *Gender, Violence and Popular Culture: Telling Stories*. London and New York: Routledge.

Spivak, Gayatri Chakravorty. 1985. 'Three Women's Texts and a Critique of Imperialism'. *Critical Inquiry* 12 (1):243–261.

Street, John. 2004. 'Celebrity Politicians: Popular Culture and Political Representation'. *The British Journal of Politics & International Relations* 6 (4):435–452.

Takacs, Stacy. 2015. *Interrogating Popular Culture: Key Questions*. London and New York: Routledge.

Thorogood, Joe. 2016. 'Satire and Geopolitics: Vulgarity, Ambiguity and the Body Grotesque in *South Park*'. *Geopolitics* 21 (1):215–235.

van Munster, Rens, and Casper Sylvest. 2015. *Documenting World Politics: A Critical Companion to IR and Non-Fiction Film*. London: Routledge.

Vucetic, Srdjan. 2011. *The Anglosphere: A Genealogy of a Racialized Identity in International Relations*. Stanford, CA: Stanford University Press.

Weber, Cynthia. 2005. *Imagining America at War: Morality, Politics and Film*. London and New York: Routledge.

Weldes, Jutta. 1999. 'Going Cultural: Star Trek, State Action, and Popular Culture'. *Millennium: Journal of International Studies* 28 (1):117–134.

Weldes, Jutta. 2003. 'Popular Culture, Science Fiction, and World Politics'. In *To Seek Out New Worlds: Exploring Links between Science Fiction and World Politics*, edited by Jutta Weldes, 1–22. Houndsmills, UK: Palgrave Macmillan.

Wood, Matthew, Jack Corbett, and Matthew Flinders. 2016. 'Just Like Us: Everyday Celebrity Politicians and the Pursuit of Popularity in an Age of Anti-Politics'. *British Journal of Politics & International Relations* 18 (3):581–598.

Part I

Mapping the (inter)discipline of popular geopolitics

1 The origins and evolution of popular geopolitics

An interview with Jo Sharp and Klaus Dodds

Jason Dittmer

Introduction

When I was a PhD student (1999–2003) studying newspaper representations of Central and Eastern Europe during NATO and EU expansion, Klaus Dodds's and Jo Sharp's work was central to the way in which I came to position my work within wider literatures. More importantly, however, when I subsequently decided to shift from 'legitimate' news media to the decidedly more vulgar study of superheroes and their imbrication in geopolitical discourse, it was their critical opening into the worlds of popular culture that gave me the courage to push the boundaries of what was acceptable to study within the field of critical geopolitics. Both had paved the way for my work, Dodds with his analyses of political cartoons and James Bond films, and Sharp with her work looking at the treatment of Russia (as a mirror for American identity) in Reader's Digest magazine. I think it is safe to say that, without these two scholars, there would either be no field of popular geopolitics or it would have taken a much different form at a much later date.

Indeed, one of the themes of the interview we had in summer 2015 is the highly contingent nature of events. I was especially interested to discover the particular collision of intellectual currents, social networks, and personal circumstances that brought about their unique formulations of popular geopolitics. A paradox can be identified in their responses to this question. First, they both acknowledge the interdisciplinary nature of the influences, from Edward Said to Michael Shapiro to Antonio Gramsci. Further, it is their wide-ranging interests in specific geographic regions—most obviously the Soviet Union (Sharp) and the South Atlantic and Antarctica (Dodds)—that fuelled their focus on popular culture and its relation to geopolitics. Nevertheless, it is the 'small world' nature of the discipline of geography—specifically the poststructuralist branch of political geography from which critical geopolitics emerged in the 1990s—that enabled popular geopolitics to take hold and become a project in its own right. The key authors of critical geopolitics were young contemporaries of one another – inviting one another to participate in panels, special issues, and other academic projects. This fertile environment enabled popular geopolitics to territorialise as a field and yet, that field has sometimes been better at reiterating these two luminaries' claims than at advancing their concepts or theories. In fact, my own early work on nationalist superheroes is susceptible to this critique, largely replicating elements of both

Klaus's and Jo's work. From Klaus I took the emphasis on the visual shorthand of comics and cartoons, and from Jo I took an interest in serialised narratives that underpin ultimately conservative geopolitical visions of the world. In the intervening years, I moved from the U.S. to London, and subsequently became good friends with both Klaus and Jo, which is likely apparent in the tone of the interview. In fact, when I first moved to London and was waiting for my furniture to arrive on a slow boat from New York City, I slept on my bedroom floor with only a pillow and duvet borrowed from Klaus to cushion me. Geography can indeed be a 'small world'.

Until recently, I think it is safe to say that popular geopolitics remained insulated from some of the broader research currents in related fields, such as cultural studies (in fact, this was the driving force behind the speakers' series that germinated the volume you hold in your hands). Nevertheless, the field is now in a period of rapid change, with a new generation of young scholars taking an interest and importing their own interdisciplinary concepts and intellectual inspirations. It is for this reason that I wanted to ensure that the early history of the field was documented, as it has an historical and geographical context that is crucial to understanding the conceptual development of the field. As Jo Sharp notes below, however, the future is not beholden to the past, and that is a good thing.

The following has been edited and abridged, with references inserted where appropriate.

Jason: Thank you very much for agreeing to do this. As the two people perhaps most foundational to popular geopolitics in geography, I was wondering if you both might describe the landscape of critical geopolitics[1] when you came to it. Klaus, why don't you go first?

Klaus: Thank you for inviting me. My first exposure to something that might be called critical geopolitics came entirely by accident. That was when I alighted upon an article published by Simon Dalby, I think in 1989 or 1990, that talked about geopolitics in ways that struck me as rather interesting, but at the same time reminded me of something by Gearóid Ó Tuathail three or four years earlier, which I had perhaps not taken as much notice of as I should have, on the language and nature of US-Salvadoran foreign relations[2]—I can't remember the title (presumably Ó Tuathail 1986). In both cases, I came away with a sense in which geopolitics could be something rather different than what I had been taught as an undergraduate, which had been a fairly conventional diet of the Great Men associated with geopolitics, and a straightforward periodisation of how ideas came and went, but one which was not informed by social theory. So, I was excited by the work of Simon Dalby and Gearóid Ó Tuathail, both of whom were interested in discourse and language and were also reasonably new to academia. You could associate with them—these were people who represented an entirely different generation of political geographers distinct from Peter Taylor, or John Agnew, or whomever.

Jason: How about you, Jo?

Jo: Well, critical geopolitics was quite new, because I came to it in 1990, I suppose, so the words had not long been used. Obviously, there was Gearóid Ó Tuathail and John Agnew's first piece (Ó Tuathail and Agnew 1992). But the piece that first captured my imagination was Simon Dalby's work and his book *Creating the Second Cold War* (Dalby 1990),[3] which I found fabulous and really brought a lot of things into focus. I didn't really come to geopolitics as such. Well, I came to it in a sort of roundabout way because I didn't have any intention of studying critical geopolitics. I think the specific institutions that I moved through have been quite instrumental in the ways that I came across and engaged with particular academic approaches. So, I had come to Syracuse because I wanted to work with John Agnew, not because of the critical geopolitics article, but because of his book *Place and Politics* (Agnew 1987), which I had used for my undergraduate dissertation to look at identity in the Shetland Islands. I had then done a course with the late Graham Smith on—well it started off at the beginning of the year as the Geography of the Soviet Union and by the end of the year it was the Geography of Russia. It was a fabulous time to be doing that course. It was a course that looked at the changing geographies of the Soviet Union and how that drove the end of the Cold War; it wasn't America winning, or any of these simplistic arguments, but that it had a very different society, particularly in the ethnic fringes. Nationalism drove the end of the Cold War. I found that argument very persuasive and very interesting; I wondered what it would be like to combine some of his ideas and some of John Agnew's ideas, but look at *Russian* nationalism as dominant rather than marginal. So that was the plan; very old-school. Two things happened which then brought me to critical geopolitics. I hope you weren't expecting any brief answers [laughter].

Jason: Never.

Jo: So, the first was that I had to take courses since I was doing my Master's and PhD in the United States, so I did everything I could on Russia and the Soviet Union. So that took me out of geography and into political science, and I couldn't understand why the person teaching the course didn't use Graham's explanation of what happened. So, he was in there saying—it was basically Sovietology, the idea that you could understand change in the Soviet Union by looking at where people were standing in funerals, and who was carrying the coffin, and these sorts of things. And I said, 'No, no!' It was in no way an arrogant interjection because I was just saying 'I've come across this other guy's ideas which are so much better!' [laughter] And I just couldn't convince the professor that this was a better explanation. This was the first time that I think I had really been hit by the idea that there were very different explanations for something. I perceived that what he had was a very old-fashioned, simplistic view; I had been exposed to a much more nuanced political-cultural-geographical explanation of what was going on. We were just ... like ships in the night. At the same time as that was happening, I was exposed to postmodernism for the first time. So, I took a course with Jim Duncan that was very interdisciplinary—he ran it with someone from landscape architecture—and I had never come across anything like this before. It was brilliantly taught, and also it introduced me to the idea of discourse, the politics of language, the politics of knowledge, all these sorts of things. So, at the same time that I was seeing this happen in the course about Russia, I was getting the conceptual grounding in the postmodernism course. So those two things really came together in my head and lead me to a Master's dissertation that was the beginnings of *Reader's Digest* as an example of this, drawing together those two experiences. Then I came across the early incarnations of critical geopolitics and thought, 'Oh yes, that is what I am trying to do.' So, it wasn't so much that I read critical geopolitics and thought 'this is what I want to do', but that I—and I suppose others came to it the same way—was experiencing competing narratives of a major historical world event at the same time as I was exposed to post-structuralism. And I wonder, also, whether there is any significance to the fact that some of the earliest proponents of critical geopolitics were from the fringes of Britain, but were at the time of working in North America. So, you have Simon and Gearóid from Ireland, John Agnew from Cumbria (which is almost its own country in Britain), myself from Scotland—Klaus is the outlier coming from the heart of Empire—but I wonder if we were much more conscious of that politics of language—of the use of 'British' versus 'Scottish', 'Irish', a sense of 'Northern Englishness'[4]—and that poststructuralist approach gave us an intellectual language for something we've been thinking about anyway. I've never spoken to the others about this; I suspect it isn't coincidence.

Jason: It would be interesting, sometime when we are all sitting down for a beer, to float that idea. I think I've heard Simon say something similar about himself and Gearóid.

Jo: I suppose there is the introduction to *Critical Geopolitics* (Ó Tuathail 1996); Gearóid grounds himself in the introduction when he starts off with a bit of Irish history, saying, 'this is where I am coming from'.

Jason: In your story, you say you started *Condensing the Cold War* (Sharp 2000) by combining these two courses, but what actually made you decide to do what we would now call Popular Geopolitics? That was a form of discourse that hadn't been studied and which was in fact excluded from a lot of the early critical geopolitics.

Jo: I think it was because I didn't come to it as critical geopolitics. I came into it wanting to study Russia; I had been doing this course that made me realise there was more than one narrative going on. I had some of the language from the postmodern/poststructural course to explain that. Edward Said's *Orientalism* also influenced me (Said 1978).[5] I thought,

'Oh, I could do something like that for my Master's!' It seemed then like a neat kind of project, and I really wasn't thinking about it anymore than that. I wanted to take Orientalism from the nineteenth/early twentieth century and see how it works out if you look at America and communism. It was a chance encounter with John Western; he always reminds me that it was his idea. He said, 'Why not look at the *Reader's Digest*?' It was meant to be a short project to look at … I think it was a term paper initially … just to look at the concept of Othering and using the Other to reflect on the Self. It was looking at that that I then came across some of the critical geopolitics literature and I realised it wasn't simply about representation and difference, but that it was doing something else in terms of American political culture, American identity, the sense of manifest destiny. At some point—I can't be sure at what point—I must have been influenced by Gramsci and the idea that the hegemonic cultural values were important. So, because I didn't come through the tradition that said, 'this isn't something to be included', it came as a surprise to me when I discovered that not everyone thought that looking at popular culture made sense in terms of International Relations or geopolitics. Because you're right, there is a huge tradition that says there is absolutely no role for popular understandings, and the first time I gave a seminar in the department, one member of staff said, 'well, this is all very interesting, but it is irrelevant, it has no influence at all on the political process'.

Jason: We'll come back to the reception of your ideas in a moment. How about for you, Klaus? What led you specifically to popular geopolitics?

Klaus: One of the other areas I was always very interested in was public geographies, specifically some of the work that people like Susan Smith did at the start of her career, Jacquie Burgess and Jo Gold, around media geographies, geography in the popular media, that kind of thing: how geography was put to work in a variety of contexts. I think through their work—as well as Simon, Gearóid, and John Agnew's work—it wasn't too much of a stretch for me at least to think about how this kind of formal, academic sort of geopolitics might actually have something else attached to it, which was a kind of popular dimension. Or at least I could think about how ideas migrate forwards and backwards in different kinds of contexts. I remember my supervisor, Les Hepple, saying to me, 'well, actually, that is a relationship that you could productively think about in your own PhD work', which was about Britain and Argentina and the way the two countries had managed their difficult relationship over the disputed South Atlantic territories.[6] It was obvious to me doing that research in Britain, Argentina, and the United States that the popular domain was hugely important. So, for instance, if Argentina believed that the Malvinas were an integral part of the Republic—they talked about them as the Little Lost Sisters—that the gendering of territory, the representation of territory, and the role of public education was hugely important in ensuring that Argentinian citizens—by the time they were even five or six—were inculcated with the sense that the Malvinas were theirs. Likewise, you had the paradox of Britain perhaps—a context where British citizens didn't have a strong sense of what the Falklands were, let alone where they were located, until of course April 1982 when things changed. So, then you had the interesting case, which you could think through more theoretically: how did this apparently obscure place suddenly loom large in the popular geographical imagination? So, it was those kinds of happenstances that led me to popular geopolitics. But at the same time, I should stress how supportive Simon and Gearóid were at the start of my career. I think a lot of this stuff partly comes out of personal relationships. In both cases here a willingness to look at my work, and to encourage me.

Jason: I know from reading the papers you published from your PhD that your PhD work would be categorised as formal and practical geopolitics, so to the best of my knowledge your first work in popular geopolitics was the Steve Bell cartoons. Is that right?

Klaus: Sort of. What happened was that I had Steve Bell's cartoons that he had produced in 1982 and I thought they were—of all the cartoons I had seen in Britain and in Argentina—the most interesting.

Jason: So you had intentionally examined popular geopolitics of political cartoons in your PhD, you didn't just publish it? I didn't realise that.

Klaus: I think with probably everyone's publication history, you can only tell so much about what they were doing and thinking at the time. People will say to me, 1996 was the first time you published something on popular geopolitics (Dodds 1996). And my reaction is, actually, I thought about popular geopolitics a lot earlier on in the midst of my PhD, some of which was in the PhD but not published more widely. Some of what I collected when in Argentina, including cartoons of Margaret Thatcher as a pirate, was rather X-rated and focussed on what Argentina wanted to do to Thatcher as a pirate. I wouldn't necessarily want to publish this or reproduce those images. Not for young audiences. So, I had this material, and it was probably thanks to someone like Peter Taylor who said, 'Oh, I love Steve Bell—if you have this stuff why not do something with it?' So, I had the opportunity to pull that stuff out of my box file and write it up. I think it appeared in the same special issue that Jo was in (Sharp 1996). For me that was the perfect opportunity. The other thing that hardened my decision to develop this stuff was meeting Jo for the first time at the AAG in 1993 in Atlanta, when she and I were on the same panel, chaired by John Agnew with Simon and Gearóid in the audience. There was only about an audience of six, with this enormous bowl room as I remember. I think I then in a sense was given an impetus—and probably the confidence—to do something more with this material, rather than just sit on it.

Jason: That's quite a panel, in retrospect. Those people are all big names today. Were there any particular models that you specifically were trying to avoid as you undertook popular geopolitics?

Klaus: I think what drove my thinking was to avoid trivialising the popular. In a sense, I'm not sure I had a model, but I was very concerned about—and I think Jo's work on *Reader's Digest* was a great help because she demonstrated in that 1993 piece that you could write very well on something that some people might have thought was banal and mundane (Sharp 1993). She showed that precisely because it was banal and mundane, and extremely popular and circulating around the world, that it was worthy of consideration. So, like everyone who is starting their career, what you crave are people to reassure you that you are doing substantial work. I think if I had a concern it was really about not being seen to trivialise anything I was doing, but at the same time to be mindful of the fact that I was concerned about public reception, however big or small that public was. I think most people will attest to the fact that there is a sense of insecurity that is particularly acute when starting off.

Jason: It is so different when you have the job security that enables you to say what you want, but usually when you are starting off you lack that.

Klaus: In 1995 I was still in a temporary lectureship at Royal Holloway, which became permanent in 1996. In retrospect I wonder if I held back on publishing some of that stuff on cartoons because perhaps I was concerned that it might be seen as a little frivolous.

Jason: It's one thing to go to Jo and Simon Dalby and tell them your ideas about popular geopolitics. If there is anyone who will go along with it, it is them. But how did your non-fellow travellers respond?

Klaus: In Bristol, where I did my PhD, the direction of traffic was very much towards cultural and economic geography,[7] so I think political geography was seen to be—well we know Brian Berry's quote about the moribund backwater (Berry 1969)—there was a sense in which political geography was not the cutting edge. Indeed, the biggest names in the British context were probably Peter Taylor, who was doing world-systems analysis, and the ever-prolific Ron Johnston, who was best known for his work in electoral geography. Certainly, geopolitics as we would understand it now was not seen as something terribly avant-garde, even if when I started the Cold War ended and we had the Gulf War crisis. It would appear to have been a golden opportunity for geopolitics to reassert itself, but it didn't feel that way in the hothouse that was the Bristol geography department at that time.

Jason: So you think that in your particular context, the shift to popular geopolitics was aided by the way in which it brought the new cultural geography to political geography and gave it that aura of the avant-garde?

Klaus: That's exactly how I would see it. I think as popular geopolitics began to establish itself, I am quite certain it was aided by the cultural turn—people like Peter Jackson, Denis Cosgrove, and others who were interested in language, discourse, and representation. That gave it a boost and an intellectual legitimacy.

Jason: Jo, how did your initial work in popular geopolitics go down? As your PhD research was explicitly about popular culture in a way that Klaus's wasn't, you must have been extra worried about the way it would be perceived.

Jo: Within geography it was no problem at all, and I was very lucky that I had a fabulous advisor in John Agnew. I even had a moment, two years into the PhD, where I thought it was ridiculous and I was going to give up. I had come up with another topic and spent a week working on it, and John just dragged me into his office and told me not to be stupid and to just get on with it. Which was of course very good advice. I also got to know Gearóid and Simon quite early on. They invited me to present at a session they were doing at the AAG—the same one Klaus is talking about—so to be a part of that group early on was fantastic. Simon, in particular, has always been incredibly supportive of me, even driving through blizzards to come to my viva, which was great. So, within political geography, I didn't feel I had to prove myself. And they were involved with the journal *Political Geography* so quite early on I got something published from my PhD, in 1993 (Sharp 1993). It was a special issue they were editing, so I got very supportive comments from them. Publishing outside the discipline with the *Reader's Digest* project was much more difficult; I never really managed that. I tried a number of different journals over the years and didn't get anywhere. I tried an American political history journal and it was a nightmare, partly because it was a terrible review process. They sent it to one reviewer and about a year later I got comments that it was too theoretical. I emphasised the empirical in the re-write and they sent it to a different reviewer, who a year later said it wasn't sufficiently theoretical. The review process for *Condensing the Cold War* was very interesting because they sent it to two reviewers initially, one in geography and one in—I guess—IR or political science. The geographer loved it and the other person had real problems with it. I made considerable revisions and it got sent out to different geographer and a different IR person and the same pattern emerged.

Jason: I realise that it's a lot to ask of you to remember specific reviews, but what was it that the presumably IR person didn't like?

Jo: I don't recall precisely. I suspect it was to do with where this fitted into the political process, because I wasn't proving influence. There have always been comments in the book reviews—it was reviewed quite broadly—the thing that most often came up is that I haven't said enough about the individual writers of the articles, and that came particularly from people with historical backgrounds who were interested in the individuals. Because I was looking at such a long time-period, I was arguing that there was a particular voice of the *Reader's Digest* that gave the seal of approval.

Jason: It's the editor, right? The articles are all condensed.

Jo: Yes. But also, my argument was that the type of person was more important; they weren't academics. That's the knowledge that the *Reader's Digest* didn't like because it was silly, complicated, too ambivalent. They were people with empirical, hands-on experience: retired generals, those kinds of people. It was that fact, plus that the editors had chosen them as a voice worth repeating, that I said gave them their authority. Some of the reviewers didn't like that. Also, that I had no evidence from audiences. I didn't do any research on audience reception, which I don't see how you can do over seventy years and across the whole US. I was arguing that the rate of re-subscription was a suggestion that people were accepting particular arguments. They weren't violently rejecting them. Why would you continue to get something that you didn't find useful or agree with? It's a surrogate for audience work. I appreciate that it's not unproblematic.

Jason: I won't push you on that [chuckles]. Since to my mind you both have helped to create the field, how have you witnessed the field unfold over the roughly 25 years in which there has been popular geopolitics? Klaus?

Klaus: I would say that a lot of things have changed. First, the sheer number of people engaged with popular geopolitics; in 1996–97 if you'd asked me how many people are working in this field, I wouldn't have gotten much past five fingers. That's what it felt like to me. I think—to personalise it for a bit—when I came across your work for the first time, it's an odd thing to say, but I suddenly thought, 'Okay, so there are more people out there.' Also, I thought more strongly about North American contributors, whereas most of the people I knew—maybe reflecting my own insularity—had been British-based. Clearly in retrospect there were others interested, broadly speaking. So, I think the increase in numbers is one change. Second, I think clearly over 25 years interest in language and representation has been hugely complicated by non-representational theory, by materiality, by interests in objects, by affect—things that I don't think 25 years ago anyone was talking about. I don't remember anything about affect and politics in geography. So, I think the mainstream was very much the politics of representation and it's not surprising that if you look back 25 years ago the work has that particular feel to it. Third, I think consumption and audiences are being thought about far more—in your own work with me and in others' work—and that is a very different iteration of popular geopolitics than existed 25 years earlier. I think the fourth thing—I'll finish on this one—is that geographically it's a lot more diverse. We have wonderful work being done in the former Soviet Union, Africa, Asia, Latin America—it has a far more global scope to it. I think it feels like a far more cosmopolitan project, whereas previously it had been very niche or U.S.-this or U.S.-that. I think it's a lot more exciting now.

Jason: My sense from talking to you outside of this interview, Jo, is that you are a bit more sceptical?

Jo: I think the first things were very exciting because there was a sense of doing something that hadn't been done before. A lot of the ideas that have become completely mainstream and a bit old-hat and no longer at all fashionable were only just emerging: our engagements with Foucault (it's difficult to think of that now as new and exciting but it was), really thinking of alternative ways of representing the world and how that might be linked into the construction of individuals' political subjectivity, national identity, all those sorts of things. Conceptually, it was a really exciting time and my feeling is that in those early years—this is maybe a kind of grumpy person's looking back, whatever the opposite of nostalgia is—the exemplars, whether they be your comic books, my *Reader's Digest*, the range of other things that were being looked at—that was being used to drive forward particular conceptual positions. To talk about not just these particular representations, but what they allowed us to understand about the politics of scale, the ways the scales were linked together. The reason I moved towards feminist geopolitics[8] is that I think what I was drawn to popular geopolitics for was the way it linked the global and the individual. That is something that is so often overlooked in what I tried to do with the *Reader's Digest*; I tried to understand how people identified themselves as political subjects. The reason I drifted away was because I got bored. There was a tendency of people to go off and find their favourite film, book, whatever, and just do another study of it. I didn't feel that there was the same amount of conceptual drive. We had a lot of good examples, but it wasn't doing anything very different with it. That, plus another institutional move, which opened up other opportunities for me, meant that I drifted away from popular geopolitics. It has only been more recently that I think the field has become more exciting again because … I'd like to think it has been responding to the feminist geopolitics challenge but also the things Klaus just mentioned about how scales are linked together and how popular culture drives political process rather than being simply another representation. I'm probably being overly critical, but that's the way it seems to have happened.

Jason: Can you give an example from your own recent return to popular geopolitics of how the field now is doing innovative things on the topics that matter to you?

Jo: Well, I really like your *Popular Geopolitics 2.0* (Dittmer and Gray 2010).

Jason: This part is definitely getting into the final edit [laughter].

Jo: Because I think it's responding to that conceptual problematic of scale. However, there is a danger as always in academic circles, and I think this has been the case with first the cultural turn, which went so far away from the material in its efforts to ignore it, and now there is the danger on the cutting edge of geography that we go so far away from the representational, everything has to be in the pre-cognitive and visceral and that sort of thing. I don't think in politics that works because the representational remains important. But to understand, as Martin Müller's (2008) work has, how discourse is performed and not simply the words on the page … feminist geopolitics has been trying to get at that too. And I think your work on *Popular Geopolitics 2.0* was also trying to find a way through—it was quite conceptual, I think—how we balance those different approaches. In Hayden Lorimer's words, it is the 'more-than-representational' (Lorimer 2005). I would be very uncomfortable with a critical geopolitics or popular geopolitics that abandoned the representational. So, I think we have kind of swung that way and are coming back to different ways of incorporating the material. I'm not sure I can separate critical geopolitics, popular geopolitics, and feminist geopolitics in any clear way that I perhaps could have in the early 2000s. So, I'm not sure that your work is purely popular geopolitics—or only popular geopolitics—particularly as you look towards the diplomatic side of things that obviously is linking back towards the core concerns of critical geopolitics (e.g. Dittmer 2015, 2016) even as the feminist geopolitical impetus has been responded to—I think popular geopolitics is thinking of these things more and more. I think Chih-Yuan Woon's work is clearly a response to both.[9]

Jason: How about you, Klaus? You and Jo have both kind of come and gone from critical geopolitics. For neither of you is it the only thing you've ever done, far from it. What made you decide to pick up the baton of popular geopolitics at various times, or put it down?

Klaus: I really think that you are on to something here. So, I think like a number of people who have had quite varied careers, I struggled to stay focussed on one thing for terribly long. That presents both opportunities and dangers. If I had spent the last 20 years consistently working on popular geopolitics, I am quite certain I would have had a big monograph out that was *Everything You Need to Know About Popular Geopolitics*. I just didn't have that consistency in me. If I'm being honest, I think the Arctic and Antarctic—in terms of regional work—was always my greater passion. What's interesting, I think, is the way in which strands occasionally cross over and hybridise somewhat. I have found it perfectly possible to do things that might appear quite disconnected from other core areas of business. Now, I follow the careers of other people who I admire and respect, and occasionally I look at their websites and publication history. Jo is a good example. When Jo teamed up with [Africanist] John Briggs for example, and this whole vista opened up of East African research—and I wondered how that was going to relate to popular geopolitics. Then eventually you see the papers she was doing on Tanzanian newspapers and the War on Terror [e.g.,(Sharp 2011a). In my own work on the Arctic and Antarctic I began to see opportunities to take the insights of popular geopolitics and look again at some of the material that perhaps I'd ignored because of my interest in governance. I realised, 'You know what, there's a lot going on here.'

Jason: This is a side question unrelated to the previous, but how did you decide to do a study of James Bond?

Klaus: A lot of it comes down to opportunities or suggestions or encouragement. When you look back on a career—how incredibly haphazard it appears! In 2002 David Newman said to me, 'I am amazed someone of your interests in popular culture hasn't written about James Bond', and I replied that I was a big James Bond fan, but I had never thought about writing something. He was the editor of *Geopolitics* and he asked me to write something. So that's when I decided to write about my favourite James Bond film, *From Russia With Love*, and to think about how the James Bond films have been extremely significant, given their popularity and longstanding appeal (Dodds 2003). *From Russia With Love* was a wonderful text to get your teeth into because of its very strong iconography revolving around Istanbul and Turkey. It appealed to me and the referees liked it, and people were very positive about it after it was published. There might be people who didn't like it, but the people who spoke to me spoke warmly about it and that reinforced my determination to do more.

Jason: You have one paper (Dodds 2005) based on archival research of the adaptation of a Bond book into a Bond film. I love that paper because it is unique within popular geopolitics, based on archival research about production, something that most popular geopolitics ignores because it is so focussed on the image or text itself. Where did that idea come from?

Klaus: That paper was a happenstance. I wrote the paper on *From Russia With Love*, and I knew the film extremely well as a result. As a teenager I had read the James Bond novels, and it happens that the transition from the novel to the film is a more faithful transition than some of the others, for instance, *You Only Live Twice*. What I began to think about was the question of who undertook this creative process, of re-working Fleming's really well-known novels into the screenplays and subsequent films? I discovered by chance that one of the most important Bond screenwriters was someone named Richard Maibaum. He had been involved in thirteen Bond films, and so if there was one person to start with it was surely him. An extraordinary discovery was that he was a graduate of the University of Iowa and he had given his old university all his screenplay materials. I got into contact with their archive and they had it all, and nobody had ever used it before. I got everything photocopied and what you saw was this amazing record of how Maibaum started with the novel, did the screenplay, and went back and forth arguing with the studio. So, you can see how the screenplay takes shape, especially the role of place in the story. You're absolutely right that popular geopolitics hasn't talked an awful lot about production. Jo's work hints at a bit of that with regard to the *Reader's Digest*, but it is hard to get a hold of a lot of that stuff methodologically.

Jason: I love that paper because it hints at a completely different world of popular geopolitics that might-have-been and still-could-be. What do you think is missing from popular geopolitics? What are the strengths and weaknesses of that kind of research?

Klaus: I think one of the stronger elements of popular geopolitics now is the way in which there is more explicit talk about the everyday, the way in which things—objects, images, texts—play a part in everyday lives. I think the work on things like affect try to get at the transitory nature of these things. They have a capacity to take over everyday life at times but also at times to be part of the background, hardly commented on, subsumed into lives. I like that sense in which the popular is seen as less box-like, more diffuse, more scattered through everyday life. So, you aren't looking at a single film as it is disconnected from everything else. I am guilty of that, so I say this knowingly. I think one of the areas that I still feel needs a lot of work is the production side—the consumption side is a work in progress of which I am more optimistic. But the way in which things get made, get produced, circulate (or not), and how they fall apart, or fail is …

Jason: You can't study a movie that was never made.

Klaus: Exactly. And yet *why* that movie wasn't made—or why *Dr. No* wasn't something otherwise—is actually interesting. What is perceived to work commercially, visually, geopolitically? For instance *Dr. No* was at one point going to be much more about the Panama Canal and Cuba, but it was a difficult time for US-Cuban relations and it was decided that it was best to avoid that topic to avoid generating controversy. So, I think the production side of popular geopolitics needs further encouragement. I think the consumption stuff is a work in progress in terms of audiences, how they do or don't coalesce, and how we can write intelligently about things that are very difficult to grasp. I think methodologically those are also—whether dealing with affect or not—incredibly tricky things to investigate.

Jason: How indebted is popular geopolitics to the disciplinary project of geography? To what extent is it worthwhile to think of it as a sub-field of geography, or is it something else?

Jo: I think the difference with political science and IR is about the connections between scales, which geographers are more attuned to. I don't think it can be seen as a purely disciplinary endeavour because if you go back to the origins of critical geopolitics, the inspiration was not from geographers. It was [Michel] Foucault, it was [Edward] Said, it was [James] Der Derian, [Michael] Shapiro, and those kind of guys: the cool, critical IR types, whom we basically used to take apart our founding fathers in geography. I think we do things differently in the sense that human geography has a very good, 'anything goes' approach—a tolerant attitude to theoretical and methodological diversity. We are much less angsty about the proper way of doing things than some of our cognate disciplines. I think part of that is because we don't have founding fathers that we have to kowtow to. Having said that, because we've had a long history of thinking about scale, I think we are much more attentive to that in geopolitics, although some feminist IR people are thinking a little bit in terms of the way that the domestic and the international are dependent on one another, someone like Cynthia Enloe obviously. Really the majority of these divisions are more entrenched, whereas in geography what sets us apart is a greater sense of the tension between the existence of scales established through institutions and regulations and such. At the same time, we realise that these are porous and creative and interdependent in all sorts of interesting ways. If there is a distinction that's where I'd like to think it is.

Jason: I think you're saying that there is something in popular geopolitics that has emerged out of this geographic sensibility, and that's not to say that other people don't do it or think about it, but that there is a legacy of its origins in geography.

Jo: I think part of this openness to approaches in geography is reflected in how critical geopolitics has developed within the discipline. In fact, it has got to the stage where critical geopolitics—which was a term—has become an approach. The words have been so rammed together that they have lost their critical purchase. Originally it was supposed to be an oxymoron, two terms sat together that should make us very uncomfortable. Now, of course, it is not a provocation, but has become mainstream. This is great, because we have a discipline that can allow within twenty years something that was such a provocation initially to become a very accepted way of doing things. Only five years ago Simon Dalby was able to say that the term had lost its critical meaning (Dalby 2010). When I introduced *subaltern geopolitics*, I was very clear in the introduction of that special issue of *Geoforum* (Sharp 2011b) that this wasn't simply about coming up with another label, but it was an attempt to get back to the jarring provocation. It is really interesting to compare that with the way that feminist and postcolonial IR seem to sit within their discipline. Critical and feminist geopolitics sessions get big audiences at the AAG; they are seen as kind of cool and interesting. I've been to a couple of ISA meetings and there have been huge turnouts for people doing old-school geopolitics and then at the ISA meeting in San Francisco a few years ago I went to hear some of my idols and there were more people on the panel than in the audience. In geography, that wouldn't happen. Those rooms would be the ones packed out. The intellectual trajectories of these things that are very close in many respects have been very different. That says a lot in a very positive way about geography.

Jason: What about you, Klaus? How important is the geo- in popular geopolitics for you?

Klaus: The geo- is very important for me. Recently there has been a trend amongst environmental geographers to come back to geopolitics emphasising the agency and vitality of the earth itself, in contrast to the discursive focus of critical geopolitics. The role of ice, wind, air, and so on in shaping the politics of earth. Until recently I saw the geo- as being about the representations of the earth, whereas now I realise there are a range of ways to conceptualise the geo-, from the discursive to the material. This is the influence of some of the assemblage work, thinking about how things come together or not.

Jason: How should we think about popular geopolitics through this materialist lens of the geo-?

Klaus: I think one of the things we might think about is the genre of disaster films and ask, 'How can we turn this around?' The focus of these films is human-centric: who survives? Which city gets destroyed? But we might really need to think about the way in which the geo-physical has been visualised as a vital force in political life. That would be a very different conversation about what a popular geo-politics might entail.

Jason: What do you think is the future of popular geopolitics as an endeavour? Do you have a preferred course you would like to see it go down?

Klaus: I think the future of popular geopolitics might well be about reconciling lots of different interests. I would like us to try to think about the production, circulation, and consumption of the popular more holistically. Often, we view media as if it is only in one of those boxes. In everyday life, however, all three apply. The things we study often defy the very ways in which we study them. Perhaps we also need to re-think the forms in which we analyse popular culture. I have started to think about the novel and why we ought to turn to the novel as a way to think through popular geopolitics. There is a lovely novel by Dave Eggers called *Zeitoun*, which is about the aftermath of the flooding of New Orleans in 2005. I think it deals brilliantly with the geo- (in this case the agency of water in the overwhelming of the city) and brings up the production of race, inequality, violence, the urban, and the War on Terror, in ways that are incredibly affecting as a reader – something that academic papers rarely achieve. Eggers shows how the everyday is utterly shaped by these broader structural forces. I still think popular geopolitics needs to be better at accounting for the scalar complexities of what we study.

Jason: What do you think about the future of popular geopolitics, Jo? How do you see that future, and how would you like to see that future?

Jo: If I could shape it I'd like to not be able to anticipate what comes up. I'd like it to be challenging me, and to be challenging some of the things that the three of us have taken for granted and written in our work. I'd like for people to tell us that we're wrong. Although theoretically we're very good at talking about how those scales connect together, I don't think empirically it has been done perfectly or consistently. I think there is still work to unpick institutions; maybe your diplomacy work will do it.

Jason: I doubt it. [laughter]

Jo: I would like for us to be thinking about how these geopolitical knowledges are being circulated, consumed in different ways, and again not just about interesting representations; for people to be using all these case studies of representation to *explain* political identity, how people engage with political process, and so on. It is more and more important as we see all sorts of non-formal, non-traditional political routes gaining in significance. I saw that in last year's [2014] Scottish Independence referendum. It wasn't about parties; the ideas were coming from a loose organisation of civil society groups that became the Yes campaign. This was all being communicated through social media and other forms that we as political geographers are not very good at understanding. We see this across Europe and the Middle East and getting our heads around some of that would be a good idea. I think we romanticise some of these groups, so it needs to be about understanding the process, rather than just capturing the representation and saying, 'yea, that's lovely'. I think the last point—probably obvious and a bit stodgy—but America is being challenged as the hegemon and we have ignored China to some extent. Is it the case that American cultural imperialism has been worn down?

Jason: What ties together the things you are saying—to me—is our need for a better understanding of power in popular geopolitics, both in the relations between scales and also in a topological sense of the nodes where these relations tend to cluster, places like Al Jazeera. Maybe Hollywood is still important, but what is going on in China? Where are the loci of representational power today? That seems to me like a general theme in what you are saying.

Jo: And that comes then to the notion of discourse being performative as well, understanding where it happens and what different representations allow us to do, and allow to happen. These were the provocations right at the beginning, and I think the early stages of critical and popular geopolitics have been horribly misrepresented in some of the critiques that have suggested they were all about language, which I don't think was ever the case. It was about discourse in that Foucauldian sense, I think. So, I think it would be nice if perhaps popular geopolitics disappeared as a term, that we don't separate it from critical geopolitics more generally. When you think about how the [2015 UK] General Election was won or lost through the media;[10] so much wasn't about policy but about personality. The extent to which it makes sense to separate the formal political sphere and the sphere of the popular—not just the media but everything—these lines are becoming ever more blurred, if they ever were distinct.

Jason: It's interesting too, because popular geopolitics is the only one of those three terms that has become a project of sorts. You don't hear people saying, 'I'm doing formal geopolitics', or 'I'm doing practical geopolitics'. They say, 'I'm doing critical geopolitics'. It is popular geopolitics that gets hived off and imagined as something distinct. Which gets at your original schism between Gearóid Ó Tuathail and John Agnew setting it out, and then you and Klaus adding in something different. In a way, that binary still exists—just not rigidly.

Jo: You're right. There is a bit of a hierarchy in some people's minds between proper geopolitics (even critical) and the popular side of things. But as I say, I'm not convinced that distinction makes sense. It's the same thing with feminist geopolitics, which persists as having a label although a lot of the things that feminist geopolitics was arguing for has become mainstream (although not often cited). I wonder if things have become less distinct in that there is a general range of approaches to geopolitics that includes popular geopolitics. There was a paper in 2010—I can't recall who wrote it—that argued we need to move away from media in popular geopolitics [this is a joking reference to Dittmer and Gray 2010]. Does that mean that if we are looking at other forms of political activity and how that ties in to other scales, would that include someone like Kye Askins, who I don't think would ever call herself popular geopolitics, but who is studying non-formal geopolitics? I think initially the label was to say that geopolitics doesn't just happen in the hallowed halls of Westminster and the White House. Again, this is probably a provocation that is no longer needed, as there is more of an acceptance that all of these other spaces count. Maybe geography has embraced the idea that politics happens everywhere.

Jason: Thank you very much for your time. I have enjoyed this immensely.

Klaus: Thank you.

Jo: Thank you.

The interview thus ends on an uncertain note. If the historical details of popular geopolitics and its origins were complicated but relatively clear in hindsight, the contemporary boundaries of popular geopolitics—disciplinary or otherwise—remain unstable, in flux, and contested. If popular geopolitics can refer to the classic analysis of geopolitical representations, to the affective and discursive politics of social media, to the geopolitics of everyday life sans media, to novelistic accounts of non-human agency, to audience practices of consumption, then what is it? And how is it that the term itself is more used than ever, even as its meaning becomes more contested and multiple?

I am of course not interested in offering a definition, or in eliminating the term. But what I would like to see is the development of rigorous, sustained debate, drawing from various conceptual and theoretical traditions. For too long popular geopolitics has been a realm of part-time scholarship, and therefore it has not developed with any sense of direction or purpose. The challenges are real: how should we make sense of the eclipse of mass media by social media and the internet? How can we decolonize our popular geopolitical knowledges? How can we imagine a popular geopolitics inclusive of, but not limited to, media? These are big questions, but they deserve answers. Inspiration abounds in neighbouring disciplines, but answering them will require sustained empirical inquiry and conceptual debate. It is my belief that by engaging in that inquiry and debate, a new generation of scholars will make the definition and utility of popular geopolitics clear for the next 25 years.

Notes

1 Critical geopolitics is a constellation of approaches to the study of geopolitics that typically share a poststructuralist concern with representation and an anti-imperialist ethos. For reviews, see Dodds 2001; Jones and Sage 2010.
2 The papers referenced here are the first efforts to frame geopolitics through the lens of discourse, and they presage the formation of what became known as critical geopolitics. For a review of this early work, see Dodds and Sidaway 1994.
3 The paper and book referenced here are both analyses of discourses through which the Cold War was framed; Ó Tuathail and Agnew's classic paper is an analysis of George Kennan's framing of Russian collective psychology at the start of the Cold War while Dalby's book (based on his doctoral dissertation) is of post-détente discourse.
4 The political and cultural geography of the United Kingdom is very complex, shaped by the historical and contemporary political dominance of London and the Home Counties ('Southern England') over a range of nations (the Irish, the Welsh, the Scottish) as well as the (formerly) industrial heartland of northern England. Sharp is here referring to a sense of distancing by these often-marginalised groups from the official discourse of geopolitics coming from London.
5 Edward Said's *Orientalism* is a landmark volume in literary studies, highlighting a mode of literary engagement with 'the East' that was common in British and French accounts of empire in the eastern and southern fringes of the Mediterranean. Said argues that this discourse enables colonialism rather than simply reflecting it.
6 Here Klaus is referring to the events leading up to, and including, the 1982 Falkland Islands conflict. He published this work in *Pink Ice* (2002).
7 The University of Bristol in the 1990s served as a hotbed of what came to be known as the 'new cultural geography', which highlighted identity and practice. These same insights were used to revitalise economic geography (see Thrift 2000).
8 Feminist geopolitics refers to a critique that both highlights the role of women in geopolitics but also contests the top-down scalar imagination of both classical and critical geopolitics. For a review, see Dowler and Sharp 2001 and Hyndman 2001, as well as Caso's essay in this volume.
9 Woon's work is marked by an interest in both state action and a feminist approach. See, for instance, Woon 2013, 2015.
10 In this election the Conservative Party won a majority, allowing David Cameron to remain Prime Minister. Despite being a parliamentary democracy (people only vote for their local MP), the media discourse was widely critiqued as being fixated on the personality of the party leaders, especially Labour Party leader Ed Miliband. Of particular interest was the ability of party leaders to eat bacon sandwiches in view of the public.

References

Agnew, John. 1987. *Place and Politics: The Geographical Mediation of State and Society*. Boston: Allen & Unwin.

Berry, Brian. 1969. 'Review of *International Regions and the International System* (by B. Russett), '*Geographical Review* 59 (3):450–451.

Dalby, Simon. 1990. *Creating the Second Cold War: The Discourse of Politics*. London: Pinter.

Dalby, Simon. 2010. 'Recontextualising Violence, Power and Nature: The Next Twenty Years of Critical Geopolitics?' *Political Geography* 29 (5):280–288.

Dittmer, Jason. 2015. 'Everyday Diplomacy: UKUSA Intelligence Cooperation and Geopolitical Assemblages'. *Annals of the Association of American Geographers* 105 (3):604–619.

Dittmer, Jason. 2016. 'Theorizing a More-than-Human Diplomacy: Assembling the British Foreign Office, 1839–1874'. *The Hague Journal of Diplomacy* 11 (1):78–104.

Dittmer, Jason, and Nicholas Gray. 2010. 'Popular Geopolitics 2.0: Towards Methodologies of the Everyday'. *Geography Compass* 4 (11):1664–1677.

Dodds, Klaus-John, and James Sidaway. 1994. 'Locating Critical Geopolitics'. *Environment and Planning D: Society and Space* 12 (5):515–524.

Dodds, Klaus. 1996. 'The 1982 Falklands War and a Critical Geopolitical Eye: Steve Bell and the If... Cartoons'. *Political Geography* 15 (6):571–592.

Dodds, Klaus. 2001. 'Political Geography III: Critical Geopolitics after Ten Years'. *Progress in Human Geography* 25 (3):469–484.

Dodds, Klaus. 2002. *Pink Ice: Britain and the South Atlantic Empire*. London: I.B. Taurus.

Dodds, Klaus. 2003. 'License to Stereotype: Popular Geopolitics, James Bond and the Spectre of Balkanism'. *Geopolitics* 8 (2):125–156.

Dodds, Klaus. 2005. 'Screening Geopolitics: James Bond and the Early Cold War Films (1962–1967)'. *Geopolitics* 10 (2):266–289.

Dowler, Lorraine, and Joanne P. Sharp. 2001. 'A Feminist Geopolitics?' *Space & Polity* 5 (3):165–176.

Hyndman, Jennifer. 2001. 'Towards a Feminist Geopolitics'. *The Canadian Geographer* 45 (2):210–222.

Jones, Laura, and Daniel Sage. 2010. 'New Directions in Critical Geopolitics: An Introduction'. *GeoJournal* 75 (4):315–325.

Lorimer, Hayden. 2005. 'Cultural Geography: The Busyness of Being "More-Than-Representational"'. *Progress in Human Geography* 29 (1):83–94.

Müller, Martin. 2008. 'Reconsidering the Concept of Discourse for the Field of Critical Geopolitics: Towards Discourse as Language and Practice'. *Political Geography* 27 (3):322–338.

Ó Tuathail, Gearóid. 1986. 'The Language and Nature of the "New" Geopolitics: The Case of US-El Salvador Relations'. *Political Geography Quarterly* 5 (1):73–85.

Ó Tuathail, Gearóid. 1996. *Critical Geopolitics*. Abingdon, UK: Taylor & Francis.

Ó Tuathail, Gearóid, and John Agnew. 1992. 'Geopolitics and Discourse: Practical Geopolitical Reasoning in American Foreign Policy'. *Political Geography* 11 (2):190–204.

Said, Edward. 1978. *Orientalism*. New York: Vintage Books.

Sharp, Joanne P. 1993. 'Publishing American Identity: Popular Geopolitics, Myth and *The Reader's Digest*'. *Political Geography* 12 (6):491–503.

Sharp, Joanne P. 1996. 'Hegemony, Popular Culture and Geopolitics: *The Reader's Digest* and the Construction of Danger'. *Political Geography* 15 (6–7):557–570.

Sharp, Joanne P. 2000. *Condensing the Cold War: Reader's Digest and American Identity*. Minneapolis: University of Minnesota Press.

Sharp, Joanne P. 2011a. 'A Subaltern Critical Geopolitics of the War on Terror: Postcolonial Security in Tanzania'. *Geoforum* 42 (3):297–305.

Sharp, Joanne P. 2011b. 'Subaltern Geopolitics: Introduction'. *Geoforum* 42 (3):271–273.

Thrift, Nigel. 2000. 'Performing Cultures in the New Economy'. *Annals of the Association of American Geographers* 90 (4):674–692.

Woon, Chih-Yuan. 2013. 'Precarious Geopolitics and the Possibilities of Nonviolence'. *Progress in Human Geography* 38 (5):654–670.

Woon, Chih-Yuan. 2015. '"Peopling" Geographies of Peace: The Role of the Military in Peacebuilding in the Philippines'. *Transactions of the Institute of British Geographers* 40 (1):14–27.

2 Popular geopolitics and popular culture in world politics

Pasts, presents, futures

Kyle Grayson

Introduction

There is a long-standing interest shared in political geography and international relations about how our ideas and understandings of the world are formed. Moreover, it has been argued that our imaginations about politics—what is politically possible, what is politically legitimate, and what is politically unpalatable—arise from dynamics that are located beyond sites of formal political activity like legislatures, negotiating tables, or, in some cases, the streets. In expanding notions of where politics takes place and what can count as political, over the past three decades political geography and international relations have begun to recognise the role played by popular culture in political socialisation, hegemony, and contestation. Of particular note in this regard has been the identification of popular culture as an important aspect of world politics, not because it represents '...a superstructure that reflects a political base', but because 'popular culture *makes* world politics what it *is*' (Grayson, Davies, and Philpott 2009, 157).

The recognition of the co-constitution of popular culture and politics has thus led to the development of two related fields of inquiry: popular geopolitics in political geography and popular culture and world politics (PCWP) in international relations (IR). In this chapter, I identify and assess the intersections between popular geopolitics and PCWP. I argue that historically both have been concerned with demonstrating how cultural artefacts are productive of discursive formations and (geo)political imaginations that shape world politics and IR. I demonstrate that this research has traditionally encompassed three forms. The first has been to focus on the inter-textualities linking formal, practical, and popular geopolitical discourses. Within this body of research, the aim has been to show how particular representations, ideas, and forms of knowledge become politically embedded as common sense. The second has been to identify how cultural artefacts are an allegorical resource that can be mobilised to make the geopolitical sensible (or at least intelligible) to a wide audience. The third, and less prominent, has been to illustrate how popular culture itself is mediated through practical geopolitics such that dynamics in world politics shape the production, circulation, and consumption of cultural artefacts. While all three strands have proved insightful, a friendly critique is then provided. The critique highlights the ocular-centrism, focus on narrative, and neglect of reception that can be found in much of this literature. I suggest that being attentive to the interdisciplinary aspects of popular geopolitics and PCWP can help to address or reframe these short-comings.

I then turn to more recent developments in popular geopolitics and PCWP that are extending the interdisciplinary reach of the field in exciting new directions. The first is the aesthetic turn in IR and political geography, which extends from approaches that advocate a conceptualisation of political theory which borrows heavily from art criticism, to those that wish to move beyond textual representation to how bodily sensations themselves are constitutive of the political. The chapter argues that a foregrounding of the politics of aesthetics, or the aesthetics of politics, provides a means of engaging with other senses that are also central to the construction of geographical imaginations in popular geopolitics. It also offers conceptual framings that position cultural artefacts as theories and practices of geopolitics in their own right, rather than reflections or representations of a separate 'real' world of (geo)politics. The point is that popular culture *does* politics by providing narratives, images, sounds, smells, tastes, and feelings that *produce* (geo)politics, while (geo)politics itself can tell us much about how popular culture is made, circulated, and consumed. The second is recent work on reception that uses novel techniques and emerging sites of audience engagement to provide better analytic purchase on how artefacts are read and received. The chapter concludes by suggesting that while the direct connections between popular geopolitics, PCWP, and concrete geopolitical behaviours will always remain open to contestation, a sensitivity to embodiment and the importance of audience suggests that a turn to assemblages may provide new insights into popular geopolitics.

Popular geopolitics: the awakening

Beginnings can often be messy things. At best, there are points of rupture, points at which what once was unquestioned no longer convinces. The emergence of critical geopolitics in the late 1980s and the concurrent rise of the so-called 'third debate' between positivist and post-positivist approaches in the cognate field of international relations mark such ruptures. Uniting these movements was a deep suspicion of claims to universalism, objectivity, and explanatory power being made by analysts of world politics as they engaged in scholarship that was argued to be limited, unreflexive, overly mechanistic, and politically reactionary (e.g., see the critique provided by Ashley 1984). By drawing on concepts, methodologies, and fields of problematisations from history, literary theory, cultural studies, and sociology, both critical geopolitics and critical international relations challenged the 'sight, sites, and cite(s)' through which the geopolitical becomes instantiated (Ó Tuathail 1996, 43 and 71). With strong leanings towards post-structuralism, this early work was heavily influenced by Foucauldian discourse analysis and Derridean deconstruction. Perhaps more important for the development of popular geopolitics, Edward Said's (1978) *Orientalism,* with its focus on the role of imaginative geographies in the textual and material practices that make the world, provided a conceptual apparatus to begin to understand how ideas about the world were produced, circulated, and absorbed as common sense. More generally, texts, and language were no longer to be viewed as referential markers, but rather served as differential dynamics through which political possibilities were forged. The result, to paraphrase James Der Derian and Michael J. Shapiro in their Preface to *International/Intertextual Relations*, was to call into question the Enlightenment foundations of geopolitics and to 'pry into and at the death grip of a candle power doctrine on nuclear power times' (1989, ix). But beyond the epistemological and ontological unmasking of geopolitics, it also expanded where geopolitics could be said to be located. From Said, geography transformed from being conceptualised as a set of material realities that regulate political action to sets of aesthetic, ideational, and discursive components used to make claims about how geography shapes political possibilities. As he argued:

> Just as none of us is outside or beyond geography, none of us is completely free from the struggle over geography. That struggle is complex and interesting because it is not only about soldiers and cannons, but also about ideas, about forms, about images, and imaginings.
>
> (Said 1994, 6)

The geopolitical, thus, was a differential discourse that mobilised spatial logics and imaginaries for processes of identity construction, while presenting itself as merely reporting on the final product of these processes. As Said remarked, '...there is no doubt that imaginative geography and history help the mind to intensify its own sense of itself by dramatising the distance and difference between what is close to it and what is far away' (Said 1978, 55). Moreover, within these processes, Said also drew attention to the relations of power that ultimately shaped how identities were produced, and by whom, with the statement that 'the imaginative geography of the "our land-barbarian land" variety does not require that the barbarians acknowledge the distinction' (Said 1978, 54). Thus, world politics and geopolitical agency could, in part, be conceived of as the imposition of 'our' imaginative geographies onto 'others'.

Political geography and strands of critical international relations thus began to take the productive power of discourse seriously (see, for instance, Ashley 1984; Walker 1986; Dalby 1988; Weber 1990; Doty 1993). Initially, while examining its role in enabling the practices of the Cold War, Gearoid O'Tuathail and John Agnew (1992) made the distinction amongst two forms of geopolitical discourse: formal (i.e. academic) and practical (i.e. policy-making). While the distinction inside the classificatory scheme could be porous, the focus was generally on elites and the transmission of geopolitical power-knowledge, as well as the entrenchment of its corresponding regime of truth. Transmission and entrenchment were primarily conceptualised as top down processes orchestrated by the 'greats' of world politics: presidents, prime ministers, public intellectuals, and practitioners. In response to the limitations identified with this approach, a third category was soon added: popular geopolitics.

Jo Sharp's (1992, 1993) initial work on the role of the *Reader's Digest* in producing the geographical imaginations in the American public, which fuelled the Cold War and contributed to an eschatological politics of identity, would prove highly influential to expanding where geopolitical discourses could be located. It also foregrounded the importance of the everyday to geopolitical practice. More specifically, Sharp, in mobilising a Gramscian conception of hegemony, drew attention to the following elements in the construction of geographical imaginations. First, it was important to expand the sites of geopolitical discourse analysis, because 'popular sources of information can enrich… by providing the context within which elite geopolitical texts are received but also in which they are produced' (Sharp 1993, 491). Second, Sharp identified a core aporia at the heart of geopolitical discourse. Despite its strong desire to do so, she noted that 'geography cannot… provide a mimetic presentation of that which it seeks to describe' (Sharp 1993, 493). In bringing this aporia to the fore, Sharp not only revealed the inevitable space between representation and the real in geopolitical discourses, but also that commonly accessible forms of mediation are necessary to convey and establish geographical imaginations by collapsing this space (e.g., maps or popular magazine articles).

Popular geopolitics thus should be seen as contributing to an 'inter-textual frame of reference' that contributes to the production of mythologies that present political outcomes as 'dehistoricised, natural, and eternal, facts' (Said qtd. in Sharp 1993, 494). Sharp's conceptualisation of popular geopolitics thus supplemented and refined O'Tuathail and Agnew's (1992) identification of policy elites as 'bricoleurs'. Her point was that, rather than engaging in top down transmission, 'geopoliticians have to draw upon discourses already granted hegemonic social acceptance. These discourses are reproduced within culture… if geopolitics were to be consistently created independently of the negotiated reality of its readership, it would face an insurmountable crisis of representation' (Sharp 1993, 493). At the same time, while popular geopolitics might reproduce understandings that were of political benefit to specific constituencies, this did not mean reducing theses understandings to the outcomes of a knowing manipulative intent. Practitioners, pundits, and commenters might be aware of the terms of discourse they were using, and the meaning given to them within the discursive field, but they too were shaped by the rules of the discourse itself, and the ways in which its language also shaped their thinking. Thus, Sharp argued that popular geopolitics should be conceived of as 'a historically constituted way of understanding the world… [that provides] narrative closure, in the form of reference to commonly accepted truisms'.

Similarly, critical international relations began its own cultural turn, 'Similarly, critical international relations began its own cultural turn, as exemplified by David Campbell's *Writing Security* (1992), James Der Derian's *Anti-Diplomacy* (1997), Michael J. Shaprio's *Violent Cartographies* (1998), and Jutta Weldes et al.'s *Cultures of Insecurity* (1999). Within this body of work, one can see similar processes of revealing how discourse is central to the construction of identities, geographies of evil, and common-sense understandings of what is politically possible within the 'international'. Thus, like early work on geopolitical discourses, this work stressed the importance of inter-textuality, performativity, and expanding conceptualisations of where and how world politics should be understood and the sites through which its relations of power circulate. Moreover, it established the channels for research streams that were to follow.

Similar streams/converging currents?

Across both political geography and critical international relations, the volume of work examining popular geopolitics and PCWP began to increase, slowly over the course of the 1990s and early 2000s, with significant growth from 2005 onwards. While sharing similar concerns over the reproduction of common sense and the instantiation of power-relations central to the production of geopolitical imaginations at the heart of hegemony, identity, violence, and security, each field has produced a slightly different topography in approaches taken to the incorporation of popular cultures (Table 2.1).

One can clearly identify (at least) three streams of work in popular geopolitics. The first is research that focussed on the role of representations and how these are both produced by, and productive of, geopolitics. Artefacts subject to examination have mainly been visual media, with a particular emphasis on films, reflecting that film criticism has a scholarly pedigree, going back to the founding of the Moscow Film School in 1919. And with film there was a considerable amount of attention initially paid to war and espionage genres, with this later also including superhero films—understandably so, given their commercial popularity as well as the degree to which their themes and focal points transpose themselves onto geopolitical practices (e.g. Dodds 1996, 2003, 2005, 2008; Black 2004; Dittmer 2005; Crampton and Power 2005; Ó Tuathail 2005; Dalby 2008). Similarly, other visual mediums like photography, newspapers, art, novels, comics, television series, and cartooning have been subject to investigation (Kirsch 1997; Kolossov 2003; Robison 2004; Falah, Flint, and Mamadouh 2006; Hughes 2007; Dodds 2007; Manzo 2012; Williams 2014; Saunders 2012; Dunnett 2009; Holland 2012; Mawdsley 2008; Glynn and Cupples 2015; Davies 2010). Interestingly, cultural mediums such as music (e.g. Gibson 1998; Dunbar-Hall and Gibson 2000; Dell'Agnese 2015), videogames, sport (e.g. Philpott 2016), toys (e.g. MacDonald 2008), advertising (e.g. Rech 2014), humour (e.g. Purcell, Scott Brown, and Gokmen 2010; Dodds and Kirby 2013; Thorogood 2016), and

Table 2.1 Strands of research in popular geopolitics

	Representation	*Mobilisation*	*Object centred*
Key question	How are representations productive of geopolitics?	How does popular culture enable geopolitical mobilisation?	How are popular objects constitutive of geopolitics?
Focal points	Inter-textuality	Production, circulation, and consumption of popular geopolitical sentiments	Materiality and the geopolitical

food (e.g. Evered and Evered 2016) have not featured as prominently, though videogames have more recently received increasing attention (e.g. Power 2007; Robinson 2012, 2015; Young 2015; Salter 2015; Ciută 2015). Where other mediums have been brought in such as music, the focus has been on that which can easily be interpreted as text, such as lyrics, rather than the rhythmic or notional composition. Regardless of the medium, a shared purpose held across this work is to show the inter-textual linkages amongst formal, practical, and popular geopolitical discourses in spoken, written, and visual languages. More recently, an exciting new direction has been to begin to examine how audiences themselves interpret and identify the inter-textual linkages connecting fields of representation—a development that will be returned to below.

The second stream has been work that has focussed on the popular as a site of (geo)political mobilisation. Echoing critical geopolitics' initial concern with how discourse influences possibilities available for action, this stream engaged with mediums, modes of expression, and scalar relationships that produce, circulate, and shape geopolitical practices (e.g. Dahlman and Brunn 2003; Mamadouh 2003; Semati 2012; Pinkerton and Benwell 2014). From the roles of radio and newspapers (written in the vernacular) to the provision of venues for the popular articulation of geopolitical preferences and understandings (e.g. Pinkerton and Dodds 2009; Sharp 2011) to aesthetics and affect (Debrix 2008; Ingram 2011; Shaw and Sharp 2013), to the ways in which conspiracy theories (e.g. Jones 2012) cultivate both a particular geopolitical view as well as channels for their wider distribution, these competing information streams demonstrate how technologies, mediums, and expressive modes embody geopolitics. As in media studies, this stream of research has challenged traditional understandings of audiences as passive spectators. It has also opened up audiences to more fine-grained analytic delineations, through data capture opportunities made possible by new communications technologies and platforms (e.g. Dodds 2006; Dittmer and Gray 2010).

The third stream has examined the ways in which objects, be it those shared in common, widely held, or popularly consumed, are constitutive of geopolitics and its imaginaries (Salter 2015, 2016). More importantly, drawing upon the literatures in new materialism and object-oriented ontologies, contributions to this literature have stressed the importance of material objects, as agential assemblages or actants, to the shaping of geopolitical practice (Meehan, Shaw, and Marston 2013). Thus, homes (Brickell 2012), food (Law 2001), mobile phones (Saunders 2009), digital technologies (Deibert 1997), and even postage stamps (Raento 2006) have been identified as central to the reproduction of geographical imaginations that designate what Dalby (1988) referred to as 'our' space from 'their' space and the constitution of identities that are used to make these designations appear natural. Thus, the purpose of this work has been to show how common (cultural) objects are themselves constitutive of the geopolitical and the imaginaries underpinning it through their materialities. Literature in this stream has not only brought in analytic insights and methodological innovation from broader work on material culture, but has also contributed to a rethinking of what elements constitute the 'popular' agency of popular geopolitics and where the popular is located by taking the role of objects seriously (Table 2.2).

Although critical international relations has traversed many of these same currents, a more sectarian, methodologically conservative, and openly hostile disciplinary space has created some different tributaries (and potentially, at least one stagnant pool) in the incorporation of popular culture. The first strand in PCWP has been research that has sought to use cultural artefacts as simple allegorical resources to convey (or teach) what are often assumed to be the more complex realities of international relations. In other words, the artefact (as a familiar object) becomes a way of explaining something less familiar that is taken unquestioningly to be a part of world politics or the discipline of international relations. On the one hand, the attraction of this view of PCWP has been influenced by the ways in which practitioners have utilised references from popular culture, from Reagan's designation of the Soviet Union as the 'Evil Empire' to China and Japan adopting the terms of the Harry Potter universe to call one another 'Voldemort' and label cultural sites as 'horcruxes' (New York Times 2014). A focus on the allegorical has also been an approach used to great effect pedagogically, for example in multiple editions of Cynthia Weber's (2013) *International Relations Theory: A Critical Introduction*. In this book, Weber, drawing upon the work of Roland Barthes, uses film as a way of revealing that which is taken to be true as an article of faith, or what is able to 'go without saying' at the heart of international relations theory, which then makes these theories appear as plausible explanations for agent behaviour in world politics (Weber 2013, 5). Weber's approach is considered, drawing attention both to the diverse cultural sites through which the logics of international relations are (re)produced (very few of the films she uses would be classified as focusing on 'formal' politics) and the means by which we can reveal the politics concealed by allegorical transposition. Thus, the key is not just what is revealed by the allegorical properties of pop culture artefacts or how 'accurately' a metaphor like 'Voldemort'—or a theory like political realism—explains dynamics in world politics. Rather, by drawing upon work in linguistics (e.g. Lakoff and Johnson 2008), our attention should be directed to what needs to be taken for granted for the allegory/metaphor to work and what other factors are lost in the process of reducing issues or actions to the properties of the metaphor. Unfortunately, reflecting an unwillingness to critically

Table 2.2 Strands of research in PCWP

	Allegorical	*Productive*	*Inverse*
Key question	How does popular culture reflect world politics?	How does popular culture constitute world politics?	What can world politics tell us about popular culture?
Focal points	Mimesis and realism of depictions	Conditions of possibility revealed through popular culture	Forces that shape the production, circulation, and consumption of popular culture

engage with core assumptions defining a provincial, yet predominant, view of the discipline and its *raison d'etre*, a great deal of work in this area does little more than reproduce hegemonic understandings of international relations as a discipline and the versions of world politics it perceives as relevant (e.g. Ruane and James 2008; Drezner 2011). Thus, rather than illuminating any of the issues raised in popular geopolitics or PCWP about the creation of imaginations, processes of identity construction, or delineation of space, much of this literature serves as empirical examples of these processes in motion within the discursive formation of international relations.

A second strand in PCWP has taken the form of work undertaken in popular geopolitics that stresses the world-creating processes of discourses (e.g Weldes 2003). Thus, work in this area has sought to explore the inter-textual linkages between formal (i.e. academic), practical (policy), and popular discourses in the production of common sense understandings of world politics. An objective of such approaches is to determine how popular culture is a part of world politics with the assumption that both are mutually constitutive (Grayson, Davies, and Philpott 2009). The difference between this approach and allegorical approaches is that the focus on inter-textuality does not assume that there is some *a priori* world to which popular culture holds a mirror. Rather, it takes the view that popular culture itself is composed of texts that make us and our worlds, in much the same way that previous work produced by the first cultural turn argued that IR theory is both an everyday and world-creating practice (e.g. Enloe 1989; Walker 1991). Popular culture is thus a means through which subjectivities are produced, identities are constructed, knowledge is shared, threats are identified, norms are established, and discipline is imposed, such that particular courses of action become not only possible to undertake by particular subjects, but are also recognised by publics as having political legitimacy (e.g. Shepherd 2013; Åhäll 2015). To put this claim into the terminology of Jacques Rancière (2004), popular culture is a site where political consensus is established and where the boundaries of what can be legitimately recognised are policed. Popular culture thus becomes a conduit for: 1) introducing particular subjectivities, ideas, and problem framings; 2) establishing these as the *entire* range of options available for consideration; 3) leaving alternatives unarticulated through popular mediums, whether by practices of commission (i.e. actively seeking to prevent their expression) or omission (i.e. not even being aware of their presence). In these ways, beyond the directly representational, popular culture produces our political worlds by establishing the horizons within which these are perceived and interpreted.

The third pillar has sought to reverse the line of inquiry of the first two from 'what can popular culture tell us about world politics?' to 'what can world politics tell us about popular culture?' Often such work engages with notions of intellectual property rights, piracy, commodity chains, media ownership, cultural production, and capitalism itself as well as new organisational forms that have emerged such as Der Derian's (2001) 'Military Industrial Media Entertainment Complex' (Lisle and Pepper 2005; Johns 2006; Kaklamanidou 2013; Griffin 2015; Pusca 2016). The contribution of this work is how it problematises what popular culture itself is, what might count as popular, how the popular is generated, and the material processes through which popular culture is produced, circulated, and consumed. Yet, this is surprisingly under-represented in relation to other strands noted above, and as Ciută (2015) argues, answering these key questions requires more sustained engagement with media and cultural studies.

Given the identification of these strands of research and their strengths, in what ways might they be improved? In particular, how might popular geopolitics and PCWP draw more fruitfully from their interdisciplinary composition and work being undertaken in other fields? This is the focus of the next section in this chapter.

Engaging with the interdisciplinary: a friendly critique

While the research strands noted above have provided important insights into the production, circulation, and consumption of geographical imaginations, common sense, and their associated relations of power through forms of discursive representation and materiality, they are not without their limitations. In part, these limitations stem from an uneven harnessing of their inherent interdisciplinary with certain components being deployed more than others. Thus, if we look at research in popular geopolitics that has been undertaken thus far, there are three general critiques that can be applied to the field: the presence of ocular-centrism, a primary focus on narrative/text, and a neglect of reception. These will be covered in turn.

Within popular geopolitics and PCWP, when it comes to the selection of artefacts to be analysed, there has been a manifest ocular-centrism in terms of focusing on practices or objects for whom our primary sense of perception is our sight: art, film, television, literature, and even colour (e.g. Guillaume, Andersen, and Vuori 2015; Andersen, Vuori, and Guillaume 2015), rather than things we hear, taste, or otherwise 'feel'. While visual studies (e.g. Foster 1988; Jay 1988; Mitchell 2002) has argued that Western cultures in the modern era are primarily visual cultures, the absence of the other senses is a puzzling development, given the role that taste, or sound might play in the ordering of space. Moreover, even in the visual analysis of artefacts, there is often a neglect of placing both the objects and subjects of analysis and the ways in which they are being analysed within broader scopic regimes that render these object/subjects visible in certain ways as well as shaping our own means for interpreting them. According to Allan Feldman, scopic regimes:

> prescribe modes of seeing and object visibility and… proscribe or render untenable other modes and objects of perception. A scopic regime is an ensemble of practices and discourses that establish truth claims, typicality, and credibility of visual acts and objects and politically correct modes of seeing.
>
> (1997, 30)

An awareness of how scopic regimes condition our visual registers is important because just like any other regime of truth, scopic regimes (re)produce relations of power that are central to making sense of our worlds. They condition not only what comes to be understood as common sense, but they also condition a sense (i.e. vision), which is to be held in common. Both work to structure the conditions of possibility for what can be recognised and what must escape notice in our accounts of worlds. Thus, while many detailed, rich, and enlightening readings of photos, films, cartoons, and television shows have been undertaken, the conditions of possibility for alternative readings have not been examined. Some notable exceptions here would include Campbell and Power (2010) on the scopic regime governing the visual geopolitics of Africa and Fraser McDonald's (2006) exploration of the visual spectacle of weapons of mass destruction and their framing within public mediascapes.

The second shortcoming in the literature has been an analytic preference to treat artefacts as transposed narratives (i.e. textual data like the written word) regardless of the senses through which an artefact is being received. While this reflects a post-structural impulse to interpret everything as text, this impulse does not need to treat every form of expression as the same kind of (narrative) text. For example, music, when examined in popular geopolitics and PCWP has far too often been reduced to an overwhelming emphasis on lyrics rather than the musical elements themselves (e.g. Boulton 2008; Street 2013), though Marianne Franklin's (2005) edited collection is a notable exception. While lyrics are certainly important, so too is the musical score of which lyrics are but one component part. Leaving musical composition aside potentially misses key dynamics in the production and reproduction of key elements central to geographical imaginations. In this regard, even a discursively weighty form like hip hop can provide many illustrations of the potential importance of musical composition in addition to lyrics in the production of geographical imaginations. For example, does the sampling of the melody from Egyptian artist Hossam Ramzy's (1957) 'Khusara Khusara' by Jay Z (2000) in his chart topping 'Big Pimpin' contribute to longer standing Orientalist understandings of the Middle East that circulated around the exotic delights of the harem? Conversely, does Gangstarr's (1991) sampling of Maceo and Macks (1974) 'Party', for their song 'Whose Gonna Take the Weight?' about their own Islamic faith, prove disruptive of these very same imagined geographies, by recoding where the religion is spatially located and the subject positions of those whose reside within those spaces?

The third, and perhaps reflecting the post-positivist influences on international relations (political geography has been much freer in this regard), there has been a neglect of audience and reception. Too often, very sophisticated readings of artefacts are offered but without any indication of how audiences may actually be interpreting them. Are understandings held by audiences hegemonic, negotiated, or counter-hegemonic? How does this differ amongst audiences, artefacts, and contexts? What kinds of temporalities do these forms of reading have and in relation to what kinds of events, discourses, or social formations? Emerging work in popular geopolitics has been particularly pioneering in its approaches to film audiences, offering methods for analysing audience reactions as well as how new communicative mediums—such as chatrooms in the Internet Movie Database

(IMDB)—might allow us to identify and study audience reception in less costly ways than the large-N surveys reminiscent of those in marketing and communication studies. (Dittmer 2008; Dittmer and Dodds 2008; Dittmer and Gray 2010; Woon 2014).

Given the current state of the art, what next then for popular geopolitics and PCWP? How might promising areas of research be cultivated with a renewed interdisciplinary impetus? In the next section, the ways in which the 'aesthetic turn' has facilitated new directions for research is explored.

Incorporating the aesthetic turn

The concerns raised above about these tendencies in popular geopolitics and their possible limitations are not new. Over the past two decades, one reaction to this has been to engage in forms of analysis that seek to investigate the (geo)politics of representation differently (i.e. to go beyond a concern with inter-textuality) as well as to move beyond a focus on representation full stop. This reflects in part another turn, an aesthetic turn, which is taking place in both political geography and international relations more widely. Engaging with representation differently started with Bleiker's (2001) *Millennium* article, where he outlined a form of aesthetic analysis, that is one where the focus is on the politics generated by the inevitable space that exists between signifier and signified; this has since developed through the work of people like Anca Pusca (2016), Cerwyn Moore and Laura Shepherd (2010), and Michael Shapiro (2010, 2013) in international relations and Alan Ingram (2011, 2012) and Alison Williams (2014) in political geography. This space, the ways in which it is recognised (or ignored), attempts to collapse it, and its impact on individual bodies, and bodies politic, has thus become a focal point for examination. This has led to two major developments. The first has been to open the door to engaging with other senses central to the development of geographical imaginations in popular culture, in part by borrowing from other fields (e.g. Cusick 2006; Hayes-Conroy and Hayes-Conroy 2008). And a part of this, to paraphrase W.T.J. Mitchell's (2002, 170) dictum on visuality, is a concern not just with how the geopolitical conditions our senses, but how our senses also condition the geopolitical. In direct reference to the etymological origins of aesthetics, forms of embodiment are thus coming to the forefront in popular geopolitics and PCWP (Table 2.3).

The second, and for the moment the more predominant, has been to position cultural artefacts as theories and practices of politics in their own right, rather than reflections or representations of a separate 'real' world of geopolitics. And this has been a move that has led to the development of new methodological directions, such as Shapiro's (2010) 'aesthetic subject'. The aesthetic subject is not an innate form of subjectivity, but rather is an approach to reading artefacts, such that the search for underlying psychological motivations for character behaviour are abandoned in lieu of analysing experiential terrains and modes of mobility for characters across them. By examining the conditions of possibility for movement in accounts of fictionalised worlds, one can also reveal relations of power

Table 2.3 Research strands in the aesthetic turn

	Representation	*Productive*	*Embodied*
Key question	What (geo)political work is done by the gap between representation and the real?	How do cultural artefacts act as (geo)political theories and practices?	What kind of (geo)politics emerges from the embodied effects of cultural artefacts
Focal points	Texts, images, and theories as representations	Aesthetic subjects and relations of power	Feelings and sensations experienced by individual bodies and bodies politic

absent from predominant explanatory frameworks within the formal and practical geopolitical literature. For example, Shapiro (2010, 11–16), by concentrating on the bodily comportment of the character Easy Rawlins in the film *Devil in a Blue Dress*, shows how postures, gaits, and body positionings demonstrate a politics of race in the U.S. and the ways in which its corresponding spatial divisions are navigated by those subject to systemic forms of discrimination. Importantly, this puts forward an embodied geopolitics of structural racism that is missed by behaviouralist approaches favoured by urban politics literature.

Third, as aesthetic methodologies continue to be introduced and refined, they also provide opportunities to return to questions of audience and reception noted above. For example, with an explicit emphasis on the phenomenology of gaming, the work of James Ash (Ash 2009, 2010) has sought to reveal the affective dimension of videogames by observing how gamers actually play them and how gaming makes them feel (see also Bos' essay in this volume). In doing so, Ash has been able to uncover the forms of geographical knowledge generated, the ways of seeing that structure gameplay, and how reward systems contribute towards the generation of physiological feedback loops that encourage gameplay. Such work not only provides the potential to better understand how artefacts work (geo)politically, but also the geopolitics of how particular artefacts are received more widely by identifying contradictions between what artefacts do to particular audiences and how they do things to audiences—whether they be gamers, film viewers, or restaurant patrons.

Popular geopolitics as assemblage

What has become clear is that popular geopolitics as a field is no longer solely focussed on tracing inter-textual linkages with formal and practical geopolitical discourses. Rather, research is demonstrating not just what representations are present in artefacts, but also how representational and non-representational elements are then embodied by audiences, and how forms of embodiment are affected by physiological capacities of audiences and the material means through which they are delivered. The concern is thus with machinic assemblages, a collection of what

Levi Bryant (2009) has called 'heterogeneous elements or objects that enter into relations with one another', in which a complicated set of factors operating in linear and non-linear ways produce effects. Machinic assemblages can be anything from the cultural, material, ideational, discursive, and affective elements enabling practices of targeted killing (Grayson 2016) to forms of artificial life and intelligence that are developing in conjunction with cultural redefinitions of what it means to be human and advances in neuro-science. It is here where notions of subjectivity, the somatic, the semiotic, the cultural, and environmental are being problematised and critically engaged that opportunities arise for new insights into popular geopolitics and the roles of geographical imaginations. The work of John Protevi (2008, 2009, 2010) has been exemplary in raising new sets of questions for popular geopolitics.

Protevi's work has sought to understand the connections between bodies and bodies politic through forms of cultural production and physiological pre-subjective orientations. His concern is with how human perception, the human body, the human mind, and socio-political stimuli assemble to produce particular kinds of reactions in individuals, but also in collectivities. At the heart of the analysis is how the cultural, physiological, and political produce behaviours, whether it be the generation of rage states that lead to violence, or forms of empathy that produce duties of care.

Protevi's (2009: location 46) starting point for analysis is that 'subjectivity can be studied both in its embodied affectivity and in terms of the distribution of affective cognitive traits in a population'. But subjectivities are not pre-determined by nature or the result of atomism. Rather Protevi argues for the importance of body politics to the formation of the subject:

> [the] concept of bodies politic is meant to capture the emergent-that is, the embodied/embedded/extended character of subjectivity, or in other words, the way the production, bypassing, and surpassing of subjectivity is found in the interactions of somatic and social systems.
>
> (2009, loc. 54)

Thus, similar to Shapiro's (2010) rejection of *homo oeconomicus* as the default subject position for (geopolitical) analysis, Protevi argues that:

> Positing an abstract subject neglects the way in which culture is the very process of the construction of bodies politic, so that access to certain cultural resources and to the training necessary to acquire certain forms of affective cognitive capacities—once again, not simply technical training for cognitive capacities in a restricted sense but also the training necessary for acquiring positive and empowering emotional patterns, thresholds, and triggers—is distributed along lines analysable by political categories.
>
> (2009, loc. 504)

The implications are that our subjectivities, their spatial relations, and the catalysts and reactions to them (which may also be embodied) are produced through a complicated assemblage in which the social and somatic interact. Protevi's point is not that flying into a rage and committing heinous acts of violence given a set of stimuli is natural. His argument is that while the capacity for rage may be within certain parameters established through the physiology of the human body, triggers must be perceived and processed to initiate action, and these triggers are often socio-culturally produced. For example, it is often culture that codes the kinds of social actions that ought to trigger fear or anger, as well as determining what are the appropriate means for managing these feelings. And in with the generation of affect modulated by cultural expectations, the somatic and social can generate feedback loops that eventually lead to pre-subjective reactions. Thus, the role of popular geopolitics takes on renewed centrality, not only by producing imaginative geographies that contain a moral grammar about what is to be done, but by establishing a range of conditions (often through process repetitions) that increase the probability of generating (violent) pre-cognitive responses in individual bodies and bodies politic. The importance here is broad, including the potential to change how we conceptualise any ethics of responsibility, if for example, we find that certain stimuli in a given set of social conditions increase the probability of anti-social or violent actions through non-subjective reflexes in an audience. It also raises questions about the circulation of virulent imaginative geographies and who can (or should) be held accountable for the affective cultures that produce them (e.g. Debrix 2008). It also speaks to the importance of moving beyond narrative and exploring how visual presentation (e.g., quick cuts), sound, smell, touch, and taste contribute to the generation of sensory feedback loops that create conditions of possibility for subject positions, pre-subjective reactions, and bodies politic.

Conclusion

In this chapter, I have made an original contribution to the study of popular geopolitics through the identification and assessment of the intersections between popular geopolitics and PCWP. Three streams of research have been classified in each and common lines of critique across them have been forwarded. I have argued that the aesthetic turn, a renewed interest in audiences, and conceptualising imagined geographies, subjectivity, and bodies politic as a machinic assemblage provide exciting new directions for inquiry for popular geopolitics. In pursuing these new areas of research, it will be important to recognise that there is no need to reinvent the wheel with regards to modes of inquiry, concepts, and methods. Many of the cognate fields for popular geopolitics in the humanities and social science have already developed the means for analysing music, moving images, taste and touch. Similarly, there are rich veins to mine conceptually from sociology, media studies, cultural studies, and literary studies. Moreover, research in the sciences and the philosophy of the mind has begun to suggest how we might conceptualise the interactive dynamics formed by physiological and neurological elements in relation to outside stimuli as well as means by which to capture these interactions and their geopolitical implications. Thus, popular geopolitics is entering an exciting new period where longer standing concerns with the representational and inter-textual aspects of geographical imaginations are being deepened while emerging concerns with embodiment, subjectivity, and audience are being opened up to new forms of investigation.

At the same time, it will be important to remain attentive to the geopolitical dynamics that make popular culture what it is by constituting the worlds in which we produce, distribute, and consume artefacts. For example, how do practices of spatial governance (from borders to new forms of territoriality) shape popular geopolitics? In doing so, there is also a need to think carefully about the 'popular' of popular geopolitics: what does this term mean in a time of increasingly segmented audiences, a contracting public commons through the enclosure of intellectual property, and the framing of political disagreement as 'culture wars' at the local, national, and global level? Thus, there is still much for popular geopolitics to do and develop by embracing its strengths, including an interdisciplinary horizon, and continuing to pursue answers to questions that are located across the 'popular culture-world politics continuum' (Grayson, Davies, and Philpott 2009).

References

Åhäll, Linda. 2015. *Sexing War/Policing Gender: Motherhood, Myth and Women's Political Violence* London: Routledge.

Andersen, Rune, Juha Vuori, and Xavier Guillaume. 2015. 'Chromatology of Security: Introducing Colours to Visual Security Studies'. *Security Dialogue* 46 (5):440–457.

Ash, James. 2009. 'Emerging Spatialities of the Screen: Video Games and the Reconfiguration of Spatial Awareness'. *Environment and Planning A* 41 (9):2105–2124

Ash, James. 2010. 'Architectures of Affect: Anticipating and Manipulating the Event in Processes of Videogame Design and Testing'. *Environment and Planning D: Society and Space* 28 (4):653–671.

Ashley, Richard K. 1984. 'The Poverty of Neorealism'. *International Organization* 38 (2):225–286.

Black, Jeremy. 2004. 'The Geopolitics of James Bond'. *Intelligence & National Security* 19 (2):290–303.

Bleiker, Roland. 2001. 'The Aesthetic Turn in International Political Theory'. *Millennium* 30 (3):509–533.

Boulton, Andrew 2008. 'The Popular Geopolitical Wor(l)ds of: Post-9/11 Country Music'. *Popular Music and Society* 31 (3):373–387.

Brickell, Katherine. 2012. 'Geopolitics of Home'. *Geography Compass* 6 (10):575–588.

Bryant, Levi. R. 2009. 'Deleuze on Assemblages'. *Larval Subjects*, available at https://goo.gl/ihBj4D [last accessed 12 July 2015].

Campbell, David. 1998. *Writing Security: United States Foreign Policy and the Politics of Identity (2nd ed.)*. Minneapolis: University of Minnesota Press.

Campbell, David, and Marcus Power. 2010. 'The Scopic Regime of Africa'. In *Observant States: Geopolitics and Visual Culture*, edited by Fraser MacDonald, Rachel Hughes and Klaus Dodds, 167–197. London and New York: I. B. Tauris.

Ciută, Felix. 2015. 'Call of Duty: Playing Video Games with IR'. *Millennium - Journal of International Studies*: DOI: 0305829815601488.

Crampton, Andrew, and Marcus Power. 2005. 'Reel Geopolitics: Cinemato-graphing Political Space'. *Geopolitics* 10:193–203.

Cusick, Suzanne G. 2006. 'Music as Torture/Music as Weapon'. *TRANS: Revista Transcultural de Musica*, available at http://www.sibetrans.com/trans/a152/music-as-torture-music-as-weapon [last accessed 12 July 2015].

Dahlman, Carl, and Stanley D. Brunn. 2003. 'Reading Geopolitics beyond the State: Organisational Discourse in Response to 11 September'. *Geopolitics* 8 (3):253–280.

Dalby, Simon. 1988. 'Geopolitical Discourse: The Soviet Union as Other'. *Alternatives: Global, Local, Political* 13 (4):415–442.

Dalby, Simon. 2008. 'Warrior Geopolitics: *Gladiator, Black Hawk Down* and *The Kingdom of Heaven*'. *Political Geography* 27:439–455.

Davies, Matt. 2010. '"You Can't Charge Innocent People for Saving Their Lives!" Work in *Buffy the Vampire Slayer*'. *International Political Sociology* 4 (2):178–195.

Debrix, François. 2008. *Tabloid Terror: War, Culture, and Geopolitics*. London and New York: Routledge.

Deibert, Ronald J. 1997. *Parchment, Printing and Hypermedia: Communication in World Order Transformation*. New York: Columbia University Press.

Dell'Agnese, Elena. 2015. '"Welcome to Tijuana": Popular Music on the US-Mexico Border'. *Geopolitics* 20 (1):171–192.

Der Derian, James. 1992. *Antidiplomacy: Spies, Terror, Speed, and War*. Oxford: Blackwell.

Der Derian, James. 2001. *Virtuous War: Mapping the Military-Industrial-Media-Entertainment Network*. Boulder, CO: Westview Press.

Der Derian, James, and Michael J. Shapiro. 1989. *International/Intertextual Relations: Postmodern Readings of World Politics*. Lanham: Lexington Books.

Dittmer, Jason. 2005. 'Captain America's Empire: Reflections on Identity, Popular Culture, and Post-9/11 Geopolitics'. *Annals of the Association of American Geographers* 95 (3):626–643.

Dittmer, Jason. 2008. 'The Geographical Pivot of (the End of) History: Evangelical Geopolitical Imaginations and Audience Interpretation of *Left Behind*'. *Political Geography* 27 (3):280–300.

Dittmer, Jason, and Klaus Dodds. 2008. 'Popular Geopolitics Past and Future: Fandom, Identities and Audiences'. *Geopolitics* 13 (3):437–445.

Dittmer, Jason, and Nicholas Gray. 2010. 'Popular Geopolitics 2.0: Towards Methodologies of the Everyday'. *Geography Compass* 4 (11):1664–1677.

Dodds, Klaus. 1996. 'The 1982 Falklands War and a Critical Geopolitical Eye: Steve Bell and the If… Cartoons'. *Political Geography* 15 (6):571–592.

Dodds, Klaus. 2003. 'License to Stereotype: Popular Geopolitics, James Bond and the Spectre of Balkanism'. *Geopolitics* 8 (2):125–156.

Dodds, Klaus. 2005. 'Screening Geopolitics: James Bond and the Early Cold War Films (1962–1967)'. *Geopolitics* 10 (2):266–289.

Dodds, Klaus. 2006. 'Popular Geopolitics and Audience Dispositions: James Bond and the Internet Movie Database (IMDb)'. *Transactions of the Institute of British Geographers* 31 (2):116–130.

Dodds, Klaus. 2007. 'Steve Bell's Eye: Cartoons, Geopolitics and the Visualization of the 'War on Terror'. *Security Dialogue* 38:157–177.

Dodds, Klaus. 2008. 'Hollywood and the Popular Geopolitics of the War on Terror'. *Third World Quarterly* 29 (8):1621–1637.

Dodds, Klaus, and Philip Kirby. 2013. 'It's Not a Laughing Matter: Critical Geopolitics, Humour and Unlaughter'. *Geopolitics* 18:45–59.

Doty, Roxanne Lynn. 1993. 'Foreign Policy as Social Construction: A Post-Positivist Analysis of US Counterinsurgency Policy in the Philippines'. *International Studies Quarterly* 37 (3):297–320.

Drezner, Daniel W. 2011. *Theories of International Politics and Zombies*. Princeton and Oxford: Princeton University Press.

Dunbar-Hall, Peter, and Chris Gibson. 2000. 'Singing about Nations within Nations: Geopolitics and Identity in Australian Indigenous Rock Music'. *Popular Music and Society* 24 (2):45–74.

Dunnett, Oliver. 2009. 'Identity and Geopolitics in Hergé's *Adventures of Tintin*'. *Social & Cultural Geography* 10 (5):583–599.

Enloe, Cynthia. 1989. *Bananas, Beaches, and Bases: Making Feminist Sense of International Politics*. London: Pandora Press.

Evered, Emine Ö., and Kyle T. Evered. 2016. 'A Geopolitics of Drinking: Debating the Place of Alcohol in Early Republican Turkey'. *Political Geography* 50:48–60.

Falah, Ghazi-Walid, Colin Flint, and Virginie Mamadouh. 2006. 'Just War and Extraterritoriality: The Popular Geopolitics of the United States' War on Iraq as Reflected in Newspapers of the Arab World'. *Annals of the Association of American Geographers* 96 (1):142–164.

Feldman, Allen. 1997. 'Violence and Vision: The Prosthetics and Aesthetics of Terror'. *Public Culture* 10 (1):24–60.

Foster, Hal. 1988. *Vision and Visuality*. Seattle: Bay View Press.

Franklin, Marianne I. 2005. *Resounding International Relations: On Music. Culture, and Politics*. Basingstoke: Palgrave Macmillan.

Gibson, Chris. 1998. '"We Sing Our Home, We Dance Our Land": Indigenous Self-Determination and Contemporary Geopolitics in Australian Popular Music'. *Environment and Planning D: Society and Space* 16 (163–184).

Glynn, Kevin, and Julie Cupples. 2015. 'Negotiating and Queering US Hegemony in TV Drama: Popular Geopolitics and Cultural Studies'. *Gender, Place & Culture: A Journal of Feminist Geography* 22 (2):271–287.

Grayson, Kyle. 2016. *Cultural Politics of Targeted Killing: On Drones, Counter-Insurgency, and Violence*. London and New York: Routledge.

Grayson, Kyle, Matt Davies, and Simon Philpott. 2009. 'Pop Goes IR? Researching the Popular Culture-World Politics Continuum'. *Politics* 29 (3):155–163.

Griffin, Penny. 2015. *Popular Culture, Political Economy and the Death of Feminism: Why Women are in Refrigerators and Other Stories*. London and New York: Routledge.

Guillaume, Xavier, Rune Andersen, and Juha Vuori. 2015. 'Paint It Black: Colours and the Social Meaning of the Battlefield'. *European Journal of International Relations* 22 (1):49–71.

Hayes-Conroy, Allison, and Jessica Hayes-Conroy. 2008. 'Taking Back Taste: Feminism, Food and Visceral Politics'. *Gender, Place and Culture* 15 (5):461–473.

Holland, Edward C. 2012. ''To Think and Imagine and See Differently': Popular Geopolitics, Graphic Narrative, and Joe Sacco's "Chechen War, Chechen Women"'. *Geopolitics* 17 (1):105–129.

Hughes, Rachel. 2007. 'Through the Looking Blast: Geopolitics and Visual Culture'. *Geography Compass* 1 (5):976–994.

Ingram, Alan. 2011. 'Making Geopolitics Otherwise: Artistic Interventions in Global Political Space'. *Geographical Journal* 177 (3):218–222.

Ingram, Alan. 2012. 'Experimental Geopolitics: Wafaa Bilal's Domestic Tension'. *The Geographical Journal* 178 (2):123–133.

Jay, Martin. 1988. 'Scopic Regimes of Modernity'. In *Vision and Visuality*, edited by Hal Foster, 3–23. Seattle: Bay Press.

Johns, Jennifer. 2006. 'Video Games Production Networks: Value Capture, Power Relations and Embeddedness'. *Journal of Economic Geography* 6 (2):151–180.

Jones, Laura. 2012. 'The Commonplace Geopolitics of Conspiracy'. *Geography Compass* 6 (1):44–59.

Kaklamanidou, Betty. 2013. *Genre, Gender and the Effects of Neoliberalism: The New Millennium Hollywood Rom Com*. London: Routledge.

Kirsch, Scott. 1997. 'Watching the Bombs Go Off: Photography, Nuclear Landscapes, and Spectator Democracy'. *Antipode* 29 (3):227–255.

Kolossov, Vladamir. 2003. '"High" and "Low" Geopolitics: Images of Foreign Countries in the Eyes of Russian Citizens'. *Geopolitics* 8 (1):121–148.

Lakoff, George, and Mark Johnson. 2008. *Metaphors We Live By*. Chicago: University of Chicago Press.

Law, Lisa. 2001. 'Home Cooking: Filipino Women and Geographies of the Senses in Hong Kong'. *Cultural Geographies* 8 (3):264–283.

Lisle, Debbie, and Andrew Pepper. 2005. 'The New Face of Global Hollywood: *Black Hawk Down* and the Politics of Meta-sovereignty'. *Cultural Politics* 1 (2):165–192.

MacDonald, Fraser. 2006. 'Geopolitics and 'The Vision Thing': Regarding Britain and America's First Nuclear Missile'. *Transactions of the Institute of British Geographers* 31 (1):53–71.

MacDonald, Fraser. 2008. 'Space and the Atom: On the Popular Geopolitics of Cold War Rocketry'. *Geopolitics* 13:611–634.

Mamadouh, Virginie. 2003. '11 September and Popular Geopolitics: A Study of Websites Run for and by Dutch Moroccans'. *Geopolitics* 8 (3):191–216.

Manzo, Kate. 2012. 'Earthworks: The Geopolitical Visions of Climate Change Cartoons'. *Political Geography* 31 (8):481–494.

Mawdsley, Emma. 2008. 'Fu Manchu versus Dr Livingstone in the Dark Continent? Representing China, Africa and the West in British Broadsheet Newspapers'. *Political Geography* 27 (5):509–529.

Meehan, Katharine, Ian G. R. Shaw, and Sallie A. Marston. 2013. 'Political Geographies of the Object'. *Political Geography* 33:1–10.

Mitchell, W.T.J. 2002. 'Showing Seeing: A Critique of Visual Culture'. *Journal of Visual Culture* 1 (2):165–181.

Moore, Cerwyn, and Laura J. Shepherd. 2010. 'Aesthetics and International Relations: Towards a Global Politics'. *Global Society* 24 (3):299–309.

New York Times. 2014. 'Latest China-Japan Spat: Who is Voldemort?' *New York Times*, available at http://sinosphere.blogs.nytimes.com/2014/01/09/latest-china-japan-spat-whos-voldemort/?_r=1 [last accessed 12 July 2015].

Ó Tuathail, Gearóid. 1996. *Critical Geopolitics*. Abingdon, UK: Taylor & Francis.

Ó Tuathail, Gearóid. 2005. 'The Frustrations of Geopolitics and the Pleasures of War: *Behind Enemy Lines* and American Geopolitical Culture'. *Geopolitics* 356–377.

Ó Tuathail, Gearóid, and John Agnew. 1992. 'Geopolitics and Discourse: Practical Geopolitical Reasoning in American Foreign Policy'. *Political Geography* 11 (2):190–204.

Philpott, Simon. 2016. 'Planet of the Australians: Indigenous Athletes and Australian Football's Sports Diplomacy'. *Third World Quarterly* 38 (4):862–881.

Pinkerton, Alasdair, and Klaus Dodds. 2009. 'Radio Geopolitics: Broadcasting, Listening and the Struggle for Acoustic Spaces'. *Progress in Human Geography* 33 (1):10–27.

Pinkerton, Alasdair, and Matt Benwell. 2014. 'Rethinking Popular Geopolitics in the Falklands/Malvinas Sovereignty Dispute: Creative Diplomacy and Citizen Statecraft'. *Political Geography* 38:12–22.

Power, Marcus. 2007. 'Digitized Virtuosity: Video War Games and Post-9/11 Cyber-Deterrence'. *Security Dialogue* 38 (2):271–288.

Protevi, John. 2008. 'Affect, Agency and Responsibility: The Act of Killing in the Age of Cyborgs'. *Phenomenology and the Cognitive Sciences* 7 (3):405–413.

Protevi, John. 2009. *Political Affect: Connecting the Social and the Somatic*. Minneapolis: University of Minnesota Press.

Protevi, John. 2010. 'Rhythm and Cadence, Frenzy and March: Music and the Geo-Bio-Techno-Affective Assemblages of Ancient Warfare'. *Theory & Event* 13 (3):189–211.

Purcell, Darren, Melissa Scott Brown, and Mahmut Gokmen. 2010. 'Achmed the Dead Terrorist and Humor in Popular Geopolitics'. *Geoforum* 75:373–385.

Pusca, Anca M. 2016. *Post-Communist Aesthetics: Revolutions, Capitalism, Violence*. London: Routledge.

Raento, Pauliina. 2006. 'Communicating Geopolitics through Postage Stamps: The Case of Finland'. *Geopolitics* 11 (4):601–629.

Rancière, Jacques. 2004. *The Politics of Aesthetics*. London and New York: Continuum.

Rech, Matt. F. 2014. 'Be Part of the Story: A Popular Geopolitics of War Comics Aesthetics and Royal Air Force Recruitment'. *Political Geography* 39:36–47.

Robinson, Nick. 2012. 'Videogames, Persuasion and the War on Terror: Escaping or Embedding the Military–Entertainment Complex?' *Political Studies* 60:504–522.

Robinson, Nick. 2015. 'Have You Won the War on Terror? Military Videogames and the State of American Exceptionalism'. *Millennium – Journal of International Studies* 43 (1):450–470.

Robison, Bridget. 2004. 'Putting Bosnia in Its Place: Critical Geopolitics and The Representation of Bosnia in the British Print Media'. *Geopolitics* 9 (2):378–401.

Ruane, Abigail E., and Patrick James. 2008. 'The International Relations of Middle-earth: Learning from *Lord of the Rings*'. *International Studies Perspectives* 9:377–394.

Said, Edward. 1978. *Orientalism*. New York: Vintage Books.

Said, Edward W. 1994. *Culture and Imperialism*. New York: Vintage Books.

Salter, Mark B. 2015. *Making Things International: Circuits and Motion*. Minneapolis: University of Minnesota Press.

Salter, Mark B. 2016. Making Things International 2: Catalysts and Reactions. Minneapolis: University of Minnesota Press.

Saunders, Robert A. 2009. 'Wiring the Second World: The Geopolitics of Information and Communications Technology in Post-Totalitarian Eurasia'. *Digital Icons: Studies in Russian, Eurasian and Central European New Media* 1:1–24.

Saunders, Robert A. 2012. 'Undead Spaces: Fear, Globalisation, and the Popular Geopolitics of Zombiism'. *Geopolitics* 17 (1):80–104.

Semati, Mehdi. 2012. 'The Geopolitics of *Parazit*, the Iranian Televisual Sphere, and the Global Infrastructure of Political Humor'. *Popular Communication* 10 (1–2): 119–130.

Shapiro, Michael J. 1997. *Violent Cartographies: Mapping Cultures of War*. Minneapolis: University of Minnesota Press.

Shapiro, Michael J. 2010. The Time of the City: Politics, Philosophy, and Genre. New York: Routledge.

Shapiro, Michael J. 2013. *Studies in Trans-Disciplinary Method: After the Aesthetic Turn*. New York: Routledge.

Sharp, Joanne P. 1992. '"The Russians Acted Like the Russians": The "Othering" of the Soviet Union in the *Reader's Digest*, 1980–90'. *disClosure: A Journal of Social Theory* 1 (1):1–5.

Sharp, Joanne P. 1993. 'Publishing American Identity: Popular Geopolitics, Myth and *The Reader's Digest*'. *Political Geography* 12 (6):491–503.

Sharp, Joanne P. 2011. 'A Subaltern Critical Geopolitics of the War on Terror: Postcolonial Security in Tanzania'. *Geoforum* 42 (3):297–305.

Shaw, Ian G. R., and Joanne P. Sharp. 2013. 'Playing with the Future: Social Irrealism and the Politics of Aesthetics'. *Social & Cultural Geography* 14 (3):341–359.

Shepherd, Laura J. 2013. *Gender, Violence and Popular Culture: Telling Stories*. London and New York: Routledge.

Street, John. 2013. 'The Sound of Geopolitics: Popular Music and Political Rights'. *Popular Communication* 11 (1):47–57.

Thorogood, Joe. 2016. 'Satire and Geopolitics: Vulgarity, Ambiguity and the Body Grotesque in *South Park*'. *Geopolitics* 21 (1):215–235.

Walker, R.B.J. 1986. 'Culture, Discourse, Insecurity'. *Alternatives: Global, Local, Political* 11 (4):485–504.

Walker, R.B.J. 1991. 'State Sovereignty and the Articulation of Political Space/Time'. *Millennium – Journal of International Studies* 20 (3):445–461.

Weber, Cynthia. 1990. 'Representing Debt: Peruvian Presidents Belaunde's and Garcia's Reading/Writing of Peruvian Debt'. *International Studies Quarterly* 34 (3):353–365.

Weber, Cynthia. 2013. *International Relations Theory: A Critical Introduction*. Abingdon, UK: Taylor & Francis.

Weldes, Jutta. 2003. *To Seek Out New Worlds: Exploring Links between Science Fiction and World Politics*. Houndsmills, UK: Palgrave Macmillan.

Weldes, Jutta, Mark Laffey, Hugh Gusterson, and Raymond Duvall. 1999. 'Introduction: Constructing Insecurity'. In *Cultures of Insecurity: States, Communities, and the Production of Danger*, edited by Jutta Weldes, Mark Laffey, Hugh Gusterson and Raymond Duvall, 1–33. Minneapolis: University of Minnesota Press.

Williams, Alison J. 2014. 'Disrupting Air Power: Performativity and the Unsettling of Geopolitical Frames through Artworks'. *Political Geography* 42:12–22.

Woon, Chih-Yuan. 2014. 'Popular Geopolitics, Audiences and Identities: Reading the 'War on Terror' in the Philippines'. *Geopolitics* 19 (3):656–683.

Young, Robert. 2015. 'Going Fifth Freedom: Fighting the War on Terror in the *Splinter Cell: Blacklist* Video Game'. *Critical Studies on Terrorism* 8 (1):147–162.

3 Towards a new paradigm of resistance

Theorising popular geopolitics as an interdiscipline

Vlad Strukov

Introduction: speaking in metaphors, or the problem of self-definition

As a relatively new discipline popular geopolitics is obsessed with providing self-definitions: virtually every publication begins with a section that critically surveys the history of popular geopolitics, its scope and methodological apparatus. For example, Jason Dittmer (2005, 627) claims that popular culture 'is one of the ways in which people come to understand their position both within a larger collective identity and within an even broader geopolitical narrative, or script'. Similarly, Thomas McFarlane and Iain Hay (2003, 211) note that '"Popular geopolitics" demands that attention be given to examination of the role of the media in the construction and perpetuation of dominant geopolitical understandings'. Robert A. Saunders (2017, 3), writing a decade after these authors, provides a more nuanced definition of popular geopolitics:

> In the current era of postmodern geopolitics defined by globalisation, deterritorialization, and cultural fragmentation, mass media's role in shaping geographical imagination and making sense of the geopolitical order, is steadily increasing, making *popular geopolitics* as important in international relations as its *elite* and *academic* counterparts.

Whilst these authors diverge in their accounts of the realm of popular geopolitics, they aim to ascertain its academic validity as a discipline.

In this type of disciplinary surveys researchers aim to account for the relationship of popular geopolitics to its sister-disciplines of critical geopolitics, international relations and cultural geography. My objective is different. I aim to trace the evolution of popular geopolitics as an interdiscipline in its relation to cultural studies. Below I demonstrate how the two fields of enquiry share a methodological and theoretical base. In fact, I wish to put forth the claim that cultural studies and popular geopolitics have followed the same analytical trajectory—what I define below as first, second and third phases—however, they have developed at different epistemological paces. Whilst cultural studies has been developing from the 1960s (or even from the mid-1950s as some would claim), popular geopolitics emerged already in the twenty-first century however has rapidly taken centre stage.[1] This analytical evolution accounts for the changes in the scope, focus and objectives of popular geopolitics as an interdiscipline.

Another critical difference that I examine here is that from its inception cultural studies has been a socio-political project. In addition to providing a new paradigm for thinking about culture, especially in terms of the relationship between the cultural centre and multiple peripheries, cultural studies has supplied an emancipatory framework. In other words, cultural studies has always aimed to liberate ideas as well as people through an exploration of cultural agency. The title of the seminal volume in cultural studies *Empire Writes Back* (Ashcroft, Griffiths, and Tiffin 1989), made an ironic reference to the classic of popular culture Hollywood's franchise *Star Wars: The Empire Strikes Back*, thus drawing the reader's attention to the fact that struggle for freedom is ongoing.[2] The volume was also a political manifesto insofar as it supplied tools for critical engagement with the existing cultural and institutional order—in the form of canon, educational practice and research methodology—as well as tools for advancing its liberating project. Indeed, the achievement of cultural studies as an emancipatory project is evident in university curricula (for example, the inclusion of black literature in the Francophone canon), employment law (for example, the struggle against workplace implicit bias), cultural arena (for example, gender and sexual equality in sports), and other areas.

By contrast, popular geopolitics has not evolved into a socio-political project. In a way, rather than challenging institutional and cultural hierarchies, popular geopolitics predominantly seems to be at ease with the current neoliberal order.[3] I do not wish to claim popular geopolitics promotes the current structure of capitalist dominance, or is complacent about it. Rather, as I will show below, popular geopolitics has proposed a rationale that has the potential to advance an entirely new emancipatory project. It is possible this new project will aim at re-negotiating global information flows and re-appropriating subjectivity with its liberating agency beyond the paradigm of identity and centre-periphery relationships that underpinned the project of cultural studies. Thus, I titled this chapter 'towards a new paradigm of resistance' in order to reveal—through critical examination and theoretical calibration—the potential and potentialities of popular geopolitics and its future project of resistance.

Cultural studies and popular geopolitics are unusually broad and increasingly varied fields of enquiry. In addition, both of the disciplines deliberately aim to blur the boundaries of academic, political and social discourses, which makes it impossible to conduct an exhaustive analysis of all of their trends and sub-flows. Instead, I centre on their complexities and ambivalences, which help me reveal the capabilities of knowledge as it is construed in theoretical discourse. I approach the two disciplines as a series of events, each of which reveals the unrepresented and pinpoints the attributes within the intellectual situation. To paraphrase the title of Alain Badiou's *Being and Event* (2013 [1988]), these events *exist* as unfolding of new forms of intelligence and new social systems. In light of the current state of the art research in popular geopolitics and cultural studies, my next section discusses the evolution of the two disciplines by focusing on the transformations in their scope and methodologies. The following three sections examine three interrelated concerns—the discourses of power, space and identity, respectively—wherein I trace and theorise the relationship between cultural studies and popular geopolitics as two meta-disciplines. I conclude the chapter by outlining a new paradigm of resistance, which is to be developed by popular geopolitics in its third, post-critical phase.

Twinning knowledge, or the third phase of popular geopolitics

Cultural studies is an elusive discipline because it aims to explore culture and, through constantly re-defining what it means by the term 'culture', cultural studies re-considers and re-invents its own objectives. From a critique of culture as a domain of arts created by the upper classes, cultural studies has shifted its attention to the culture and cultural practices of other social groups. By problematising the regimes and modes of cultural production, cultural studies has supplied an extremely useful arsenal of intellectual tools, which have been used in the broader field of humanities. However, during this same process of decentring culture and challenging social attitudes to culture, cultural studies has obscured its own object. Thus, nowadays cultural studies functions as a meta-discipline for which culture is a universal form of producing knowledge and a means to exercise power and control. Similarly, popular geopolitics appeared in a response to the poststructuralist turn of the late twentieth century, by finding its object—the spatial *qua* politics—not only on the arena of 'high politics', such as the institutions of the state and international, inter-state relations, but also in the realm of popular imaginings and popular culture. (Here, popular geopolitics has been dependent on how culture is defined by cultural studies and its various parallel projects such as postcolonial studies.) By expanding the notion of the political, whereby the relations between the agents of the political are always spatial, popular geopolitics—or for that matter, geopolitics of the popular—has transformed our understanding of how popular culture and politics really work. The profound impact of popular geopolitics has been in identifying how popular culture impacts citizens' political action. While advancing this extremely valuable project, like cultural studies, popular geopolitics has been concerned about the legitimatisation of its discourse and own academic institutionalisation.

In institutionalised form,[4] cultural studies emerged in the 1970s as a school of cultural materialism; or at least this is how it was conceived in the British cultural studies, whereby scholars such as Stuart Hall (1973, 1986, 1992), Raymond Williams (1974, 1977), Paul Gilroy (1982, 1993) and Richard Dyer (1986, 2002) conceived culture as a material, embodied extension of social processes, and originally, those related to the working class (see, for example, E. P. Thompson's 1963 *The Making of the English Working Class*). The class logic, or rather the class lens—that would eventually evolve into the race lens, the gender lens, the sexuality lens, and so on— meant that the British cultural studies viewed social processes as dialectical manifestations, each leading to a negation of its own antecedent and seeking a resolution by forging new identities (see the discussion of identity politics below). In its second, critical, or poststructuralist phase, cultural studies rejected dialectical materialism in favour of theories that promoted non-antagonistic, open-ended and in-exhaustive modes of cultural exchange, such as Homi Bhabha's (1994) notion of cultural hybridisation, mimicry and ambivalence, or Gilles Deleuze's (1994) concept of the fold. Deconstruction—both of the liberal (Jacques Derrida) and conservative (Paul de Mann) types—contributed to the cultural studies of the second phase by vitiating it with post-realist doctrines, that is, the belief that concepts such as class, gender, intelligence, security, and so on, are cultural constructs. The influence of the French thought—especially Jean Baudrillard, Jean-François Lyotard, and Michelle Foucault, which have been 'rediscovered' in the Anglophone tradition of cultural studies—was profound in so far as it altered the very notion of human cognition. The performative, gender-centred approach to the study of culture—such as the work of Judith Butler—emerged as a sub-discipline of cultural studies (see Chapter 4 of this volume for a discussion of this strand of research).

At the same time, cultural studies began to concern itself with its own theoretical apparatus, interpretative freedom and the ontology of culture (see, for example, Agger 1992; Alexander 1990; During 2005; Ferguson and Golding 1997; Mukerji and Schudson 1991). In other words, over the course of three decades, cultural studies as a discipline progressed to investigate articulations of meaning, not meaning per se. These articulations—speech acts, enunciations, pronouncements, iterations, performances, gestures, mediations, and so on—began to constitute its own, separate field of investigation for a number of cultural studies scholars, including: Auslander (2003); Balme (2004); Blundell et. al. (2013); Hammer and Kellner (2009); and Phelan and Lane (1998). As cultural studies moved from one phase to another, there also occurred a shift in the scope of its conceptualisations, particularly, from considerations of macro- to micro-contexts, which was as a result of the perceived imperative to produce historically and socially specific studies. To use a spatial metaphor, in this period cultural studies shifted to the margins of its own discourse because the discipline believed that meaning was to be produced and to be revealed on the peripheries, or when crossing borders, and within the folds of meaning.[5]

As I already mentioned above, on one level, popular geopolitics returns to the concerns of British cultural studies, namely, its interest in the popular, or what then was labelled as the mass, working-class culture. Or, in more abstract terms, popular geopolitics affirms its interest in a type of culture that purposefully opposes itself to that of the dominant social group, and which opposes itself to the kind of culture that adheres to an established canon, or institutionalised tradition such as theatre.[6] For example, in his analysis of a graphic narrative about the war in Chechnya, Edward Holland uses the concept of the anti-geopolitical eye to unsettle 'the dominant narratives that justify hegemonic geopolitical visions and their resultant policies' (2012, 105).

On another level, popular geopolitics is infused with the problematics of the second critical, poststructuralist phase, constantly and self-obsessively defining and redefining its own realm, and seeking to investigate phenomena in their micro-contexts, whilst producing macro-value. For example, Raffaella Coletti writes that 'while popular geopolitics were basically aimed at understanding *whether* popular culture impacts our geopolitical imagination, geographical imaginaries, identities and imagined communities, the discipline has now moved to concentrate on *how* this happens' (2017, 3). In the end, popular geopolitics has been framed as a discipline concerned with symbolic articulations, whereby its two principal constituent parts—space and politics—have been reimagined as cultural articulations. For example, in his article on *Captain America*, Dittmer defined popular geopolitics as 'the construction of scripts that mould common perceptions of political events' (2005, 626). This is specific to the first phase because political events appear to be separated from popular imaginings. I argue that in the second phase—due to the borrowings from cultural studies[7]—popular geopolitics considers these very imaginings to be political events, thus eschewing the distinction between different realms of culture, 'real' and 'imaginary', the script and the action, and so on. In terms of how popular geopolitics works with space, during the first phase of popular geopolitics, Dittmer described it as a process of attaching 'symbolic meaning to territory' (2005, 626). In my conceptualisation, during the second phase, popular geopolitics worked with territory as symbol, that is, space appears to be filled with symbols and to be a symbol itself, thus, completing the cycle of the critical auto-reflection, which was characteristic of cultural studies of the second phase.

Finally, like cultural studies, popular geopolitics is in the process of shifting its attention onto audiences, or, to be precise, on constructing prosumers (producer + consumer) of geopolitical content as a type of actual geopolitical content. As a result, the prosumer emerges not only as an agent of geopolitical relations, but as an articulation—a realm of—geopolitical meaning, that is, popular geopolitics has designated the self as a spatial practice of political relation. As speakers at various conferences on popular geopolitics have confirmed, work in this direction has been slow, which is perhaps due to the complexity of methodological tools required for the estimation of prosumer activities. However, some initial claims and theoretical frameworks have been presented in Dittmer and Gray (2010); Dodds (2006); Woon (2014).

Whilst some scholars of popular geopolitics have tasked themselves with delivering a taxonomy of popular geopolitical articulations (see, for example, Dittmer 2010; Saunders 2017), I believe future work will focus on how we can conceive of subjectivities in geopolitical terms. These disciplinary developments will be prompted by advances in artificial intelligence, automatics and cloning, whereby subjectivities emerging thanks to these technologies will reveal their own geopolitical concerns which, like during the second phase, will be interrogated in opposition to the dominant discourse and in opposition to tradition. In other words, the critique will unfold from a perspective of post-humanism, according to which being human will be considered a type of tradition, or a contextual ontology (see, for example, Badmington 2000; Braidotti 2013; Herbrechter 2013; Nayar 2014). These concerns, including considerations of humanity as a tradition, are evident in recent televisual productions such as Laeta Kalogridis' 2018 *Altered Carbon* (Netflix, ten episodes).

The imagination of power versus the power of imagination

Popular geopolitics is an interdiscipline, which is characterised by its critical relationship with its own predecessors. However, its lineage is nothing but a misnomer: the connection of popular geopolitics to geopolitics is in the name, but not in essence or purpose (like a link between astrology and astronomy). Rather, popular geopolitics is a product of a critical turn in international relations, more specifically, it is an outcome of the transition from positivist to post-positivist paradigms. On one level, this transition is based, of course, on acknowledging the difference in epistemology; on another, it reveals a scepticism about rationality per se, which has been manifested in the rejection of the rational choice theory and similar 'realist' approaches (see, for example, a discussion of the threat which rational choice theory poses to democracy in Petracca 1991).

From the impact perspective, unlike international relations, which is interested in exploring the power of institutions and regimes, popular geopolitics is concerned with examining the power of texts. Popular geopolitics considers institutions as texts, too, thus shifting the focus of our attention from structure and action to discourse (although, following the logic of poststructuralist theory, it treats structure and action as elements of discourse, too). In this regard, popular geopolitics belongs to a group of disciplines which privilege the ideas of social constructivism. For them, the structure-versus-agency debate is primarily a debate about validity and scope of supposition, and about the distinction between belief and knowledge. By engaging with the notion of the popular, popular geopolitics expands its remit to include cultural construction, too. To confirm, popular geopolitics aims to identify agency outside the established discourse; in other words, it analyses the popular realm as an alternative, a projection, or at least as a duplication of the official realm. Just like cultural studies, popular geopolitics fluctuates between different realms of agency and modes of production with the single objective of identifying, critiquing and theorising power.

In its apprehension of power and discourse, popular geopolitics borrows explicitly from poststructuralist theory—and also from how this theory has been applied in cultural studies—including Michel Foucault's notion of power (see, for example, Debrix 2008; Dittmer 2010; Purcell et al. 2010; Shepherd 2013; Shepherd and Hamilton 2016). His concept of 'dispositif', often translated from French as 'apparatus' and 'deployment', defines institutional, physical and administrative mechanisms and knowledge structures, which are utilised to exercise and advance power within a social entity, be it a group, a class or the state. Foucault theorises the function, structure and impact of dispositif as biopower, which, for him is a form of (discursive) technology employed to control people. Biopower reconsiders the body as a product of power relations and social construction. In other words, the body emerges as a discursive category. For example, George Yúdice's *The Expediency of Culture: Uses of Culture in the Global Era* (2003) repositions culture in connection with governmentality and biopower. And Joanna Zylinska's *The Ethics of Cultural Studies* (2005) conceptualises new forms of post-humanities such as migration, experiments in plastic surgery and digital communication, simultaneously from the point of view of biopower and ethics.

We find this style of thinking about power in popular geopolitics, too. For example, following Paasi (1991, 1996), Dittmer believes that the project of popular geopolitics is in concentrating 'on how bounded territories and identities are constructed and policed' (2005, 626). At the same time, other scholars (see, for example, Kitchin and Kneale 2001; Pierce 2002; Warf 2002), have identified the purpose of popular geopolitics in exploring alternative geographies of power. Writing about tabloid representations of war, François Debrix notes that biopower of media encompasses a specific structuring of discourses, the result of which is that the reader is expected to take such representations at face value. 'The facts, events, maps, figures, statistics, and images displayed through the tabloid medium are what they are, so called "purely objective realities" that are not supposed to be contested and do not even require active recognition or acquiescence' (Debrix 2008, 45). When analysing the genre of science fiction, Charlie Carpenter et. al. pay attention to how power is distributed among different characters, spaces and modes of production. For example, they note that in *Battlestar Galactica* Roslin and Adama, the two lead characters, split power so that the former is charged with the civilian fleet and personnel and the latter with military vessels and crew (Carpenter, Cvijanovic, and Mason 2013, 146). Here Carpenter et al. connect Foucault's idea of administration and production of power with post-Marxist ideas about labour-related organisation of society. Thus, popular geopolitics extends the project of cultural studies by focusing on the imaginative life of citizens and by providing it with political power such as knowledge about world politics, its various logics and structures, including the legitimacy of military government, and so on.

Popular geopolitics reappropriates Foucault's understanding of culture to account for local, trans-local and global spatial relations. For Foucault, culture is 'a hierarchical organization of values, accessible to everybody, but at the same time the occasion of a mechanism of selection and exclusion' (2001, 173). These modes of organisation and exclusion and inclusion reveal structures and agents of power in national and international contexts. In international relations these agents are instrumentalised in the discourse about power and governance, most notably in various regime theories that have been reinvigorated since the rise in the 2000s of global powers, such as Russia and Turkey (see, for example, Keohane 2002; Krasner 2009). For example, the discourse of culture and values such as 'the conservative values', or 'traditional values', has been used to account for the compatibility of interests and (un)willingness to compromise in certain political contexts (in this case the war in Syria wherein different parties and among them the U.S., Russia and Turkey have utilised political propaganda to account for their military action). The cultural regime approach has been employed in post-hegemonic contexts—in the West, the idea of 'resurgent Russia', or 'nationalistic Turkey'—which depend on reciprocity of discourses (see especially Laruelle 2016; Lin 2016), thus forming feedback loops of agency in transnational context (Saunders and Strukov 2017).

Conversely, popular geopolitics seeks for popular manifestations of power, such as the power of brands and stereotypes. By way of example, Saunders (2017) provides a typology of representations of Russians in Hollywood films:

categorised as hired killers, mad scientists, or other villainous archetypes. These cinematic tropes of Russianness reveal U.S. cultural hegemony, on one level, and on another they structure cultural contexts as brands, or, to paraphrase Foucault, they function as 'a mechanism of selection and exclusion'.[8] It is via Foucault that popular geopolitics shifts from the concerns of cultural studies—in this regard, the critical examination of narratives of orientalisation and oppression—to its own concerns as an interdiscipline: How are cultural discourses re-tooled in the era of globalisation? What is the politics of symbolic interactions? How does it affect the realm of political agency (participatory democracy or authoritarian rule, or a hybridisation of both)? Just like Foucault's work, which is both historical in nature and a-historical, or rather, 'beyond-history' in its theoretical scope and impact, popular geopolitics is both a politically charged project and a system that extends well beyond the world of politics. Eventually, popular geopolitics begins to operate with concepts and systems that had been disregarded in classical international relations, such as desire, imagination and performance, thus re-visiting the seminal works of cultural studies, including applications of Lacanian psychoanalysis, Judith Butler's performance theory and cultural economy of Stuart Hall. However, unlike cultural studies of the poststructuralist phase, popular geopolitics works with systems that reveal coherent organisations and transparent environments. Therefore, the world of popular geopolitics is the world to be comprehended and apprehended by means of analytical exploration of metaphors of control and power, that is, popular geopolitics blurred the boundaries between different concepts of the real, different ontologies and modes of imaginative production.

For Foucault, discipline is a metaphor, which is often used to define the ways in which the behaviour of individuals is regulated in the social body (for example, architecture, timetables, drills and posture organise social interaction, including a person's body). Hence the social body exists as a disciplinary society, because it employs disciplinary institutions such as prisons, hospitals, asylums, army barracks, and so forth. Popular geopolitics builds on this understanding of power and discipline insofar as it is concerned with the *disciplinary nature of imagination*, especially its ability to construct an image of the other. As I demonstrate below, the relationship between these different subjectivities, or, as a matter of fact, between these constructions of the 'I' and its opposing 'I', requires 'disciplinary disciplining' to avoid methodological and existential discontinuities. To confirm, popular geopolitics wishes to respond to the imperative to move away from essences towards fluidities; however, at the same time popular geopolitics critically re-interprets those same categories of effect, progress and tradition. In this regard, popular geopolitics employs the epistemological strategy of cultural studies, which challenges categories such as class, ethnicity and race by tracing their history and examining their impact, or disciplinary power.

Where the two disciplines de-converge is the conceptualisation of space. For both of the fields of enquiry, space has the capacity to reveal power relations; in fact, space produces power (in fact, Foucault's metaphors of power are spatial:

prisons, hospitals, and so on). In *Discipline and Punish*, Foucault conceptualises space and knowledge as modes of disciplining: the ever-visible inmate always 'the object of information, never a subject in communication': 'He who is subjected to a field of visibility, and who knows it, assumes responsibility for the constraints of power; he makes them play spontaneously upon himself; he inscribes in himself the power relation in which he simultaneously plays both roles; he becomes the principle of his own subjection' (1975, 202). Therefore, for cultural studies space is a type of perspective (or what Foucault labelled as panopticism), whereas for popular geopolitics space is a type of connective (or what I would call 'panopticallity'). Let me provide an example from cultural history to elucidate my theoretical point. Cultural studies challenges the hegemony of the canon of English literature by including texts produced in the former British colonies. Here a scholar must adopt a different spatial perspective to achieve a different construction of the subject, which now includes one from the imperial centre and one from the colonial periphery.

To examine power, knowledge and subjectivity as texts, popular geopolitics employs all kinds of connectives—temporal, affective, ideological, and so forth—and produces a view of power as self-generating matter. To use a biological metaphor, for popular geopolitics power is a kind of connective tissue in the human body: just like the three types of connective tissue—epithelial, muscle and nervous tissue—power envelops and transcends social and political institution. Popular geopolitics aims to map power as a system of spatial relations, where power is akin to the nervous system, which consists of nodes and membranes. Therefore, in its critique of power popular geopolitics goes further than the task of de-stabilising the centre-periphery relations, which is specific to cultural studies and especially its sister-discipline of postcolonial studies.

Contested spaces: popular geopolitics and postcolonial studies

In cultural studies and popular geopolitics, the question of power is linked to the question of the gaze, or, to be precise, its orientation, hence panopticism, as outlined above. In *Orientalism* (1978), Edward Said argued that the subject produces the other through the work of the gaze, that is, through looking at the other the subject emerges as a coloniser of meaning and controller of identities. At the same time in his *Discipline and Punish: The Birth of the Prison* (1975), Foucault posited the power of the gaze of the self to represent/speak on behalf of the other. He linked the French verb 'voir', meaning 'to see', with 'savoir'—'knowledge'—and 'pouvoir'—power. In his system seeing—as the gaze of the authority—produces knowledge, which, in its turn, produces power. With the help of the notion of panopticism, or the power-gaze (a way to observe, control and discipline societies, which do not have to be other), Foucault identifies a type of gaze that can be directed back at the self and results in the discipline of the mind. Both for Said and Foucault, the position of the observer is extremely important as it relates to the self that is fixed in a single point enjoying the gravity of the discourse. The observed is always distanced from the observer in his/her attempt to elude and detract the observer's gaze. Therefore, the observed, or the other, is constructed as an appearance and as a spatial relation, that is, as someone separated from the self by the gaze. The gaze reveals the relationship of power and its direction affirms the presence of the hegemonic discourse.

Popular geopolitics is interested in how subjectivities, power and discourse are spatialised; conversely, it tends to interpret the modes of domination *through* the gaze, imagination and spatial interactions, hence panopticality, as outlined above. In this regard, popular geopolitics re-affirms the spatial politics of culture examined in Homi Bhabha's seminal work *The Location of Culture* (1994). As a matter of fact, neither Dittmer (2010) nor Saunders (2017) makes explicit references to Bhabha. However, in many ways, they—and other scholars in the field of popular geopolitics—re-affirm Bhabha's key concerns and especially his ideas about the construction of space as a political realm. Dittmer defines popular geopolitics in opposition to formal geopolitics as a phenomenon 'immersed in everyday' (2010, 14). He argues that the U.S. foreign policy depends on the popular constructions of other nations, whereby those who are known to the US citizens always win over those who remain obscure in the American popular imaginings. For Bhabha, this is a debate about majorities and minorities, which he interprets not as basic arithmetic correlations, but as speculative categories that define the actual and the imaginary. Most importantly, he argues that our perceptions about minorities and majorities rely on the notion of difference, which, like Said's and Foucault's gaze, has an *orientation* insofar as it establishes power relations and accounts for dominance. Thus, for Bhabha, difference is a spatial category; difference is primarily a question about the location of culture, that is, about *where* it is located and *why* it has been placed there. However, difference does not equate the binary antagonisms of cultural materialism, and this is how Bhabha escapes the debates of the first phase of cultural studies (see the discussion of concomitant work of Appadurai below).

Bhabha considers difference from the perspective of the third space. That space is occupied by the migrant narrator who is simultaneously a cultural translator (the agent of panopticality). His/her purpose is to re-write and re-tell narratives, and to challenge and to obfuscate meaning. The migrant narrator, who is neither the self nor the other—or the 'I' and the opposing 'I', as we mentioned above—inhabits a third space where difference is revealed, oppositions are deconstructed, and narration is reversed:

> Any change in the statement's conditions of use and reinvestment, any alteration in its field of experience or verification, or indeed any difference in the problems to be solved, can lead to the emergence of a new statement: the difference of the same.
>
> (1994, 22)

This logic of 'the difference of the same' underpins many phenomena including the stereotype. Bhabha devotes a whole chapter to stereotypes in *The Location of Culture*:

> The stereotype, which is its [colonialism's] major discursive strategy, is a form of knowledge and identification that vacillates between what is always 'in place', already known, and something that must be anxiously repeated.
> (1994, 66)

Popular geopolitics explores stereotypes as a form of knowledge (similar to Bhabha) and as a form of agency (dissimilar from Bhabha). For example, Saunders (2017) provides a schema of stereotypical representations of the post-Soviet states in American popular culture which, he argues, exist as a system of positive and negative nation brands. Here, Saunders extends Bhabha's project by critically surveying images of 'the post-Soviet bogeyman' and revealing the colonial, hegemonic nature of the Western discourse. He also includes these images in his discussion of the post-Soviet nation brands; in other words, he accounts for them from the perspective of neoliberal commodification of discourse whereby 'nation-branding is often more focussed on affectual content and other sorts of non-tangibles, a recognition of postmodern sensibilities, which privilege the sublime over the rational, affect over effect, and feeling over knowing' (Saunders 2017, 50). This definition yields several important considerations. Firstly, by de-rationalising the discourse of the other/the *stereotyped*, popular geopolitics affirms, or at least accounts for, the hegemonic discourse of the self/the *stereotyper*. To paraphrase Saunders, it affects the very discourse it wishes to effect by normalising the power structure embedded in this type of gaze which anxiously repeats what is already in place.

Secondly, while Bhabha attempted to construct culture as a transient, migratory space, always in the making and always incomplete, popular geopolitics re-appropriates this notion to produce an area of impact: when Bhabha speaks of a need for an absence of a perspective, popular geopolitics employs functionalist approaches to conceive of the relationship between discourse and politics. I posit that this is the legacy of international relations on which popular geopolitics is built. To remind the reader, when aiming to define popular geopolitics, Dittmer speaks of 'sites and modalities' (2010, 37) which, unlike Bhabha's modes, presuppose functionalities and rational choice which are operationalised as symbolic commodities.

Finally, for Bhabha the third space is a position which one can adopt when trying to escape existing discursive contraptions and constraints; for scholars of popular geopolitics, the third space is the space of popular culture per se, that is, one that transcends the realms of official, 'formal' politics and the realm of institutionalised, 'high' culture. Here popular geopolitics makes an ironic return to the concerns of early cultural studies which would consider, say, for example, comic books in opposition to novels studied in universities/the cultural canon and to the political institutions of the state that keep the canon and the university in place.

Is post-identity politics really possible?

The critical analysis of the evolution of cultural studies and popular geopolitics has revealed the dependence of the latter on the notion of identity. For example, when Saunders discusses the proliferation of stereotypes about the former Soviet republics, or when he makes use of the term nation-branding to account for the struggle of nation-states to gain a place on the global stage, or when he analyses the narratives and styles of popular representations of geopolitical entities such as parody, in actual terms he advances our understanding of how identities are formed, perceived, and consumed in the era of increased globalisation and mediation. Similarly, for Dittmer in his analysis of *Captain America* the focus is on identifying how *Captain America* is connected to American nationalism, internal organisation and foreign policy, whereby the individual defines the scale of investigation. In other words, Dittmer (2005) de-personifies agency in his analysis of the case in order to put forward abstract notions of characterisation vis-à-vis their representation. Although he talks about 'structures of expectations', for him they are manifested in 'scripts' and narratives, or, I argue, in identities which are, to paraphrase Dittmer, understandable and legitimate.

Indeed, the project of early cultural studies—the first phase—had its purpose in identifying, or making visible, and legitimising identities that had previously been obscured by other entities in the dominant discourse. In the beginning, to a large extent, this work was self-focussed, that is, the scholars worked with, for example, working class cultures within the British context. However, with the arrival of scholars from what eventually was labelled as the postcolonial strand of cultural studies, such as Homi Bhabha, whose work I discussed above, the focus of investigation shifted on the conceptualisation and the analysis of that type of Other who is immediately recognisable, visible, but remains incomprehensible and therefore 'illegitimate'. Here, the political project of cultural studies is in exposing the logic and mechanisms of exclusion in order to provide this agency with the required visibility and status. Eventually, cultural studies grew to include other disciplines such as gender studies; or rather, cultural studies evolved into a meta-discipline whose focus is on the theory and politics of identity and difference and who provides theoretical models for all its sister-disciplines. In the end, cultural studies affirmed a theoretical position in which the space of identity *is* the site of the political struggle and that differences must be disassociated from borders as spaces of contestation, not legitimisation.

In the early 1990s, cultural studies, for example, in the texts by Hall (1973, 1986, 1992), identity was conceived as a socially constructed phenomenon, thus problematising all kinds of essentialists positions and expectations. In the early 2010s, popular geopolitics considers socially constructible identities at the level of nations, not just individuals and social groups. For example, Nadia Kaneva (2012) and Saunders (2017) have demonstrated how different states are involved in marketing themselves to the global community, whereby political and communication experts are hired to construct certain images of countries and nations, often with the purpose of challenging existing preconceptions and stereotypes. This is how, for example, Estonia acquired a new brand (see Curry Jansen 2008): it is now a pioneer of digital technologies, and not a former Soviet state with an obsolete communication structure (in actual terms Estonia is no different from other Soviet states, which have contributed immensely to the development of global digital networks). Among other things, this example demonstrates that a) identities are used as commodities in the neoliberal economic system, b) identities are re-essentialised in order to separate them from competing identities (in this case, the suggestion that Estonians are somehow more digitally minded than other people), and c) identities are stratified in order to reveal their relationship to other geopolitical concerns (in this case, the location of Estonia within the European Union, and within the Baltic realm, not within 'the former Eastern Europe', and so forth) (see Saunders forthcoming).

Like cultural studies, popular geopolitics is concerned with the ways in which identities are produced, although it uses a different procedure and engages a somewhat different set of concepts derived from international relations. In this process, popular geopolitics inadvertently reiterates the identity politics of the previous era. To clarify, popular geopolitics does not celebrate sameness and does not promote essentialist readings of identities and cultural traditions; however, popular geopolitics does affirm a type of thinking that is *within* the space of identity. In this regard, popular geopolitics fails where cultural studies has failed: it does not offer—at least now—a type of consideration of the world order that would be *outside* the realm of identities, or in spaces that are not determined by the identity discourse.

The critique of identity politics was already present in the critical phase of cultural studies. For example, for Judith Butler 'identity categories tend to be instruments of regulatory regimes, whether as normalizing categories of oppressive structures or as rallying points for a liberatory contestation of that very oppression' (2006 [1991], 255). In her appraisal of Foucault's scholarship on sexuality, she argues that 'a Foucauldian perspective might argue that the affirmation of "homosexuality" is itself an extension of homophobic discourse' (Butler 2006 [1991], 260). It may sound like a paradox but an examination of structures of domination in power discourse may aim to de-stabilise power relations on one level, and on another, it may produce that same power discourse because it may approach power from that same standpoint. In fact, the resulting inclusion of otherness, also known as normalisation, does not prevent stereotyping and discrimination: these continue to occur in other situations and in other realms because the hegemonic discourse persists in its affirmative power (see chapter 8 on meta-hegemony).

To confirm, popular geopolitics has inherited this 'identity problem' from cultural studies. It is unlikely popular geopolitics will produce a paradigm of thinking which would be outside the realm of identity politics. This is the objective of philosophers such as Rancière (2004) and Badiou (2005, 2006) who have attempted to produce a critique of subjectivities, not a critique of identities, by considering the anti- or non-spaces, which are not like Bhabha's third spaces, but rather re-configurations of the self, thus allowing the self to exist in multiple forms (see, for example, my conceptualisation of Badiou's concept of one as many in relation to contemporary Russian film [Strukov 2016]). Nevertheless, I believe popular geopolitics may produce a new strategy for resisting the pressures of identity politics, which is something I wish to conceptualise in the remaining section of the chapter.

Conclusions, or an argument in favour of resistance politics

As I have demonstrated above, as disciplines, cultural studies and popular geopolitics display an overlap in terms of their genealogy, objectives and modes of enquiry. Firstly, both have emerged as a result of the interpretative turn,[9] or a renewed interest in cultural meanings, iterations and interpretations, which had previously been the domain of cultural anthropology (see, for example, Crapo 2001; Nanda and Warms 2010; Srivastava 2012). The two disciplines expand our understanding of culture as an enormous and variated repertoire of texts, practices and relations, including their visual manifestations. However, the popular geopolitics considers culture to be primarily a spatial practice, that is, its chief concern is to trace the impact of cultural practice on spaces and also to examine how space is imagined thanks to cultural practice and exchange of identities. By spatialising culture, popular geopolitics reveals its own affinity to political economy whereby all realms of culture manifest themselves as areas of production and consumption, including production of imaginings, with political relations emerging as a result of negations and control over resources such as tradition, rituals, and symbols.

Secondly, cultural studies and popular geopolitics include a reflexive turn, or what I labelled above as the second phase, insofar as both disciplines advance a critical and conscious self-examination which shifts from the issues of representation to issues of writing, doing and performing culture to cultural world-making. If cultural studies has been preoccupied with relational categories such as the question of what constitutes the centre and periphery, or what makes knowledge a relational category, popular geopolitics has focussed on the re-interpretation of the political and the popular as inter-relational categories, that is, on how they are affected, not only constructed. The affective segment of popular geopolitics has roots in the discipline of international relations and is also an independent concern which has emerged in response to the affective turn as a new configuration of bodies, technologies and thought that took central ground in the 2000s. Scholars of popular geopolitics tend to politicise the relationship between the subjective and pre-subjective as a mode of meaning production. This is where cultural studies and popular geopolitics de-converge, since the latter often views subjectivities in an instrumentalised way, that is, as an 'audience' of popular geopolitical discourse. Another important difference in this category is that cultural studies has developed gradually, moving from its materialist to critical and contemporary phase whilst popular geopolitics had advanced not in phases but rather in feedback loops whereby it has included pre-critical and critical points of focus from the very outset. In terms of its own critical discourse, popular geopolitics has been outward-looking, constantly employing new theoretical frameworks, and also incredibly inward-looking which is due to highly self-conscious self-examination.

Thirdly, cultural studies has often overlapped with postcolonial studies, which is due to their shared interest in the interpretation of spaces, especially borders, cross-border crossings and peripheries, producing multi-centric and multi-agent visions for the organisation and structuring of global spaces. Arjun Appadurai's (1996) famous five -scapes of globalisation—ethnoscape, technoscape, finanscape, mediascape and ideoscape—exemplify a simultaneous attempt to critique and systematise areas of cultural production and impact. Equally, popular geopolitics aims to produce a universal picture of inter-spatial connections determined by the conditions of economic development, local and global identities and intra-political characteristics (by this I mean the objective of popular geopolitics to assess how spaces are imagined, not only constructed, thus producing a different type of political discourse). Here, however, two key differences between cultural studies and popular geopolitics can be identified: A) If cultural studies—thanks to its engagement with postcolonial studies and area studies—has been properly internationalised, popular geopolitics has been locked-in its own Anglophone, Western-centric paradigm. This could have been avoided since as a discipline popular geopolitics had emerged in the aftermath of the transnational turn in humanities. B) Cultural studies has always been preoccupied with the question of power and equally with the question of resistance; in fact, one can argue that cultural studies is a meta-study of resistance. That can be resistance to certain forms of culture, social conditions and political regimes, all conceptualised through and with the help of the language of hegemony understood in Gramscian (2011) (2001 [1929–35]) and neo-Gramscian terms (see, for example, van Apeldoorn 2004; Cohn 2010; Williams 2014). (See chapter 7 on meta-hegemony in the Indian context.) Popular geopolitics has been less interested in the logic and rhetoric of resistance: I trust this type of discourse will evolve as popular geopolitics advances as a discipline and as new forms of resistance appear, that is, those that have characterised popular geopolitics with its focus on space from the outset.

In cultural studies, among other things, the rhetoric of resistance has produced a critique of the globalisation discourse. For example, Roland Robertson (1992) has promoted the notion of glocalisation. On one level, the term showcases how globalisation principles and processes are absorbed and re-interpreted by different social groups depending on their vantage point, historical and cultural setting, and a range of economic conditions, which impact the globalisation process. In Robertson's theoretisation, glocalisation accounts for the moments when homogenisation and heterogenisation intertwine. On another level, cultural studies would argue that these processes of differing between the imperatives of homogenisation and heterogenisation and the very fact that globalisation is interpreted at a local level suggest that local subjectivities resist the dominant logic of globalisation by engaging in its critique through thought and practice. The counterargument would be that these subjectivities are trapped in their glocal identities, and therefore unable to challenge the dominant discourse outside the framework of that same discourse. Popular geopolitics has the potential to resolve this theoretical deadlock by ascribing value to spatial imaginings and examining those not only at the level of consumption such as the consumption of media, or in fact, at the level of political engagement such as elections and activism, but also through re-examining mundane situations and assessing how in their everyday practice subjectivities respond to global challenges, that is, how they affect globalisation through their imagining.

The last is possible in a system that privileges multiplicity, such as in Shmuel Eisenstadt's notion of multiple modernities. 'The idea of multiple modernities presumes that the best way to understand the contemporary world—indeed to explain the history of modernity—is to see it as a story of continual constitution and re-constitution of a multiplicity of cultural programs' (Eisenstadt 2000, 2). Here, resistance is conceived as a type of fluctuation between different types of modernity, a type of vibration, or what Derrida would call 'a radical trembling':

> The strategic bet. A radical trembling can only come from the outside … This trembling is played out in the violent relationship of the West to its other, whether a 'linguistic' relationship (where very quickly the question of the limits of everything leading back to the question of the meaning of Being arises), or ethnological, economic, political, military, relationships, etc.
>
> (Derrida 1982, 135)

These are, indeed, the preeminent concerns of popular geopolitics which deploys its own 'radical trembling' as a means to transcend narratives since they purport hegemonic discourses and establish multi-sited forms of critical investigation. Therefore, the debate should be not about the universalisation of particularism (e.g., identities such as gender, also see Caso's chapter in this volume) and the particularisation of universalism (e.g., state sovereignty), but about how each of these reveals its own geopolitical discourse, whereby the notion of the popular should be critically examined as connoting hegemony in its own right. The discipline of popular geopolitics should advance as a discipline of contestation: that is, a field of knowledge that queries and problematises globe-oriented ideologies available in popular culture, popular politics or popular imaginings.

Finally, the resistance project of cultural studies has been in *radical contextualisation* – working class men, diaspora artists, transgender athletes, and so on – whereby their difference in terms of cultural association and practice has been labelled as liberating thanks to its ability to forge new identities. The analysis of subordinate groups and their production of freedom through cultural practice has revealed how the popular is not a neutral realm of mundane activities, but a strategy of resistance. Popular geopolitics should re-visit the debates promoted by John Fiske (1989), who argued that a deconstruction of popular texts such as music videos reveals the incompleteness and inconsistencies of media discourse, thus enabling forms of resistance – 'symbolic battles'. Moreover, popular geopolitics can utilise the outcome of the cultural studies' concepts of the 'local other' (see, for example, Saukko 2003), who is capable of resisting power structures outside symbolic systems. Thus, popular geopolitics can work with contexts through *radical re-contextualisation*, always speaking from the perspective of interdisciplinarity in a world characterised by over-production of meaning.

Notes

1 Some pioneering ideas appeared already in the 1990s, see, for example, Sharp 1993.
2 1980, dir. by Irvin Kershner.
3 Some critique of the dominance of the US and UK is available in various studies, including Dittmer 2010 and Dodds 2006.
4 The Birmingham Centre for Contemporary Cultural Studies opened in 1964; however, its work was consolidated in the 1970s under the leadership of Stuart Hall. Writing about the history of cultural studies, Hall (1973) traces the intellectual origins of cultural studies to the 1950s and the publication of Richard Williams' *Culture and Society* (1958).
5 See, for example, Bartolovich and Lazarus (2002); Melrose (1996); Morgan and Leggett (1996); Norquay and Smyth (2002); and White and Schwoch (2008).
6 Hence, we have Dittmer's study of comic books which, as a medium and a form of visual culture, is still perceived by many as marginal.
7 These borrowings are evident in direct engagement with cultural studies scholarship as, for example, we see in Dittmer (2005) where he uses Williams' idea of structuring feelings to account for 'the social effects of regional institutionalization' (1977, 627).
8 Dittmer's engagement with the notion of hegemony is even more direct: for example, he notes that 'at the heart of popular culture's importance to the construction of national and global geopolitical scripts is Antonio Gramsci's concept of hegemony' (2005, 627).
9 I include the so-called 'visual turn' into the broader framework of the interpretative turn.

References

Agger, Ben. 1992. *Cultural Studies as Critical Theory*. London: Psychology Press.

Alexander, Jeffrey C. 1990. *Durkheimian Sociology: Cultural Studies*. Cambridge: Cambridge University Press.

Appadurai, Arjun. 1996. *Modernity at Large: Cultural Dimensions of Globalization*. Minneapolis: University of Minnesota Press.

Ashcroft, Bill, Gareth Griffiths, and Helen Tiffin. 1989. *The Empire Writes Back: Theory and Practice in Post-Colonial Literatures*. London: Routledge.

Auslander, Philip. 2003. *Performance: Critical Concepts in Literary and Cultural Studies*. London: Routledge.

Badiou, Alain. 2005. *Handbook of Inaesthetics*. Translated by R. Brassier and A. Tosca Stanford: Stanford University Press.

Badiou, Alain. 2006. *Briefings on Existence: A Short Treatise on Transitory Ontology*. Translated by N. Madarasz. Albany: State University of New York Press.

Badiou, Alain. 2013 [1988]. *Being and Event*. London: Bloomsbury Academic.

Badmington, Neil. 2000. *Posthumanism*. London: Palgrave Macmillan.

Balme, Christopher. 2004. *Beyond Aesthetics: Performance, Media, and Cultural Studies*. Berlin: Verlag Trier.

Bartolovich, Crystal, and Neil Lazarus. 2002. *Marxism, Modernity and Postcolonial Studies*. Cambridge: Cambridge University Press.

Bhabha, Homi K. 1994. *The Location of Culture*. London: Routledge.

Blundell, Valda, John Shepherd, and Ian Taylor. 2013. *Relocating Cultural Studies: Developments in Theory and Research*. London: Routledge.

Braidotti, Rosi. 2013. *The Posthuman*. Cambridge: Polity.

Butler, Judith. 2006 [1991]. 'Imitation and Gender Insubordination'. In *Cultural Theory and Popular Culture: A Reader*, edited by John Storey, 255–270. Atlanta and Athens: University of Georgia Press.

Carpenter, Charlie, Hrvoje Cvijanovic, and Wesley Mason. 2013. 'Security, or Human Security? Civil-Military Relations in *Battlestar Galactica*'. In *Battlestar Galactica and International Relations*, edited by Nicholas J. Kiersey and Iver B. Neumann, 136–165. London and New York: Routledge.

Cohn, Theodore. 2010. *Global Political Economy: Theory and Practice*. London: Pearson.

Coletti, Raffaella. 2017. '*The Good Wife*'s (Geo)Politics between Originality and Stereotypes: A New Wine or Just a New Bottle?' *Geopolitics* 23 (1):50–66.

Crapo, Richley. 2001. *Cultural Anthropology: Understanding Ourselves and Others*. London: McGraw-Hill.

Curry Jansen, Sue 2008. 'Designer Nations: Neo-liberal Nation Branding – Brand Estonia'. *Social Identities* 14 (1):121–142.

Debrix, François. 2008. *Tabloid Terror: War, Culture, and Geopolitics*. London and New York: Routledge.

Deleuze, Gilles. 1994. *The Fold: Leibniz and the Baroque*. Minneapolis: Minnesota University Press.

Derrida, Jacques. 1982. *Margins of Philosophy*. Translated by Alan Bass. Chicago: Chicago University Press.

Dittmer, Jason. 2005. 'Captain America's Empire: Reflections on Identity, Popular Culture, and Post-9/11 Geopolitics'. *Annals of the Association of American Geographers* 95 (3):626–643.

Dittmer, Jason. 2010. *Popular Culture, Geopolitics, and Identity*. Lanham, MD: Rowman & Littlefield.

Dittmer, Jason, and Nicholas Gray. 2010. 'Popular Geopolitics 2.0: Towards Methodologies of the Everyday'. *Geography Compass* 4 (11):1664–1677.

Dodds, Klaus. 2006. 'Popular Geopolitics and Audience Dispositions: James Bond and the Internet Movie Database (IMDb)'. *Transactions of the Institute of British Geographers* 31 (2):116–130.

During, Simon. 2005. *Cultural Studies: A Critical Introduction*. London: Psychology Press.

Dyer, Richard. 1986. *Heavenly Bodies: Film Stars and Society*. New York: St. Martin's Press.

Dyer, Richard. 2002. *The Culture of Queers*. London: Routledge.

Eisenstadt, Shmuel. 2000. *Multiple Modernities*. New York: Transaction Publishers.

Ferguson, Marjorie, and Peter Golding. 1997. *Cultural Studies in Question*. London: Sage.

Fiske, John 1989. *Understanding Popular Culture*. London: Unwyn Hyman.

Foucault, Michel. 1975. *Discipline and Punish: The Birth of the Prison*. New York: Pantheon Books.

Foucault, Michel. 2001. *L'hermeneutique du sujet. Cours au Collège de France, 1981–1982*. Paris: Gallimard Seuil.

Gilroy, Paul. 1982. *The Empire Strikes Back: Race and Racism in '70s Britain*. London: Routledge.

Gilroy, Paul. 1993. *The Black Atlantic: Modernity and Double Consciousness*. London: Verso.

Gramsci, Antonio. 2001 [1929–1935]. *The Prison Notebooks*. Translated by Joseph A. Buttigieg and Antonio Callari. New York: Columbia University Press.

Hall, Stuart. 1973. *Encoding and Decoding in the Television Discourse*. Birmingham: Centre for Cultural Studies, University of Birmingham.

Hall, Stuart. 1986. 'Gramsci's Relevance for the Study of Race and Ethnicity'. *Journal of Communication Inquiry* 10 (2):5–27.

Hall, Stuart. 1992. 'The Question of Cultural Identity'. In *Modernity and its Futures*, edited by Stuart Hall, David Held and Anthony McGrew, 274–316. Cambridge: Polity Press.

Hammer, Rhonda, and Douglas Kellner. 2009. *Media/Cultural Studies: Critical Approaches*. Amsterdam: Peter Lang.

Herbrechter, Stefan. 2013. *Posthumanism: A Critical Analysis*. London: A&C Black.

Holland, Edward C. 2012. 'To Think and Imagine and See Differently': Popular Geopolitics, Graphic Narrative, and Joe Sacco's 'Chechen War, Chechen Women''. *Geopolitics* 17 (1):105–129.

Kaneva, Nadia. 2012. *Nation Branding in Post-Communist Europe: Identities, Market, and Democracy*. New York and London: Routledge.

Keohane, Robert. 2002. *Power and Interdependence in a Partially Globalized World*. London: Routledge.

Kitchin, Robert, and James Kneale. 2001. 'Science Fiction, or Fiction Fact? Exploring the Imaginative Geographies of the New Millennium'. *Progress in Human Geography* 25:17–33.

Krasner, Stephen. 2009. *Power, the State, and Sovereignty: Essays on International Relations*. London: Routledge.

Laruelle, Marléne. 2016. 'Russia as an Anti-Liberal European Civilisation'. In *The New Russian Nationalism: Imperialism, Ethnicity and Authoritarianism 2000–2015*, edited by Pål Kolstø and Helge Blakkisrud, 275–297. Edinburgh: Edinburgh University Press.

Lin, Christina. 2016. 'A New Eurasian Embrace: Turkey Pivots East While China Marches West'. In *Toward Well-Oiled Relations? China's Presence in the Middle East Following the Arab Spring*, edited by Niv Horesh, 30–47. New York: Springer.

McFarlane, Thomas, and Iain Hay. 2003. 'The Battle for Seattle: Protest and Popular Geopolitics in The Australian Newspaper'. *Political Geography* 22 (2):211–232.

Melrose, Robin. 1996. *The Margins of Meaning: Arguments for a Postmodern Approach to Language and Text*. Amsterdam: Rodopi.

Morgan, Michael, and Susan Leggett. 1996. *Mainstream(s) and Margins: Cultural Politics in the 90s*. New York: Greenwood Press.

Mukerji, Chandra, and Michael Schudson. 1991. *Rethinking Popular Culture: Contemporary Perspectives in Cultural Studies*. Berkeley and Los Angeles: California University Press.

Nanda, Serena, and Warms Richard. 2010. *Cultural Anthropology*. London: Cengage Learning.

Nayar, Pramod. 2014. *Posthumanism*. London: John Wiley & Sons.

Norquay, Glenda, and Gerry Smyth. 2002. *Across the Margins: Cultural Identity and Change in the Atlantic Archipelago*. Manchester: Manchester University Press.

Paasi, Anssi. 1991. 'Deconstructing Regions: Notes on the Scales of Spatial Life'. *Environment and Planning A* 23 (2):239–256.

Paasi, Anssi. 1996. *Territories, Boundaries and Consciousness: The Changing Geographies of the Finnish-Russian Border*. London: John Wiley & Sons.

Petracca, Mark. 1991. 'The Rational Choice Approach to Politics: A Challenge to Democratic Theory'. *The Review of Politics* 53 (2):289–319.

Phelan, Peggy, and Jill Lane. 1998. *The Ends of Performance*. New York: NYU Press.

Pierce, Marge. 2002. 'He, She and It'. In *Lost in Space: Geographies of Science Fiction*, edited by James Kneale and Rob Kitchin, 74–89. London: Continuum.

Purcell, Darren, Melissa Scott Brown, and Mahmut Gokmen. 2010. 'Achmed the Dead Terrorist and Humor in Popular Geopolitics'. *Geoforum* 75:373–385.

Rancière, Jacques. 2004. 'Aesthetics, Inaesthetics and Anti-Aesthetics'. In *Think Again: Alain Badiou and the Future of Philosophy*, edited by Peter Hallward, 218–231. London: Continuum.

Robertson, Roland. 1992. *Globalization: Social Theory and Global Culture*. London: Sage.
Said, Edward. 1978. *Orientalism*. New York: Vintage Books.
Saukko, Paula. 2003. *Doing Research in Cultural Studies: An Introduction to Classical and New Methodological Approaches*. London: Sage.
Saunders, Robert A. 2017. *Popular Geopolitics and Nation Branding in the Post-Soviet Realm*. London and New York: Routledge.
Saunders, Robert A. forthcoming. ''Brand' New States: Post-Socialism, the Global Economy of Symbols, and the Challenges of National Differentiation'. In *The Future of Post-Socialism*, edited by Danijela Lugarić, Dijana Jelača and John F. Bailyn. Albany: SUNY Press.
Saunders, Robert A., and Vlad Strukov. 2017. 'The Popular Geopolitics Feedback Loop: Thinking beyond the 'Russia versus the West' Paradigm'. *Europe-Asia Studies* 69 (2):303–324.
Sharp, Joanne P. 1993. 'Publishing American Identity: Popular Geopolitics, Myth and *The Reader's Digest*'. *Political Geography* 12 (6):491–503.
Shepherd, Laura J. 2013. *Gender, Violence and Popular Culture: Telling Stories*. London and New York: Routledge.
Shepherd, Laura J., and Caitlin Hamilton. 2016. *Understanding Popular Culture and World Politics in the Digital Age*. London and New York: Routledge.
Srivastava, A. R. N. 2012. *Essentials of Cultural Anthropology*. New York: PHI Learning Pvt. Ltd.
Strukov, Vlad. 2016. *Contemporary Russian Cinema: Symbols of a New Era*. Edinburgh: Edinburgh University Press.
Thompson, E. P. 1963. *The Making of the English Working Class*. London: Vintage Books.
van Apeldoorn, Bastiaan. 2004. 'Transnational Historical Materialism: the Amsterdam International Political Economy Project'. *Journal of International Relations and Development* 7 (2):110–112.
Warf, Barney. 2002. 'The Way It Wasn't: Alternative Histories, Contingent Geographies'. In *Lost in Space: Geographies of Science Fiction*, edited by James Kneale and Rob Kitchin, 17–37. London: Continuum.
White, Mimi, and Schwoch James. 2008. *Questions of Method in Cultural Studies*. London: John Wiley & Sons.
Williams, Richard. 1958. *Culture and Society*. London: Chatto and Windus.
Williams, Richard. 1977. *Marxism and Literature*. London: Oxford University Press.
Williams, Yentyl. 2014. Shifting Between Hegemony and Dominance? A Neo-Gramscian Analysis of the EU as a Structural Foreign Policy Actor: The Singular Case of the Cariforum-EU Economic Partnership Agreement. Bruges Regional Integration & Global Governance Paper.
Woon, Chih-Yuan. 2014. 'Popular Geopolitics, Audiences and Identities: Reading the 'War on Terror' in the Philippines'. *Geopolitics* 19 (3):656–683.
Yúdice, George. 2003. *The Expediency of Culture: Uses of Culture in the Global Era*. Durham: Duke University Press.
Zylinska, Joanna. 2005. *The Ethics of Cultural Studies*. New York: A&C Black.

4 Gender studies and popular geopolitics

Indispensable bedfellows for interdisciplinarity

Federica Caso

Introduction

Gender representations in popular culture matter to politics, for they reproduce certain understandings of gender that inform the way politics is conducted. The common narrative of male heroes saving helpless female victims replicates the popular imagination of the Just Warrior saving the Beautiful Soul from the villainous other (Elshtain 1987). The association of women with the domestic and the family reinforces the imaginary of the national community as a devoted mother who, when threatened, needs protection from its (male) citizens (Steans 2010, 404). As a number of scholars have shown (see Jeffords 1994; Weber 2005; Emad 2006; Dodds 2011; Shepherd 2013; Åhäll 2015), popular culture grounds national security in gendered grammars. Others before me have made compelling cases for including gender analyses to the politics of popular culture. That being stated, this chapter will not add to the list of how gender narratives and representations in popular culture reproduce gendered imaginations of politics in the everyday. Instead, the contribution of this chapter will be a methodological and theoretical attempt at explicating the value of concepts and methods drawn from gender studies to understandings of popular geopolitics.

In the spirit of demonstrating the interdisciplinary nature of popular geopolitics, this chapter aims at outlining how the (inter)discipline of gender studies makes methodological and theoretical contributions to the study of popular geopolitics. I focus on three pillars of gender studies: gender, sexuality, and the body, to demonstrate how these are integral to understand popular geopolitics, and are requisite for any interrogation of the politics of popular culture. This is because these three concepts belong to the realm of the popular and the everyday, and they shape our understanding of place, space, and politics. Gender, sexuality, and the body are made in the everyday activities that popular geopolitics aims at unravelling, and thus inform our understandings of political space. Hence, they are epistemological vectors that popular geopolitics can neither ignore nor neglect.

I argue that, in many respects, popular geopolitics and gender studies share similar concerns, particularly the decentring of public/private dualism, the critique of epistemologies of the closeted and the silenced, and the body as locus of political subjectivity.[1] Consequently, I contend that by drawing from feminist and gender-informed critiques and epistemologies, scholars researching popular geopolitics can draw from established theories and methodologies that can help make their arguments more grounded. The goal of this chapter is to demonstrate how concepts from gender studies assist in elucidating popular geopolitics, allowing scholars to go beyond reading male-centred, triumphalist national metaphors in popular culture as a way of telling parables of masculinist politics. In the spirit of Grayson, Davies, and Philpott's recommendation that 'if the incorporation of popular culture into international relations (IR) is going to be fruitful, there must be a willingness to go beyond an engagement with illustrations of world politics' (2009, 156), I will demonstrate how gender studies helps move us towards achieving the goal of considering 'the ways in which the popular culture assemblage "leaks" into other aspects of political life' (Dittmer 2015, 48). This leakage, I will argue, comes in the form of the production of subjectivities. I maintain that popular culture is implicated in world politics insofar as it shapes political subjectivities, fostering co-optation into existing systems of oppression but also opening avenues for resistance (Grayson, Davies, and Philpott 2009, 157). I suggest that the investigation of geopolitical subjectivities would benefit from perspectives grounded in gender studies.[2]

An overview of aims and approaches

The chapter is structured as follows: First, I introduce the interdisciplinary nature of popular geopolitics as a field of research, and contextualise the political implications of popular geopolitics scholarship in the larger field of critical international relations, cultural studies, and political geography. Interdisciplinary research is extremely challenging, and the very interdisciplinarity that is intended to ground our research in many fields, inevitably tends to place such scholarship at the margins of established disciplines—a problem that many readers of this volume have likely encountered in their own experiences with academic publishing and its associated vagaries. My concerns will be directed at the marginal spaces of the discipline of IR, highlighting that this same position of marginality has been occupied by feminist research in IR. While it would be naïve to suggest that gender is now part of the IR 'mainstream', it would be similarly foolish to fail to acknowledge that it has gained epistemic territory among contemporary thinkers in the field. Currently, feminist research in IR should still be considered to be situated at the semi-periphery of knowledge of global politics, though certainly it is a 'rising power' (if we are to use world-systems terminology to discuss the structures of academe). This move has been abetted by sustained efforts to validate its methodologies and epistemologies. Consequently, I argue that popular geopolitics would benefit from its example, as well as the application of its various concepts and analytical lenses.

In the second part of the chapter, I unpack the three pillars of gender studies mentioned above. First, I address the concept of gender, understanding gender to be a social construction rather than a biological given (Butler 1999). I highlight the epistemic milieu in which masculinity and femininity have come to be associated with the public and the private respectively. These associations have been identified as extremely problematic for women and for other entities associated with femininity, to the extent that feminist scholars and activists have directed their considerable energies to deconstruct these binaries and dismantle their lines of demarcation. Additionally, this section will demonstrate that both gender studies and popular geopolitics share a common commitment to the motto 'the personal is political'. Because this aphorism is grounded in decades of feminist critique of knowledge production and politics, I contend that popular geopolitics substantively benefits from engaging with gender studies and its critique of the gendered spaces of knowledge.

The second concept I will unpack is sexuality, primarily drawing from queer approaches.[3] Queer and gender studies have engaged the politics of sexuality to deconstruct how certain positions are normalised, while others are made to be perverse. Processes of normalisation involve processes of marginalisation and silencing. Thinking along these lines, I will argue that in the discipline of IR, popular culture has been silenced on the grounds that it is not politics but a diversion from it. The implication of this silencing practice is that we are blind to a series of political subjectivities and political spaces that can be analysed only through popular culture. Queer and gender studies have elaborated epistemologies of the closeted to give a voice to marginalised positions. Thus, I will argue that popular geopolitics can benefit from an exploration of these epistemologies to validate and legitimate the political subjectivities and spaces produced in the realm of popular culture, which has been located at the margins of politics.

The final concept that I will explore is that of the body. Over the past few decades, gender studies has become concerned with the materiality of the body. The intent of this has been to demonstrate that discourses and narratives about gender are not enough to explain the sexed and gendered body. Sexed and gendered subjects are materialised in the lived experience of the body. In this last section, I will argue that a focus on the body answers those calls of frustration that see popular geopolitics engrained in discursive approaches, which grant no space to the materiality of lived experiences of popular culture (Dittmer & Gray, 2010; Dittmer, 2010). Given gender studies' proud tradition of dealing with the lived experience of the body, I contend that popular geopolitics can benefit from a 'corporeal turn'. Moreover, a focus on the body would move away from discourse-dependent approaches that dominate popular geopolitics.

Interdisciplinarity, intertextuality, and the margins of disciplines

Popular geopolitics is a field of research interested in how popular culture and everyday cultural practices shape national identities and global politics (Dittmer and Dodds 2008). It stems from the critical branches of IR and political geography, yet neither these disciplines can claim exclusive parenting over it. Some scholars think of popular geopolitics as a sub-discipline of critical geopolitics (Sharp 1993; Ó Tuathail 1996; Dittmer 2010), while others of international relations (Grayson, Davies, and Philpott 2009; Weldes and Rowley 2015)—not necessarily in an exclusionary way. And certainly, there are other disciplines which are concerned with the political dimension of popular culture and the everyday, including cultural studies, game studies, imagology, sociology, and gender studies (see Grayson's chapter in this volume). Diplomatically, Weldes and Rowley (2015) call popular geopolitics a 'sub-inter-discipline'. I rather see it as a rhizomatic field of research, in order to avoid its colonisation from and appropriation by other disciplines, and to underscore the way it draws from various and diverse disciplines without necessarily becoming them.[4] A rhizome is, in fact, interconnected, heterogeneous, and generative (Deleuze and Guattari 1987).

As several other contributors to this volume argue, popular geopolitics is an inherently interdisciplinary field of study, for it requires the researcher to be literate at least in world politics, cultural studies, sociology, media studies, gender studies, and human geography (see also Saunders and Strukov 2017). Its interdisciplinary nature makes it a demanding and challenging field of study. The researcher has to be flexible in their thinking to be able to make connections; they also have to be rigorous in their methods to make these connections stand the test of knowledge production. Connecting the dots through a mix of creative, innovative, and proven methodologies is crucial for the development of research in popular geopolitics. Drawing from concepts belonging to different and disparate disciplines allows not only for breath, but also and especially depth of the analysis. Hence the importance of this collection, which brings into conversation multiple disciplines, and aims at addressing many of the critiques of popular culture-centric analyses of IR and world politics by making explicit the parallels in our field's interdisciplinarity with similar trajectories in gender studies.

Without the pretention of arrogating popular geopolitics to the discipline of IR, as a student of this discipline I am concerned with how the research in popular geopolitics has been positioned in relation to it. In IR, popular geopolitics emerged as part of the resurgence of culture in analyses of politics and world politics facilitated by poststructuralism (Weldes 1999). Its focus on popular culture is intended to decentre power from political and cultural elites and to denaturalise structures and relations of power (Sharp 1993; Weldes 1999). In other words, popular geopolitics propels a shift in focus from 'high politics' to the everyday networks of power. Drawing from poststructuralism, popular geopolitics underscores the production of geopolitical knowledge/understanding, for popular culture is understood as a text imbricated in the production of power/knowledge about world politics. Couched in these terms, world politics is not a top-down enterprise done by politicians and elites, but a networked system of power that is produced in the spaces of the everyday, which goes up influencing elites, and is repackaged by elites and recycled into the everyday (Weldes 1999).

At the core of popular geopolitics is a commitment to showing that we are all involved in making politics, with the 'us' being politicians, academics, unemployed, nurses, police officers, women, children, men, people from the so-called Global South, migrants, etc., and we make it through our most mundane and quotidian practices. We all consume some sorts of popular culture, and even if we choose not to consume it, we can still be the subject of its representation. In these terms, even the castaway or pariah can be involved in politics, as subject of a novel for instance. Popular culture is also a connector, for it links bodies and embodied experiences. A woman living in Pakistan might have watched the same television series as has a transgender man in Italy, thus creating a moment of symbolic, real, or virtual encounter, thereby creating what can be called a 'wormhole' of common culture. A student in Japan might post a walkthrough video of the videogame *Zelda*, and this might be watched by a shop assistant in Tunisia, and, through online forums, these two fans may eventually meet and exchange ideas. Most popular culture is produced and consumed across national boundaries producing cultural homogenisation, hybridisation, cultural imperialism, or globalisation, depending on one's political views (Weldes and Rowley 2015, 16).

Foregrounded by a series of earlier interventions (Sharp 1993; Shapiro 1999; Weldes 1999), the literature on popular geopolitics proliferated in the aftermath of 9/11. This should come as no surprise given the technological revolution brought by social media, and the pervasiveness of media technologies in our everyday life. It can be argued that this technological revolution has generated a growing interest in the politics of popular culture, broadly conceived, because of the attendant emergence of the what Berlant calls the 'intimate public sphere', that is, 'the collapsing of the political and the personal into a world of public intimacy' (Berlant 1997, 1). Privacy is harder to maintain in the digital era, and people feel they can share their opinions in the public intimacy of their social media, thus creating a web of connections between ideas and people that is unprecedented. Performing our daily life increasingly draws on readings of other people's lives. Berlant argues that within this intimate public sphere, the nation depends on 'personal acts and identities performed in the intimate domains of the quotidian' (1997, 3). And this is exactly the target of popular geopolitical research. Thus, while the politics of popular culture is not new, the social, cultural, and technological milieu allows for greater purchase, wider transmission, and new platforms for contesting meanings. In this respect, popular culture is an intertext, for it draws from various scenarios that are interpreted and read next to each other, both far and wide from their sites of origin.

This, however, does not necessarily translate into the mainstreaming of the politics of popular culture. The discipline of IR is still implicitly or explicitly sceptical about the value of popular culture as data. Drawing from Agathangelou and Ling's (2004, 23) metaphor of the 'house of IR', Cait Hamilton and I have argued that the discipline of IR is well trained in practices of knowledge policing, and that it keeps popular geopolitics ever 'on the doorstep, [as] an uninvited and unwelcomed guest' (Caso and Hamilton 2015, 3). While conferences, workshops, monographs, edited volumes, journal special issues, and blogs have been established and received considerable attention from scholars across the world, research in popular geopolitics from the discipline of IR is still not entirely legitimate, especially when conducted by early career scholars (Caso and Hamilton 2015)

A same position of marginality once characterised feminist scholarship in IR. The discussion and debates between gender specialists and political and IR theorists has long been a 'dialogue of the deaf' (Enloe 2001, 661). Steans (2003) remarks that while other critical positions have found more accommodating spaces in IR, this has not been the case for feminist scholarship. She adds that mainstream IR has engaged feminist critique in a selective way, privileging those positions that work with stable categories of analysis (see also Weber 1994). This selectivity functions to reproduce the structure of 'the house of IR' through the appropriation of data which reinforces mainstream positions, and through the rejection of those which challenge them (Agathangelou and Ling 2004). The marginality of feminist scholarship in IR stems from the nature of the questions it poses, and attendant methods that these require. Feminist scholarship is not about causality, and therefore conventional methods in the social sciences are not well suited for the research questions it addresses (Tickner 2005).

Critical feminist scholarship, like popular geopolitics, has been kept at the margins of IR because of the nature (or more accurately, the positionality) of its inquiry, which relates to the construction of gendered political subjectivities, representations of gender, and everyday practices of subversion of constituted gender relations. The methods of research—and political actions—cannot be the same as conventional IR wants them to be, because both feminist positions and popular geopolitics do not deal with cause-effect relations and problem-solving (Ó Tuathail 1996). Conversely, they both deal with the material semiotic production of meanings and their political implications for everyday politics. While the dialogue between IR feminism and mainstream IR is still sometimes a 'dialogue of the deaf', it is important to acknowledge that IR feminism has gained epistemic ground in IR, thus posing some interesting lessons that popular geopolitics can draw from.

The spaces of gender and gendered spaces

Feminist scholarship and popular geopolitics share similar concerns about the spaces of gender, and how these become gendered spaces. Both aim at subverting the existing dichotomy between public and private, and are committed to a politics that affirms that the personal is political. I highlight in turn how gender studies and popular geopolitics challenge the 'proper' space of politics by focusing on the politics of the private and the intimate.

Women and femininity have historically been associated with the domestic and the private sphere. Feminist politics and gender studies have challenged this association and demonstrated that the personal, the private, and the feminine are political. The relegation of women and femininity to the private and the domestic is documented by Betty Friedan (1963) in her discussion of 'the problem that has no name'. Here she writes about the deep sense of dissatisfaction that American housewives felt because they were not able to move beyond the boundaries of domestic femininity that were imposed on them by the American society. They were led to believe that the identity of 'housewife' was the best, or at least most viable, option for women to express their femininity. In the 1950s, the perfect woman was the master of the home, 'freed by science and labour-saving appliances from the drudgery, the dangers of childbirth, and the illnesses of her grandmother. She was healthy, beautiful, educated, concerned only about her husband, her children, her home' (Friedan 1963, 15–16). While Friedan was addressing the American housewives, the association of femininity and women to domesticity was not geographically specific. Women were absent from politics. Feminist scholars like Cynthia Enloe (2000), Ann Tickner (1992), Katharine Moon (1997), and Carol Cohn (1987) have demonstrated the marginalisation of women from politics in multiple contexts. Additionally, they have revealed how such marginalisation served to grease the wheels of politics, as demonstrated, for instance, by Moon (1997) who discussed how prostitution played a significant role in shaping the U.S.–Korean relations during the Korean War (1950–53). She exposed that ensuring the provision of sex for American soldiers and officials was paramount to maintain continued American commitment to the liberation of the occupied peninsula. The creation and maintenance of gendered spaces and roles was a key determinant of the Korean War. Thus, women and femininity were confined to the private and domestic spheres, but feminist scholars have demonstrated that these spaces have always been political spaces.

The absence of (visibility of) women from politics and international affairs has deep semiotic and epistemological roots. Women have historically been associated with passivity, while men with being marked as active agents, an association that related to perceptions of the body (de Beauvoir 1949). The female body is coded into passivity in the sexual act of receiving penetration, while the male as active in the act of penetrating. The heteronormative sexual culture reproduces this association (Rich 1993; Segal 1997). Monique Wittig (1992) argues that heterosexuality is an ideology that serves to maintain the dominance of men over women and privileges masculinity over femininity. This edifice of semiotic associations between what bodies mean and what social and political power they are granted, is erected upon an epistemological base that codes public knowledge as masculine and private experience as feminine, something which is frequently reinforced through popular culture of all forms (but especially by comic books, videogames, films, and television shows). Historically, reason was a human faculty that was believed to be possessed by men only. For Aristotle, women were impotent men incapable of connecting to reason (Lloyd 1992). Furthermore, women were also associated to their bodies, which in the Christian tradition was regarded as sinful, and therefore detached from God (Lloyd 1992, 112). With the affirmation of the Cartesian method of knowledge production and the dualism between mind and body, where the former was elevated to transcendental knowledge producer, and the latter relegated to the doubts of existence, women were further marginalised from the production of knowledge. Lloyd writes,

> We now have a separation of functions backed by a theory of mind. Given an already existing situation of sexual inequality, reason – the godlike, the spark of the divine in man – is assigned to the male. The emotions, the imagination, the sensuous are assigned to women. They are to provide comfort, relief, entertainment and solace for the austerity that being a Man of Reason demands.
>
> (Lloyd 1992, 117)

Male subjectivity was thus made into objectivity (Haraway 1997). Simultaneously, the bodily and sensorial—experiences associated to women and the feminine—were neglected as mere private experience, and experience does not count as knowledge. Thus, feminists have had to fight for the subversion of a cultural framework that has positioned women outside of knowledge production because associated with the private, domestic, and sensorial domains.

Popular geopolitics, like feminism and gender studies, faces similar dismissal based on the unwritten rules and highly policed spaces of knowledge production. Politics is too often conceived to pertain exclusively to public or global affairs. On the contrary, with its claim that the popular, domestic, and everyday are spaces to investigate politics, popular geopolitics challenges the 'proper' spaces of politics. Popular geopolitics emerged from the recognition that world politics is not done solely within elite political environments. As Sharp argues,

> In order to have their texts accepted as reasonable, geopoliticians have to draw upon discourses already granted hegemonic social acceptance. These discourses are reproduced within culture.
>
> (Sharp 1993, 493)

She rules out both the idea that the contemporary media simply reflects the narrative of political elites, and that it projects bottom-up popular understandings. What she endorses instead is the complex relation that exists between citizens, political elites, and the media. Weldes (1999) makes a similar point that the study of political discourses has mostly focussed on the discourses made by political elites, thus overlooking aspects of politics that happen elsewhere.

> Just as official representations of international politics and foreign policy depend upon the cultural resources of a society, so too do the ways in which those state actors' representations of international relations are understood.
>
> (Weldes 1999, 119)

Both Sharp and Weldes propose to decentre our focus from 'high politics' and direct it towards 'low data' (Weldes 2006). This diffused understanding of politics as something that happens in the popular and the everyday does not fit with IR obsession with state-centrism and static policy documents (Weldes 2006, 178). In this respect, popular geopolitics struggles to fit with IR's definitions of the political.

Furthermore, popular geopolitics suggests that politics happens in the privacy of domestic realms where popular culture is consumed (the 'living room' or the 'den', for instance). For instance, Dodds (2008) argues that films are a particularly effective form of political communication as they demand people's attention and trigger affective responses to often preposterous scenarios. By informing geopolitical imaginations, films help to create understandings of world politics and produce national identities. They help define who 'we' are as a nation, and how we conceive of the other nations around us (Saunders 2017). Rowley argues that 'the ways in which people make sense of world politics is, in large part, via the knowledge and understanding created through interactions with the world in the realm of the popular, the mundane and the everyday' (2010, 309). Popular culture is a *magister vitae*. People rarely read academic books or policy documents, and there is an increasing tendency to rely on social media platforms to receive news and updates about current affairs. News and entertainment media are source of political information and communication. With this in mind, popular geopolitics identifies the private sphere as a space of geopolitical imagination that deserves academic attention if we are to understand contemporary dynamics in world politics. Focusing attention on the everyday, the popular, and the private sphere, popular geopolitics exceeds the boundaries of the political in IR.

A similar battle is fought by feminist scholars in IR, who have shifted attention from macro geopolitical systems to the lived experience of marginal and vulnerable people, emphasising their subjectivities and the loci where these are formed (Hudson 2005; Soreanu and Hudson 2008). Feminism has deconstructed the gendered spaces of knowledge production in order to open a window on marginalised subjectivities, so they can be visible in politics and IR. This has been done, for instance, by emphasising how matters considered to be private such as emotions, experiences of war, and feelings are political and have political implications (see inter alia Parashar 2013; Sylvester 2013; Åhäll and Gregory 2015). In order to gain purchase in IR, these positions stress what we miss from not engaging with the private and the personal. For instance, the feminist scholarship focusing on experiences of war highlights that starting from the micro, rather than the macro, we gain information about what war is, 'how it operates, who takes part, how they are affected and affecting, and what the politics of war looks like beyond the war rooms of state' that we would not gain otherwise (Sylvester 2011, 129).

While popular geopolitics recognises the private and the domestic as political, it would greatly benefit from feminist frameworks that challenge the gendered spaces of knowledge to make visible subjectivities, positions, and issues that would otherwise remain in the shadow of established epistemologies and methodologies. Feminism has given a public language for private silence (Landes 1998), stretching the boundaries of knowledge of IR. Like most feminist theories, popular geopolitics is concerned with the interplay between the private, the everyday, the public, and the political. The synergy could be made more explicit. The feminist motto 'the personal is political' perfectly fits the scope of popular geopolitics, and its analysis of how geopolitics permeates popular culture, and the 'prosumption' of popular culture produces geopolitical discourses (see Ritzer and Jurgenson 2010). In fact, popular culture does not happen in a cultural vacuum and it produces political subjectivities (Grayson, Davies, and Philpott 2009). In this respect, the 'public' of world politics, and the 'private' of the domestic consumption of popular culture interplay—if ever they existed as separate realms—in ways that reveal how the personal is political and international. For instance, 'violent films or videogames may not cause young men to go out and kill but they may provide one layer in the complex continuum that congeals into deeply seated antagonisms towards particular others' (Grayson, Davies, and Philpott 2009, 156). Thus, even though popular culture does not necessarily shape political behaviour, scholars interested in the popular culture/world politics continuum contend that it affects our geopolitical imagination. Drawing from feminist and gender scholarship, popular geographers can emphasise more how the consumption of popular culture in the domestic and private sphere is political insofar as the personal, the private, the intimate, and the domestic are political according to gender studies.

Queering popular geopolitics

Both queer approaches and popular geopolitics are concerned with what is often understood to be at the margins: queer approaches at the margins of heterosexuality, and popular geopolitics sitting at the edge of international relations/world affairs. Both aim at bringing to the fore subject positions that would otherwise remain silenced and/or unexplored. Both transgress boundaries of the normal in search for new spaces of politics and subjectivity formation. In this section, I outline how they can join forces to give voice to marginal knowledge positions, as well as produce more accurate representations of what is *really* going in the world.

Queer approaches challenge normative grammars and attendant uncontested subject positions. Queer research is quintessentially 'positioned within conceptual frameworks that highlight the instability of taken-for-granted meanings and resulting power relations' (Browne and Nash 2010, 4). It is a way of challenging through the lenses of sexuality, but is not limited to sexual normativities. The terminology 'queer theory' was coined by Teresa de Laurentis in 1991 and was conceived to be the post-gay and lesbian studies discipline, interested in processes of normalisation, categorisation, exclusion, and policing of borders of sexual identities (see Jagose 1996). Jagose (1996) argues that queer is a category of analysis to destabilise the linear relation that has historically been established between sex, gender, and desire, and that constructs heterosexuality as the normal sexuality. In virtue of this destabilisation, queer approaches make epistemic space for multiple and alternative subject positions to emerge (see also Butler 1999). Warner (1993) goes further suggesting that queer approaches are not just about the critique of identity politics, but of society itself. He contends that queer politics defines itself against what is considered to be normal, rather than against the heterosexual subject. He and Berlant use queer perspectives to challenge, stating that:

> the institutions, structures of understanding, and practical orientations that make heterosexuality seem not only coherent, but also privileged. [...] Contexts that have little visible relation to sex practice can be heteronormative in this sense, while in other contexts forms of sex between men and women might not be heteronormative.
>
> (Berlant and Warner 1998, 548)

Thus, the primary concern of queer politics is neither homosexuality nor heterosexuality, but the hierarchies of power that make certain practices seem normal and unquestionable (an entirely appropriate model for those working in the field of popular geopolitics).

Queer approaches have given sexuality an epistemological perspective, arguing that it is not a matter of private life, medicine, or minority rights, but a political issue (Foucault 1980). This epistemology shades light on the silenced, the closeted, and the perverse that is foreclosed by discourse of the normal. As Kosofsky Sedgwick (1990) argues, silence itself is a form of knowledge that produces subjects, and therefore is imbued with power. Drawing from insights by Foucault (1980), Kosofsky Sedgwick (1990), Butler (1999), Warner (1993), and Halberstam (1997) among others, it can be argued queer epistemologies are epistemologies of the silenced, the marginalised, the abject, the failed. These terms are all constructed in binary opposition to something else that is given priority in virtue of its normality. But also, they are constructed within language with material consequences for the unspeakable non-normative subject and for what is marginalised by language. This is constructed as the perverse. Thus, queer approaches are concerned with the margins of knowledge and the binary oppositions that make in language certain subject perverse with the material consequence of marginalisation and silencing.

The discipline of IR has a long history of normalising, condoning, and silencing (see Keohane 1989, for an example of border policing; and Weber 1994; Tickner 1997, for a critique). Thus, bringing in queer perspectives answers the call from IR critical scholars to pay attention to silence (Der Derian 1989; George and Campbell 1990; Grayson 2010), as well as diversity (Tickner 1997; Sjoberg 2012). Weldes (2006) argues that a way to police the borders of IR is to define certain data as 'high' (policy documents), and others as 'low' (popular culture); certain forms of politics as 'high' (war), and others as 'low' (the construction of subject positions). Often, 'low data' are appropriated by IR in ways that benefit the orthodoxy, but do not give credit for the politics of the space where the data was produced. For this reason, Agathangelou and Ling label IR a 'colonial household' (2004, 23). This is the fate that scholars of popular culture and world politics are constantly confronted with, silenced on the ground of their analyses' irrelevance to politics and international relations, and relegated outside of what politics is understood to be (Caso and Hamilton 2015).

Queer-international approaches allow to focus on those subjects that do not signify monolithically. Weber (2016) argues that queer curiosity in international relations helps us detect how the normalised can simultaneously be perverse and perverted, and how one thing or person can embody multiple phenomena and subjectivities. With its focus on sexuality, queer theorising draws attention to the regimes of desire and pleasure, shifting away from stable identities and orientations that define the subject (Wilcox 2014, 615). As such, the subject of queer politics is understood as an assemblage of heterogeneous parts (Puar 2007). The end result is that nothing is essential, and the analyst is forced to map the intersections of power, rather than rely on pre-packed normativities. This is exemplified by the work of Weber (1999) and Puar (2005), which demonstrate that the state acting in the international system is not necessarily heterosexually masculine, but queer, embracing at times the normativity of heteronormative masculinity, and at other times homonormativity and transgenderism. Gendered hierarchies and dichotomies are not absolute (Wilcox 2014, 613), and as a consequence, queer approaches build upon the concept of performativity and the idea that being is doing (see inter alia Butler 1999; Weber 1999). For the sake of summary, queer international approaches function to interrogate the ambiguous subjects, and the normativities within which these subjects are constituted.

On these grounds, a queer curiosity is well positioned to explore the space of marginality in politics that popular culture occupies. It allows us to recognise and dissect the subject of politics that is not constructed solely in foreign policy texts and established/elite discourses, but also in the everyday and mundane acts of engagement with popular culture artefacts. Contradictions can be mapped in the *either/and* logic that Weber (2016) puts forward. For instance, by queering popular geopolitics, we can embrace the dissonance that exists between fiery disputes in foreign policy circles on the so-called ‘war on terror’, and the minimal box office success that critical films on the war on terror have garnered (Philpott 2010). From an approach that does not allow for consideration of ambivalences, it would be difficult to make sense of the fact that governments and news media have vehemently campaigned in support of the war on terror, but films explicitly about the war on terror have performed poorly at the box office. Conversely, by queering popular geopolitics, we can see these as two different manifestations of the same phenomenon, just like homosexuality and heterosexuality are product of the same discourse of sexuality. Similarly, understanding the subject of popular geopolitics as a queer subject made of/by assemblages of pleasures and desires, the analyst is equipped with new tools to explore the role of popular culture in the interplay between agency and structures, between private and public, between foreign and domestic politics, between emotions and rationality. The ‘queer’ subject of popular geopolitics is constructed in the assemblage of these spaces, and in-between the boundaries constructed between them.

Embodying geopolitics

Finally, popular geopolitics might enormously benefit from gender studies in the ways it deals with the materiality of the body. Taking the body seriously in popular geopolitics will help to ground the field of research in the politics of what matters at the level of the everyday, and escape critiques about abstraction and lack of political relevance. Furthermore, a body-focussed popular geopolitics is transformative for the fields of critical geopolitics and international relations, for it shifts the narrative from abstracted macro-systems of world politics, to the micro-politics that shape world politics.

Critical geopolitics and international relations have both been accused of being abstract and detached from the human experience of global politics. Commenting on critical geopolitics, Thrift (2000) contends that geopolitics has academically been constructed as a discourse; this construction has problematic implications for the way we understand and analyse the practices of geopower. Similarly, Martin Müller (2008) expresses discontent for the way the notion of geopolitical discourse is ill-defined and too vague to the point of confusion and arbitrariness. According to him, this risks detracting epistemological rigour and missing the junctions of material geopower. In a similar vein, critical—mostly feminist—scholars have launched a critique on the ways the discipline of international relations focusses on macro-systems and patterned behaviour. The political implication of this is that the discipline misses the lived experience of world politics. IR, a field of research interested in war, does not know what war feels like (Sylvester 2011; Parashar 2013). Combined together, these critiques exhort scholars interested in world politics and geopolitics to pay attention to the materiality of the body as well as the lived experience of embodied subjectivities.

Popular geopolitics is not devoid of this critique. Dittmer and Gray (2010) point out that popular geopolitics has overlooked the body and forms of embodiment in favour of elite discourse and textual analysis. All too often a narrative approach is undertaken in order to validate the scholarly interest (Dittmer 2015). Grayson, Davies and Philpott remark that popular geopolitics should move on from simply identifying allegories, metaphors, and imaginaries of world politics in popular culture towards analyses of how popular culture constructs subjectivities (2009, 156–157). The danger of a narrative and/or discursive approach to popular geopolitics is the production of analyses that are subjective and lack relevance to real-world problems. Identifying metaphors is extremely important. This is demonstrated, for instance, by the fact that science fiction as a genre invites us to imagine different political alternatives and dystopia. As Kiersey and Neumann put it, 'One of the great virtues of science fiction is its ability to pose fictional worlds that, while cognitively coherent on their own unique terms, nevertheless inevitably maintain a link with experiences we share in our own world' (2015, 80). Thus, tracing analogies in science fiction helps us explore what politics might look like in certain contexts that do not exist, but could potentially occur. Furthermore, as Saunders notes, this genre 'has a long history of using allegory to critique the actions of political elites, at both national and international level' (2015, 151). However, while allegories, metaphors, and narratives are important to popular geopolitics, they cannot be all the field of research has to offer.

A focus on bodies allows a different perspective on popular geopolitics, one that is grounded on the materiality of bodily encounters and the materialisation of power. As Dittmer points out:

> The human body emerges as important [in popular geopolitics] because not only do traces of popular culture materialise in the body – as somatic archives of sorts – but also because the body serves as a site of affective interactions, where new forms of popular culture interact with previous ones, and with experiences of current events, as resources for political subject formation.
> (2015, 49)

Understood in these terms, the body is the locus where popular culture 'sticks' onto corporeal spaces, and thus constructs or materialises subjectivities. By this I am not suggesting that watching a war film constructs the soldiering subject. It is not a causal relation; however, something happens in the encounter between the subject and the object. An experience is formed. This experience is informed by past experiences, ideas, and values. Specific forms of knowledge and judgment emerge in the encounter, in ways that shape our perception of ourselves in the world. This phenomenological moment might be dismissed by those interested in macro-systems and patterned behaviours in global politics; however, it becomes extremely important if we shift the focus of our research on micro-politics and idiosyncrasy to account for the multiple possibilities of becoming. In order to do this, we have to also change the way we think about popular culture. Dittmer (2015, 47–48) suggests that rather than conceptualising popular culture as a 'thing' that can be analysed and unpacked, we take it as a 'doing', or otherwise as an 'assemblage'. This means that rather than trying to capture popular culture in itself, we start asking what does it do and how it connects bodies and objects. This involves tracing the links between producers, consumers, meanings, and cultural values of production and consumption—among other things. It also means that the popular culture artefacts can never be considered in isolation, but always in relation, as part of an ongoing system of repetition, regurgitation, and recycling. They become part of a system. By doing so, we are positioned to identify how popular culture 'leaks' into other aspects of politics. By taking this epistemological position, the phenomenological experience of the encounter with the popular culture artefacts becomes a way of being in the world (Merleau-Ponty 1962). As Dittmer states, 'This indicates how bodies may carry within them all sorts of political capabilities that are entirely virtual, or latent' (2015, 49). Thus, under different circumstances encounters with popular culture may produce different and idiosyncratic outcomes; yet, the common denominator of these outcomes is the way they shape our subjectivities and being in the world.

Taking the body as epistemological ground allows for consideration of two factors that so far have been mostly overlooked by popular geopolitics. First, by focusing on bodies we can unpack emotional geographies, that is, how emotions are attached to ideas of spaces and places. Dittmer and Gray (2010) note that popular geopolitics has tended to abstract pleasure, thrill, and tragedy from people's experiences of popular culture. On a practical level, this means that emotions have been abstracted, and the subjectivity of the researcher is made into scientific objectivity, thus reproducing that masculinist epistemological position that feminist scholars have long tried to dismantle (Haraway 1997). Within a framework that does not consider emotional and bodily reactions to the encounter with popular culture, the subject of the research is the subjectivity of the researcher. Second, by taking bodies and embodiment seriously, we can move away from studying popular culture artefacts on their own, and consider them in relation to the material political environment around them. Dittmer and Gray (2010) argue that by doing so we can move from a propaganda model towards an audience reception model. This audience reception model, though, ought not to be regarded as a patterned model, but as a phenomenological encounter to explore the construction of subjectivities. As they put it, '[b]ecause audiences have differing degrees and varieties of cultural capital, audiences create their own systems of meanings within a text, consciously or unconsciously, which may or may not overlap or reflect that which it was originally intended to convey' (Dittmer and Gray 2010, 1669).

Having accorded due importance to the body in the field of popular geopolitics, I want to highlight how gender studies would facilitate this kind of research, in an effort to support this volume's goal of presenting popular geopolitics as an inter-discipline that links a variety of other, somewhat marginal, spaces in academe. Feminist scholarship is quintessentially positioned to inform research on the body and embodied subjectivities. Feminist scholars like Judith Butler, Moira Gatens, Elisabeth Gosz, Luce Irigaray, Donna Haraway, Rosi Braidotti and Sarah Ahmed, among many others, have brought attention to the body as the locus of material and sensible experience. From this perspective, the body is understood as the base for political subjectivity. Paradigmatically, Luce Irigaray (1977) has focussed on the female body and the embodied experience of female pleasure to restore female subjectivity, outside of a man-made epistemic world. Women have historically been defined by men as their altereity and have been confined to man-made representations of what women ought to be (de Beauvoir 1949). Women have therefore been neglected the possibility to define their own subjectivity. Irigaray argues that within a phallogocentric order, women are objects not subjects. Phallogocentrism is a space dominated by vision and phallic power, wherein female genitalia is signified as 'the horror of nothing to see' (Irigaray 1977, 26). Taking the female body as subject of action and encounter, Irigary dismantles the discursive and scopic regime of representation that has caged women, and argues that as an affective being, female subjectivity ought to be constructed away from the patriarchal schema of vision, and in relation to embodied experiences of touching.

Discursive representation is an issue Rosi Braidotti (2002) and Judith Butler (1999) take seriously too, as that power that has neglected female subjectivity. Butler asks, 'what relations of domination and exclusion are inadvertently sustained when representation becomes the sole focus of politics?' (1999, 9). Representation, visual or discursive, limits the subject of politics such that it is constrained to function within its boundaries. However, while subjects are shaped by discourses and representations, they always exceed them in their embodied forms. Representations are norms of intelligibility, but bodies cannot be reduced to them. For Braidotti embodiment is the locus of resistance. She argues that the subject is located in the ontology and materiality of the body, and its subjectivity—the awareness of itself being in the world—is shaped by the material encounters with other subjects and objects around. In these terms, subjectivity is not essential but procedural (Braidotti 2002, 199). Hence, the subject becomes a multiplicity of potentialities that are actualised in encounters. The subject is not pre-given, but the product of relations, and these relations can be multiple and contrasting. Hence, the subject is not single and monolithic, but multiple and schizophrenic.

Embodied experiences are necessarily related to bodily feelings and emotions. Thus, in order to analyse the production of subjectivities in affective encounters, these have to be taken into account. Emotions and feelings run through the liminality of the body, connecting the inside and the outside (Grosz 1994; Ahmed and Stacey 2001). Sarah Ahmed (2004) contends that emotions shape what bodies can do, increasing or decreasing their power to act.

> Emotions shape the very surfaces of bodies, which take shape through the repetition of actions over time, as well as through orientations towards and away from others. Indeed, attending to emotions might show us how all actions are reactions, in the sense that what we do is shaped by the contact we have with others.
>
> (Ahmed 2004, 4)

In these terms, emotions are determinant factors of political subjectivity, for they are inherently linked to agency. Understood as a point of perpetual interaction between the inner personal self and the social environment in ways that neither is pre-determined, but is in a constant flow of exchange and reciprocity, we can contextualise agency within structures of power.

Using this preliminary—and certainly not exhaustive—feminist framework to analyse bodies and embodied subjectivities we can formulate a popular geopolitics that centres on the construction of political subjectivities, rather than on narratives and metaphors. While examples of this approach are scarce, it is important to point at some scholarship that moves in this direction. For instance, analysing the film *Jarhead* (2005), Victoria Basham (2015) identifies the role of boredom in constructing the militarised masculine soldiering subject. The film shows how boredom in war—sitting and waiting for the action to happen—is a defining element in the process of becoming a soldier. This process is highly gendered, for boredom in war is filled with sexualised moments of heteronormativity that are conceded to men but not to women. Basham quotes an excerpt from the film:

> Suggested techniques for the Marine in the avoidance of boredom and loneliness: masturbation. Rereading letters from unfaithful wives and girlfriends. Cleaning your rifle. Further masturbation .... Discussing in detail every woman the Marine has ever fucked ... Left- versus right-handed masturbation.
>
> (Basham 2015, 129)

In this passage the film articulates how boredom constructs the male soldiers in ways that are foreclosed to the female soldier. Basham positions the heteronormative militarised masculinity presented in the film against the historical construction of the military as a masculine space. In fact, female soldiers are not allowed that sexualised space that is allowed to men without the implication of exceeding the boundaries femininity. In so doing, Basham identifies not only the construction of subject positions within the film—the 'proper' male soldier—but also the embodied experience, or subjectivity, that informs that subject position—the feeling of boredom and the heteronormative way of avoiding it.

Basham does not tackle audience reception, thus making the analysis of popular culture as an assemblage limited in some respects. Bernazzoli and Flint (2010) have produced a study that addresses this. In an attempt to unpack the militarisation of society, they focus on the everyday practices of embodiment of militarism. While not focusing on popular culture artefacts as conventionally understood, they analyse practices of the everyday, illuminating a path that can be applied to popular culture as well. They have analysed people's embodied experiences of parks, museums, shopping malls, memorial sites and other spaces of the everyday. They have collected data through participant observation, interviews, and documentaries such as photos, and triangulated them. They have focussed on embodied experiences and perceptions of gender roles in the military, sympathy for soldiers, and support for the troops and war, and have considered representations of the garrison state in popular culture. In so doing they have produced a unique study on the ways the militarisation of society is not only represented in popular culture, but also lived and embodied in the everyday practices of 'ordinary people'. The article does not give much space to emotions, but its relevance resides in the space it opens for people's experiences to be voiced, in ways that not many studies on popular geopolitics do at the moment.

Neither of these two examples that I have discussed are complete pictures of popular culture and the everyday as assemblages, but they both show best practices that move beyond the narrative and discursive approach dominant in popular geopolitics, for they take under consideration emotional responses and affective politics in popular culture, and embodied subjectivities lived in everyday spaces of politics.

Conclusion

Popular geopolitics is a relatively new field of research. This novelty poses challenges to the researcher who intends to venture into this scholarly space. This same novelty, however, is a guarantor of innovation. In this chapter, I have suggested that by drawing from gender studies and feminist positions, popular geopolitics can overcome some of its challenges and ensure such innovation. Doing so will help popular geopolitics root is research space within the private and domestic, thus taking the lead from other disciplines. In doing so, it will ultimately benefit from the exploration of embodied and interconnected subjectivities rather than stand-alone narratives.

Going back to the notion of popular geopolitics as a rhizomatic field of research that I proposed at the start of the chapter, this interdiscipline should remain open to influences from other disciplines. Its interdisciplinarity is the key of its innovation as well as source of continuity with the existing knowledge. Thus, neither gender studies nor IR should colonise popular geopolitics, but rather they should offer tools of analysis and perspectives to ground research. So too should the other (inter)disciplines presented in this edited volume.

I would argue that the unique perspective that gender studies offers to popular geopolitics is a focus on the body. Grappling with the corporeal and embodied experiences in the social sciences is not an easy enterprise, and it requires a change in our analytical frameworks. We must learn to give more importance to the micro-systems of politics and connect them to the bigger picture. The importance of doing so is the commitment to the materiality of what matters to people and real politics. Thus, as I have suggested in this chapter, gender studies might help popular geopolitics shift away from a narrative and/or discursive approach and move towards a situated embodiment paradigm. This can be done by focusing on the construction of subjectivities in the encounter with the popular culture artefact, which should then be positioned within the networks of power that surround it.

Notes

1 Ways of constructing knowledge that privilege the normative and marginalise the non-normative (see Kosofsky Sedgwick 1991).
2 Certainly, it can be argued that the reverse is also true, but that would drag us into an endeavour different from what this chapter aims to do.
3 I will avoid throughout the language of 'queer theory' in spirit of acknowledging the plurality of positions and concerns within it. I will use instead the language of 'queer approaches or studies'.
4 As many do, I take the language of rhizome from Deleuze & Guattari (1987). However, I am not trying to argue for a Deleuzian approach to popular geopolitics. Rhizome is a botanical term to refer to a subterranean stem that is ramified in various directions, so drawing from diverse sources and connecting to different spaces.

References

Agathangelou, Anna M., and L. H. M. Ling. 2004. 'The House of IR: From Family Power Politics to the Poisies of Worldism'. *International Studies Review* 6 (4):21–49.

Åhäll, Linda. 2015. *Sexing War/Policing Gender: Motherhood, Myth and Women's Political Violence* London: Routledge.

Åhäll, Linda, and Thomas Gregory. 2015. *Emotions, Politics and War*. London and New York: Routledge.

Ahmed, Sara. 2004. *The Cultural Politics of Emotions*. Edinburgh: Edinburgh University Press.

Ahmed, Sara, and Jackie Stacey. 2001. *Thinking Through the Skin*. London: Routledge.

Basham, Victoria M. 2015. 'Waiting for War: Soldiering, Temporality and the Gendered Politics of Boredom and Joy in Military Spaces'. In *Emotions, Politics and War*, edited by Linda Åhäll and Thomas Gregory. London and New York: Routledge.

Berlant, Lauren. 1997. *The Queen of America Goes to Washington City: Essays on Sex and Citizenship*. Durham: Duke University Press.

Berlant, Lauren, and Michael Warner. 1998. 'Sex in Public'. *Critical Inquiry* 24 (2):547–566.

Bernazzoli, Richelle M., and Colin Flint. 2010. 'Embodying the Garrison State? Everyday Geographies of Militarization in American Society'. *Political Geography* 29 (3):157–166.

Braidotti, Rosi. 2002. *Metamorphoses: Towards a Materialist Theory of Becoming*. Padstow, UK: Polity Press.

Browne, Kath, and Catherine J. Nash. 2010. 'Queer Methods and Methodologies: An Introduction'. In *Queer Methods and Methodologies: Intersecting Queer Theories and Social Science Research*, edited by Kath Browne and Catherine J. Nash, 1–24. Surrey: Ashgate Publishing.

Butler, Judith. 1999. *Gender Trouble: Feminism and the Subvertion of Identity* (2nd ed.). London: Routledge.

Caso, Federica, and Caitlin Hamilton. 2015. 'Introduction'. In *Popular Culture and World Politics: Theories, Methods, Pedagogies*, edited by Federica Caso and Caitlin Hamilton, 1–8. Bristol, UK: E-International Relations.

Cohn, Carol. 1987. 'Sex and Death in the Rational World of Defense Intellectuals'. *Signs: Journal of Women in Culture and Society* 12 (4):687–718.
de Beauvoir, Simone. 1949. *The Second Sex*. New York: Vintage.
Deleuze, Gilles, and Félix Guattari. 1987. *A Thousand Plateaus: Capitalism and Schizophrenia*. Minneapolis: University of Minnesota Press.
Der Derian, James. 1989. 'The Boundaries of Knowledge and Power in International Relations'. In *International/Intertextual Relations: Postmodern Readings of World Politics*, edited by James Der Derian and Michael J. Shapiro, 3–10. Lanham: Lexington Books.
Dittmer, Jason. 2010. *Popular Culture, Geopolitics, and Identity*. Lanham, MD: Rowman & Littlefield.
Dittmer, Jason. 2015. 'On *Captain America* and "Doing" Popular Culture in the Social Sciences'. In *Popular Culture and World Politics: Theories, Methods, Pedagogies*, edited by Federica Caso and Caitlin Hamilton, 45–50. Bristol, UK: E-International Relations.
Dittmer, Jason, and Klaus Dodds. 2008. 'Popular Geopolitics Past and Future: Fandom, Identities and Audiences'. *Geopolitics* 13 (3):437–445.
Dittmer, Jason, and Nicholas Gray. 2010. 'Popular Geopolitics 2.0: Towards Methodologies of the Everyday'. *Geography Compass* 4 (11):1664–1677.
Dodds, Klaus. 2008. 'Hollywood and the Popular Geopolitics of the War on Terror'. *Third World Quarterly* 29 (8):1621–1637.
Dodds, Klaus. 2011. 'Gender, Geopolitics, and Surveillance in *The Bourne Ultimatum*'. *Geographical Review* 101 (1):88–105.
Elshtain, Jean B. 1987. *Women and War*. New York: Basic Books.
Emad, Mitra C. 2006. 'Reading Wonder Woman's Body: Mythologies of Gender and Nation'. *Journal of Popular Culture* 39 (6):954–984.
Enloe, Cynthia. 2000. *Maneuvers: The International Politics of Militarizing Women's Lives*. Berkeley, Los Angeles: University of California Press.
Enloe, Cynthia. 2001. 'Interview with Professor Cynthia Enloe'. *Review of International Studies* 27 (4):649–666.
Foucault, Michel. 1980. *The History of Sexuality: Volume I*. New York: Vintage.
Friedan, Betty. 1963. *The Feminine Mystique*. London: Penguin.
George, Jim, and David Campbell. 1990. 'Patterns of Dissent and the Celebration of Difference: Critical Social Theory and International Relations'. *International Studies Quarterly* 34 (3):269–293.
Grayson, Kyle. 2010. 'Dissidence, Richard K. Ashley, and the Politics of Silence'. *Review of International Studies* 36 (4):1005–1019.
Grayson, Kyle, Matt Davies, and Simon Philpott. 2009. 'Pop Goes IR? Researching the Popular Culture-World Politics Continuum'. *Politics* 29 (3):155–163.
Grosz, Elizabeth. 1994. *Volatile Bodies: Towards a Corporeal Feminism*. St. Leonards, N.S.W: Allen and Unwin.
Halberstam, Jacob. 1997. 'Who's Afraid of Queer Theory?' In *Class Issues*, edited by Amitava Kumar, 256–275. New York: New York University Press.
Haraway, Donna J. 1997. *Modest_Witness@Second_Millennium.Female Man©_Meets_OncoMouseTM*. London and New York: Routledge.
Hudson, Heidi. 2005. 'Doing' Security as Though Humans Matter: A Feminist Perspective on Gender and the Politics of Human Security'. *Security Dialogue* 36 (2):155–174.
Irigaray, Luce. 1977. *This Sex Which Is Not One*. Ithaca: Cornell University.
Jagose, Annamarie. 1996. *Queer Theory*. Melbourne and New York: University of Melbourne Press.

Jeffords, Susan. 1994. *Hard Bodies: Hollywood Masculinity in the Reagan Era*. New Brunswick, NJ: Rutgers University Press.

Keohane, Robert. 1989. 'International Relations Theory: Contributions of a Feminist Standpoint'. *Millennium – Journal of International Studies* 18 (2):245–253.

Kiersey, Nicholas J., and Iver B. Neumann. 2015. 'Worlds of Our Making in Science Fiction and International Relations'. In *Popular Culture and World Politics: Theories, Methods, Pedagogies*, edited by Federica Caso and Caitlin Hamilton, 74–82. Bristol, UK: E-International Relations.

Kosofsky Sedgwick, Eve. 1990. *Epistemology of the Closet*. Berkeley, CA: University of California Press.

Landes, Joan B. 1998. 'Introduction'. In *Feminism: The Public and the Private*, edited by Joan B. Landes. Oxford: Open University Press.

Lloyd, Genevieve. 1992. 'The Man of Reason'. In *Women, Knowledge and Reality*, edited by Ann Garry, 149–165. London: Routledge.

Merleau-Ponty, Maurice. 1962. *Phenomenology of Perception*. London: Routledge.

Moon, Katharine H. S. 1997. *Sex among Allies: Military Prostitution in US-Korea Relations*. New York: Columbia University Press.

Müller, Martin. 2008. 'Reconsidering the Concept of Discourse for the Field of Critical Geopolitics: Towards Discourse as Language and Practice'. *Political Geography* 27 (3):322–338.

Ó Tuathail, Gearóid. 1996. *Critical Geopolitics: The Politics of Writing Global Space*. Minneapolis: University of Minnesota Press.

Parashar, Swati. 2013. 'What Wars and "War Bodies" Know About International Relations'. *Cambridge Review of International Affairs* 26 (4): 615–630.

Philpott, Simon. 2010. 'Is Anyone Watching? War, Cinema and Bearing Witness'. *Cambridge Review of International Affairs* 23 (2):325–348.

Puar, Jasbir K. 2005. 'Queer Times, Queer Assemblages'. *Social Text* 23 (3–4):121–140.

Puar, Jasbir K. 2007. *Terrorist Assemblages: Homonationalism in Queer Times*. Durham, NC: Duke University Press.

Rich, Adrian. 1993. *Existence, Compulsory Heterosexuality and Lesbian*. London: Onlywomen Press.

Ritzer, George, and Nathan Jurgenson. 2010. 'Production, Consumption, Prosumption: The Nature of Capitalism in the Age of the Digital Prosumer'. *Journal of Consumer Culture* 10:13–36.

Rowley, Christina. 2010. 'Popular Culture and the Politics of the Visual'. In *Gender Matters in Global Politics: A Feminist Introduction to International Relations*, edited by Laura J. Shepherd, 361–374. London: Routledge.

Saunders, Robert A. 2015. 'Imperial Imaginaries: Employing Science Fiction to Talk about Geopolitics'. In *Popular Culture and World Politics: Theories, Methods, Pedagogies*, edited by Federica Caso and Caitlin Hamilton, 149–159. Bristol, UK: E-International Relations.

Saunders, Robert A. 2017. *Popular Geopolitics and Nation Branding in the Post-Soviet Realm*. London and New York: Routledge.

Saunders, Robert A., and Vlad Strukov. 2017. 'The Popular Geopolitics Feedback Loop: Thinking beyond the "Russia versus the West" Paradigm'. *Europe-Asia Studies* 69 (2):303–324.

Segal, Lynne. 1997. 'Sexualities'. In *Identities and Difference*, edited by Kathryn Woodward, 183–238. London and New York: Sage.

Shapiro, Michael J. 1999. *Cinematic Political Thought: Narrating Race, Nation and Gender*. Edinburgh: Edinburgh University Press.

Sharp, Joanne P. 1993. 'Publishing American Identity: Popular Geopolitics, Myth and *The Reader's Digest*'. *Political Geography* 12 (6):491–503.

Shepherd, Laura J. 2013. *Gender, Violence and Popular Culture: Telling Stories*. London and New York: Routledge.

Sjoberg, Laura. 2012. 'Toward Trans-gendering International Relations?' *International Political Sociology* 6 (4):337–354.

Soreanu, Raluca, and David Hudson. 2008. 'Feminist Scholarship in International Relations and the Politics of Disciplinary Emotions'. *Millennium – Journal of International Studies* 37 (1):123–151.

Steans, Jill. 2003. 'Engaging from the Margins: Feminist Encounters with the "Mainstream" of International Relations'. *British Journal of Politics & International Relations* 5 (3):428–454.

Steans, Jill. 2010. 'Revisionist Heroes and Dissident Heroines: Gender, Nation and War in Soviet Films of "the Thaw"'. *Global Society* 24 (3):401–419.

Sylvester, Christine. 2011. 'Pathways to Experiencing War'. In *Experiencing War*, edited by Christine Sylvester, 118–130. New York: Routledge.

Sylvester, Christine. 2013. *War as Experience: Contributions from International Relations and Feminist Analysis*. London and New York: Routledge.

Thrift, Nigel. 2000. 'It's the Little Things'. In *Geopolitical Traditions: A Century of Geopolitical Thought*, edited by David Atkinson and Klaus Dodds, 380–387. London: Routledge.

Tickner, J. Ann. 1992. *Gender in International Relations: Feminist Perspectives on Achieving Global Security*. New York: Columbia University Press.

Tickner, J. Ann. 1997. 'You Just Don't Understand: Troubled Engagements Between Feminists and IR Theorists'. *International Studies Quarterly* 41 (4):611–632.

Tickner, J. Ann. 2005. 'What is Your Research Program? Some Feminist Answers to International Relations Methodological Questions'. *International Studies Quarterly* 49 (1):1–22.

Warner, Michael. 1993. *Fear of a Queer Planet: Queer Politics and Social Theory*. Minneapolis: University of Minnesota.

Weber, Cynthia. 1994. 'Good Girls, Little Girls, and Bad Girls: Male Paranoia in Robert Keohane's Critique of Feminist International Relations'. *Millennium – Journal of International Studies* 23 (2):337–349.

Weber, Cynthia. 1999. *Faking It U.S. Hegemony in a 'Post-Phallic' Era*. Minneapolis: University of Minnesota Press.

Weber, Cynthia. 2005. *Imagining America at War: Morality, Politics and Film*. London and New York: Routledge.

Weber, Cynthia. 2016. 'Queer Intellectual Curiosity as International Relations Method: Developing Queer International Relations Theoretical and Methodological Frameworks'. *International Studies Quarterly* 60 (1):11–23.

Weldes, Jutta. 1999. 'Going Cultural: Star Trek, State Action, and Popular Culture'. *Millennium: Journal of International Studies* 28 (1):117–134.

Weldes, Jutta. 2006. 'High Politics and Low Data: Globalization Discourses and Popular Culture'. In *Interpretation and Method Empirical Research Methods and the Interpretive Turn*, edited by Dvorah Yanow and Peregrine Schwartz-Shea, 176–186. London and New York: M.E. Sharpe.

Weldes, Jutta, and Christina Rowley. 2015. 'So, How Does Popular Culture Relate to World Politics? '. In *Popular Culture and World Politics: Theories, Methods, Pedagogies*, edited by Federica Caso and Caitlin Hamilton, 11–34. Bristol, UK: E-International Relations.

Wilcox, Lauren. 2014. 'Queer Theory and the "Proper Objects" of International Relations'. *International Studies Review* 16:612–615.

Wittig, Monique. 1992. *The Straight Mind and Other Essays.* Worcester: Harvester Wheatsheaf.

5 Crossing the boundary

'Real-world' geopolitical responses to the popular[1]

Robert A. Saunders

Introduction

The myriad and ever-expanding connections between popular culture and geopolitics underpin this volume, just as they increasingly inform how citizens and polities understand the world around them. In this essay, I focus on those instances where the *popular* has triggered the *real*, specifically when certain pop culture artefacts (films, videogames, TV series, and YouTube videos) have provoked state actors into geopolitical action, thus resulting in ramifications for international relations. While depictions of the 'enemy Other' in media goes back to almost as far back as does recorded history (by way of example, Aeschylus' *The Persians* premiered in 472 BCE), the notion that popular culture can prompt conflicts between states is something that is peculiar to the current era of deterritorialised and digitalised media flows. As a consequence of globalisation, flows of power (particularly via mediated representation) that have little-to-no regard for the boundaries that have long characterised the post-Westphalian state system are increasingly the norm. Inarguably, popular culture in its many forms serves as fundamental element in shaping images of the self (*Selbstbild*) and the other (*Fremdbild*). Beginning with nursery rhymes, which so often depict children who are separated from the uterine security of hearth and home, only to encounter perfidious 'strangers' bent on robbing them of their innocence, property, and/or life, any given polity's cultural production is rooted in establishing the borders between 'us' and 'them'. As one moves up the structural and narrative hierarchies of (popular) culture production, from jokes, slogans, and songs to videogames, full-length motion pictures, and television series, the formulaic tenets of identity production are generally maintained, although often with the affective/pedagogical components being enhanced and made more complex.

While the sophistication and aesthetics of representation of the 'Other' may differ between totalitarian, authoritarian, and liberal systems and citizenries, the contours of such geopolitically informed (and informing) content remain surprisingly comparable. One society's 'messaging' might certainly be viewed as another's 'propaganda', but this does not mean that any system of popular-cultural indoctrination of world-views (*Weltbilden*) and world-formations (*Weltbildungen*) is, in actuality, substantively different from another. Produced during the height of

Stalinism, Sergei Eisenstein's *Alexander Nevsky* (1938) is thus as reflective of the esprit of contemporary Soviet popular culture, as is the unctuously patriotic U.S. musical *1776* (1969). Both of these examples portray historical events in artistic and affective ways, with tactically different yet strategically similar goals, vis-à-vis the deepening of national identity. While those nations who were depicted as the enemy other in these productions (Germans and British, respectively) might have bristled at their mediated representations, these cultural artefacts did not, in and of themselves, result in actual geopolitical action. However, since the end of the bifurcated global system based on U.S.-Soviet spheres of influence, the dramatic increase in world trade abetted by neoliberalism, and the media transformation associated with digital information and communication technologies, there has been an acute change in the ways in which popular culture impacts and interacts with world politics (Grayson, Davies, and Philpott 2009). Some have even gone as far as to argue that popular culture has become 'weaponised', with text and images now being capable of producing 'destructive outcomes' (Mirzoeff cited in Saunders 2017a, 1), as has most spectacularly occurred through Russia/Russians' intervention in the 2016 U.S. presidential election, via Facebook, Twitter, and other social media platforms (Shane 2017). In a world where image trumps capacity in international relations, a state's reputation has become one of the most important components of its power. Consequently, geopolitical popular culture—whether in the form of films, TV, videogames, or comic books—is now a major factor in shaping world affairs, whether as a system for prefiguring action on the part of citizenries through 'somatic markers' (Connolly 2002; Ó Tuathail 2003; Dittmer 2010) or, in the cases presented in this essay, as 'action' in diplomacy, military affairs, or economics.[2]

Employing Nexon and Neumann's conceptual framework of international relations approaches to popular culture, which argues for four modalities: 1) popular culture and politics; 2) popular culture as mirror; 3) popular culture as data; and 4) popular culture as constitutive (Nexon and Neumann 2006, 10), this chapter attempts to bring the first and fourth approaches into union. This is done through the critical examination of a representative set of geopolitical controversies prompted by popular culture, examining each case as part of a relatively linear development wherein 'entertainment' is increasingly presenting as 'state action', regardless of the involvement or absence of (quasi-)governmental actors in the production and distribution of such artefacts. In terms of methodology, I follow Rowley's (2010) approach to analysing popular culture, which focusses on artefacts produced by 'dominant cultures'. In that, my analysis examines 'mass culture produced in, by, and primarily for audiences in the West, and even more narrowly on dominant representations from the US and UK' (Rowley 2010, 310); however, departing somewhat from Rowley, my choice of case studies is determined by their mediatisation *qua* artefacts rather than their content per se. This chapter interrogates a selection of the most prominent examples of interventions in the popular culture-world politics continuum since 9/11 (a seminal, if not unproblematic temporal marker in popular geopolitics).[3] Using a variety of online search tools and new media platforms from Lexis-Nexis to Google to Facebook and a synthesis of Strukov's (2016) three-pronged (visual, audio, and textual) critical-discourse approach to popular geopolitical analysis, I attempt provide a holistic overview of IR 'events' in popular culture, based on the reactions of 'impugned states'.

I begin with a brief overview of the role played by 'Hollywood' in international relations and world politics, before moving on to analyse several instances where popular culture provoked international controversies, from the worldwide release of big-budget films to 'fake' documentaries and news via social media. As part of this discussion, I provide several possible explanations of why popular geopolitics has emerged as both as a practice (by cultural producers) and a field of scholarly analysis (by researchers in geography, IR, and cultural studies). While the latter is now commonly (if somewhat begrudgingly) accepted in academe, the former—i.e. the notion that 'popular geopolitics' is a thing that can be *done* as well as *studied*—is somewhat controversial. My aim is to complicate the very idea of popular geopolitics, splintering its current conceptual coherence into two separate forms: phronesis and praxis. Phronesis is most often defined as a form of practical wisdom, and this is where most academic work in the field of popular geopolitics focusses. Scholars examine a set of artefacts and then relate these to the real world, employing their various skills to convey the meaningfulness of the given representation (and interpretations thereof). I am more interested in the praxis, or the transforming activity and its intentionality (or lack thereof).[4] While I concede that this turn in popular geopolitics may be somewhat controversial, I argue that it reflects the increasingly accepted notion that popular culture informs and influences world politics and geographical imagination in ways that are unprecedented. Moreover, I contend that scholars should be ever more attentive to the role played by cultural producers (and, conversely, that filmmakers, showrunners, [graphic] novelists, and videogame programmers should realise that, borrowing a phrase from former Marvel Comics chairman Stan Lee: 'With great power comes great responsibility!')[5]

The 'Hollywood' culture industry as an interlocutor in world politics[6]

Looking back to late 2014, when U.S. President Barack Obama was drawn into an international imbroglio over the Seth Rogan and Evan Goldberg film *The Interview* and the concomitant hack of Sony Pictures Entertainment, there was a farrago of reportage about Hollywood's turbulent past with films featuring geopolitical themes. In his photo essay for the New York-based *Daily News*, Ethan Sacks (2014) listed a number of earlier American (and British) films that angered enemies of the U.S., such as *The Great Dictator* (1940), *From Russia with Love* (1963), and *Cry Freedom* (1987). Writing in the *New York Post*, Reed Tucker (2014) also tackled the issue, but instead focussed on friendly states that were perturbed by their representations in big-budget films, including the depiction of Slovaks as maniacal torturers in Eli Roth's *Hostel* (2005) and Kazakhstan's ire at its benighted rendering in Sacha Baron Cohen's *Borat: Cultural Learnings of America for Make Benefit*

Glorious Nation of Kazakhstan (2006) (the latter is explored below). Similar lists of Hollywood's run-ins with overseas governments appeared on the web sites of *NBC News* (Arkin 2014) and *The Daily Beast* (Suebsaeng 2014). In all these instances, filmmakers have trundled into the realm of geopolitics, provoking official and/or popular condemnation of their cultural products. In examining the long arc of Hollywood's engagement with world politics, there is no shortage of analysis when it comes to the political ramifications of celluloid representation of people and places beyond America's shores. In fact, many scholars attribute cinema with a powerful capacity to shape national identity (Edensor 2002; Dittmer 2005; Weber 2005; Philpott 2010), as well as being one of the most important inputs in determining attitudes towards foreign polities (Sharp 1993; Ó Tuathail 1996; Weldes 2003; Shapiro 2008; Shepherd 2013). This is particularly true when it comes to the 'enemy other', whether it be Germans and Japanese (1940s), Soviets and Chinese (Cold War), or Islamist terrorists (1970s–today) (see Saunders 2017a). Such outcomes are predictable given the highly cooperative relationship between mainstream Hollywood studios and the American military-industrial complex since the late 1930s (Der Derian 2009; Alford 2010).

In the case of the aforementioned movie *The Interview*, which employs a sitting head of state as the antagonist, the xenophobic representational paradigm of an 'Asiatic dictator' is overt, even reaching hyperbole to achieve comic effect; however, sculpting of the Other in film and other forms of popular culture is often subtler (though nonetheless important in shaping political culture and attitudes to the outside world). No genre is completely free of such generalisations; for instance, Keith M. Booker (2010) points out that Disney animated classics are rife with crass Orientalism, anti-Germanism, and Russophobia. Correspondingly, Cold War-era science fiction films (notably the alien-invasion genre) used metaphors to metastasize incipient fears of a Communist take-over, besmirching the national images of Russians, Eastern Europeans, and others in the process (Hendershot 2001). As the epitome of the spy-thriller genre, the James Bond franchise has 'educated' generations of cinema-goers about the perils of world politics, from KGB assassins to African mercenaries to Latin American drug smugglers (Dodds 2003, 2005), while also serving as a salvo in the scripting of the contemporary geopolitical world order (see Table 5.1). In the wake of the 11 September 2001 attacks on the World Trade Centre and the Pentagon, Hollywood shifted into overdrive with a host of films addressing the 'global war on terror' (GWOT), further deepening the link between cultural production and world politics (Dodds 2008; Schopp and Hill 2009; Der Derian 2010) and making use of the 'reel bad Arab' motif (Shaheen 2001). As an adjunct to this steadily thickening nexus, we are witness to a worldwide change in socialisation, wherein cultural producers are perpetuating geopolitical codes and discourses (Crampton and Power 2005) that can be called upon by politicians at a moment's notice, combined with the proliferation of countless pop-culture-based geopolitical somatic markers in the viewing/voting population that respond to affective triggers used by said politicians.

During the twentieth century, dozens of countries found something they did not like about the representation of their respective national images in American-made

Table 5.1 Potential audiences and key impact factors of popular geopolitical filmic content

	Origin country	*Fraternal allies*	*Adversarial states*	*Neutral powers*
Elites	Mechanism for mobilisation of nationalist/patriotic sentiment	Index of origin country's *Selbstbild* in relation to foreign policy actions	Confirmation of aggressive intent of origin country's foreign policy elites	Identification of geopolitical meta-narratives and their country's marginal position therein
Masses	Manipulation or reinforcement of geopolitical codes and visions; stereotyping of foreign 'Others'	Affective influence on origin country's *Fremdbild* in viewing market; invitation of emulation	Hardening of popular resentment to origin country (including via counter-readings of the artefact)	Invitation to identify with the 'hero' or 'villain' based on individual geopolitical orientations

motion pictures. Yet, until quite recently, the relationship between Hollywood and foreign governments has not been one of a direct nature. The standard response to any negative portrayal was to ban the distribution of the film (sometimes prompted by a populist 'boycott' of the film prior to its release), and in exceptional circumstances to issue an official condemnation, often confusingly directed both at the filmmakers and their 'host' governments, i.e. the White House or 10 Downing Street. However, in the new millennium, there has been a discernible shift, and one which more clearly reflects the importance of the popular within the realm of the geopolitical. Not unimportant in this new calculus is the increasing importance of alternative nodes of popular (geopolitical) culture production, e.g. Bollywood, Nollywood, etc.,[7] as well as prosumption of existing artefacts and counter-readings by attentive publics, via the ever-expanding array of digital platforms. Governments are now engaging in multivariate forms of negotiation with international publics, multinational corporations, and other transnational actors to negate or at least attenuate negative representation via film and other forms of popular media. In the next section, I explore several high-profile cases of such quasi-diplomacy, with each reflecting different power dynamics on the world stage, before moving on to 'small screen' analogues.[8]

Novel geopolitical responses to filmic pop culture: *Borat* (Kazakhstan), *300* (Iran), *Red Dawn* (China), and *The Interview* (North Korea)

Hyper-attuned to Western opinion, particularly that of economic elites in the U.S. and British fossil fuel industries, Kazakhstan's refusal to countenance Sacha Baron Cohen's parody of the Central Asian republic was not surprising (Saunders 2007). As early as 2000, embassy officials in the UK requested that then-Prime Minister Tony Blair 'ban' Borat (at the time, Borat was simply a character on a minor British skit-comedy programme, *The 11 O'Clock Show*). However, as Baron Cohen's star rose, the battle for the national image of Kazakhstan took on almost surreal proportions. Following an appearance on the 2005 MTV Europe Music Awards, Kazakhstan threatened to sue the British comedian and removed his web site from Kazakhstani servers. As the premiere of the *Borat* film loomed in 2006, Kazakhstan changed tack, actively engaging in the farce in an effort to employ the media blitz generated by Baron Cohen's antics to teach a Western public that had suddenly become aware of the fact that Kazakhstan might actually be a real country.[9] Tongue-in-cheek press interviews, YouTube videos, and calculated uses of the Borat meme to promote everything from fashion to architecture came to characterise Astana's new attitude to the Cambridge-educated trickster (Saunders 2008a). The end result: Kazakhstan became a known quantity, with tourism increasing in a measurable way. Over one decade on, the shtick still resonates (and continues to prickle citizens and government elites alike), but most Kazakhstanis now admit their country is better for the experience having seen their nation brand permanently distinguished from those of its neighbours in the region, and gaining increasingly positive associations worldwide (see Peerson 2015).

Whereas Kazakhstan represents one of the world's newest countries, one of its oldest faced a similar situation (or at least perceived it to be so) shortly thereafter, when Frank Miller's epic graphic novel *300* (1998) was adapted for the big screen. Given the moral geopolitics of popular culture's engagement with the GWOT, as well as rampant rumours that the next Middle Eastern country the U.S. sought to invade was Iran, Zack Snyder's 'monstrous othering' (Rai 2011) of the ancient Persians in the historical fantasy *300* did not go over well in Tehran. As Es (2011) points out, *300* represents a genuine shift in Hollywood's treatment of world geopolitics (though a return to the hoary fundaments established by the aforementioned Aeschylus) in that it dispenses with Cold War-era trope equating the individualistic, thalassic, cerebral Athenians with the U.S. and the authoritarian, land-bound, martial Spartans with Nazi Germany/USSR, eliding the two allies (both discernibly 'white' and coded as [northern] 'European') together against a swarthy, bestial, preternatural Oriental enemy other (i.e. the Persians). Given the highly anti-Islamic, Orientalist content of the source material, populist responses to the film[10]—buttressed by secular and theocratic condemnation in Iran—came as no surprise to Middle East experts in the foreign policy community. What proved to be most interesting was that the Iranian regime finally found common cause with the expatriate community, which took particular umbrage at the film due to the fact that most of the latter had self-identified as 'Persians' since 1979, to distance themselves from the fundamentalist government in Tehran (an ethnic legerdemain that was perniciously undermined by the film). The Iranian government, hoping to replicate the *ummah*-wide revulsion associated with the Danish cartoons controversy (see Saunders 2008b), sought to rally other Muslim countries to a common cause against the Orientalist 'Hollywood' assault on the Islamic East. While this strategy was largely ineffective, surprisingly Persian bloggers of all ideological stripes raged against the film, including engaging in 'Google-bombing', whereby searches for *300* would redirect traffic to sites lauding Persian culture (Joneidi 2007).

Like Iran, China lays claim to an ancient and influential cultural patrimony; however, the country is also a fast-rising global power with economic and political interests across the Pacific Rim and farther afield. Consequently, Chinese elites are highly attentive to their national image and its representation in international media (see Ding and Saunders 2006). So, when confronted with the prospect of a big-budget remake of the iconic 1984 anti-Communist action film *Red Dawn*, wherein an ascendant China invades the U.S.,[11] Beijing used market pressures to facilitate a critical change in the plot (Landreth 2010). When details of the script were leaked in 2010, one of China's leading state-run newspapers, *The Global Times*, railed against the film as a 'demonisation' of China. Through informal channels, Beijing was able to persuade the studio to make changes prior to release. Rather than a Chinese contingent, the occupying army was changed to a squadron of North Korean paratroopers. With the growing size of its theatrical marketplace and limits on the number of foreign films that can be screened in a given year, China has developed substantial leverage over mainstream Hollywood representations of its country, people, and culture. While in some cases, this means forcing studios into cutting 'Chinese' villains from the plot, in other instances, studios are encouraged to add scenes depicting Chinese culture in a positive light (Homewood's chapter in this volume).[12] Rather than taking the defensive posture of the Iranians, or co-opting a *fait accompli* as the Kazakhstanis did, China brandishes its economic might to manipulate the global media ecosystem, preventing 'enemy images' of the Middle Kingdom from seeing the light of day.

Given its small size and self-imposed isolation, Hollywood treatments of the People's Democratic Republic of Korea have been surprisingly common over the past two decades. Perhaps this reflects a certain form of nostalgia for the purported 'simplicity' of the late Cold War and its popular culture production modalities (see, for instance, Lipschutz 2001; Van Jelgerhuis 2015; Kirchik 2016), given that North Korea clings so furiously to its ideological opposition to the West. Alternatively, it could be because the leadership in Pyongyang seems to only become more bizarre, unpredictable, and belligerent over time. Having witnessed the insidious representation of its country in several big-budget films, including *Die Another Day* (2002), *Team America: World Police* (2004), and *Olympus Has Fallen* (2013), the North Korean Government acted swiftly and with resolve. When *The Interview* began to receive press attention prior to its release, Pyongyang went on the offensive condemning its 'gangster filmmakers', calling it an 'act of war', and emblematic of the 'wanton acts of terrorism' the U.S. engages in in the Middle East and elsewhere.

Pyongyang promised a 'merciless' response if and when the film premiered (McCurry 2014). North Korea then raised the issue with the United Nations' Secretary General, Ban Ki-moon, accusing the Obama Administration of 'sponsoring terror', and shortly thereafter entreated U.S. officials to intervene to prevent the film's then-scheduled October release (ultimately delayed to the Christmas season). On 24 November 2014, employees of Sony Pictures Entertainment (the parent company of Columbia Pictures, the studio behind *The Interview*) encountered a disturbing skeletal image on their desktops notifying them of a network-wide hack. The message stated that 'This is just the beginning', and that Sony must meet the (then-unspecified) demands of the 'Guardians of Peace'. While the corporation initially assumed the hack to be a minor one, the negative effects of the breech cascaded, resulting the release of sensitive internal documents, including salaries of company executives. Subsequent leaks would expose company in-fighting, scathing critiques of creative content, employee medical records, and hundreds of other pieces of confidential and embarrassing information. In total, Sony suffered untold damages, both in real terms (e.g. computer hardware and infrastructure) and via the impact on its reputation (online threats against movie-goers prompted Sony to withdraw the film from commercial release). Subsequently, the U.S. Federal Bureau of Investigation launched a probe, ultimately finding that Pyongyang was behind the attacks. Reflecting the seriousness of the situation, U.S President Barack Obama spent much of his 2014 year-end press conference discussing the attacks and the U.S. response.

As an 'image superpower' (Frèches, qtd. in Morley and Robins 1995) engaged in a global conflict with any and all pretenders to the throne, the U.S.—with its film, television, videogame, and new media industries—enjoys a massive capacity to create, perpetuate, and modify the national images of its friends and enemies

Table 5.2 Selected case studies in 'real world' popular geopolitics

Artefact	*Impacted country*	*Major IR/geopolitical themes*	*State reactions*
Borat: Cultural Learnings of America for Make Benefit Glorious Kazakhstan (2006)	Kazakhstan	Post-Soviet 'backwardness', economic underdevelopment, patrimony, ethnic and religious strife	Official condemnation; mediatised co-optation
300 (2006)	Iran	East-West incompatibility, democracy vs. despotism, GWOT, (Islamic) imperialism, Rimland vs. Heartland thesis	Mobilisation of Muslims and expatriate community online
Red Dawn (2012)	China	'Yellow Peril' combined with Russian revanchism, EU economic crises, NATO weakens the homeland	Economic pressure to replace China (with the DPRK) as the 'enemy other'
The Interview (2014)	North Korea	cult of personality, WMD, espionage, regime change, targeted assassination, nation-building	Cyberattacks on Sony Pictures; threats against U.S. gov't and movie-goers

alike. However, we also see from recent developments that other powers around the world now have tools at their disposal to 'speak back' to the hegemon (see Table 5.2). In each of the instances discussed above, we see national governments (and their proxies) taking novel approaches to the release and distribution of motion pictures that depict their countries in a less-than-positive way. Given that North Korea's response to *The Interview* triggered a measurable diplomatic response on the part of the U.S. government, as well as significant physical and financial damage to a major transnational corporation based in the U.S., it is apparent that the nexus between popular culture and geopolitics is deepening and becoming ever more dangerous. While film may be the most visible and enduring popular culture platform, it is just one of many pop culture formats. Increasingly, television series, videogames, and online videos are functioning as tools of geographical imagination and emerging as key planks in shaping perceptions about the world beyond one's shores; in doing so, these media are triggering a host of geopolitical issues that challenge preconceived notions about what international relations is all about.

Big world/small screens: television, videogames, YouTube, and social networking sites in world affairs

Indubitably, film has long held a commanding presence in the realm of popular cultural production, stemming from the medium's power to gather an audience into an almost temple-like space and keep them enraptured and connected for a period of time that lasts longer than many religious ceremonies. However, in recent years, the dominance of cinema has faltered due to a number of economic, social, and structural reasons. The advent of the Internet and the proliferation of personal media devices has transformed entertainment consumption practices, and in turn, come to shape production processes. While television and videogames both predate the rise of the World Wide Web, the increasingly networked structure of global society through the Internet has promoted major evolutions in both platforms since the end of twentieth century. Consequently, deterritorialisation of media has become a reality, and television shows, videogames, and short videos produced in one country are often consumed far from their country of origin, often resulting in fan bases that can be genuinely transnational (as discussed in Caso's chapter in this volume). As part of this phenomenon, pop-cultural production has now 'gotten real' on small screens as much as on big ones.

While it is quite early in the evolution of geopolitical television (Glynn and Cupples 2015), the medium is currently in a state of rapid transformation. In the worlds of international affairs specialist Daniel Drezner (2016): 'We live in a Golden Age of international relations programming on television. The proliferation of so many platforms has meant that there are now great shows that tackle the eternal dilemmas of world politics in metaphorical worlds, past metaphorical worlds, and the world we live in right now'. Unlike motion pictures, television series have the capacity to 'rip' their content 'from the headlines', often resulting in showrunners folding world events into the narrative. With the explosion of social media in the new millennium, fans have also gained the ability to influence the plot lines and character development of their favourite programs. New platforms for distribution such as DVDs, DVRs, digital downloads, and streaming-subscription services have completely transformed the viewing experience (Saunders 2017b). These changes have been magnified by novel ways of 'watching' TV, particularly via personal media devices (PCs, smart phones, iPads, etc.), thus disrupting once-accepted truths about the inflexible 'boob tube', bringing us into the so-called 'post-television era', where the viewer has left the couch to find a brave new world (Spigel and Olsson 2004; Levine 2011). With demanding series like *Lost* (2004–2010), *The Wire* (2002–2008), and *Forbrydelsen* ('The Killing') (2007–2012), television viewing has entered a new phase, and one which is increasingly defined by ongoing geopolitical issues. From *Homeland*'s fetishisation of Islamist terror, to *Game of Throne*'s ruminations on game-theory and nuclear deterrence (Moïsi 2016), world affairs is now part-and-parcel of television viewing.

While the medium may have always been infused with geopolitics, to some degree, it is now an immediate concern for many televisual dramas, as evidenced by the increased use of global themes, international characters, and overseas settings for remakes of 'classic' programmes like *Hawaii Five-O* (2010–) and *MacGyver* (2016–), which are measurably more geopolitical in their content than their precursors from the 1970s and 80s, respectively. While some shows like *24* (2001–2010) and *Borgen* (2010–2013) choose to situate events in fake countries such as 'Kamistan' and 'Turgisia', respectively, other series are more provocatively setting the action in real places. Two programmes in particularly have provoked ire from the government of the Russian Federation in recent years: the short-lived faux reality-TV competition *Siberia* (2013) and the highly successful political thriller *House of Cards* (2013–). The former, which depicts black-clad FSB agents as murderous agents of the state, drew protests from the *Siberian Times* for stereotyping the region and being 'frozen in a Cold War time warp' (Stewart 2013). The latter, which employed a rather unflattering stand-in for President Vladimir Putin in Season 3, was famously denied the ability to film at the United Nations by a Russian 'veto' on the grounds of security concerns (Lynch 2014). Russia is not alone in taking exception to its televisual representation. In 2014, North Korea informed the British Government that it must prevent the airing of the forthcoming Channel 4 thriller *Opposite Number*, or face harm to its bilateral relations with Pyongyang, also noting that the programme was being 'orchestrated at the tacit connivance, patronage and instigation by "Downing Street"' (Guardian 2014). As Netflix, Hulu, and other on-demand services internationalise television consumption, the geopolitical conceits of TV series are increasingly 'leaking' into real-world political debates, epitomising the very definition of 'popular geopolitics' (Dittmer and Dodds 2008).

While often dismissed as little more than juvenile escapism, videogames are also making their presence felt in the realm of world affairs. In the words of gaming reporter Tim Surrette: 'For most, games are a pleasant time-wasting diversion, an interactive mini-vacation from life's daily grind. For others, they are nothing but a way for governments to plant seeds into the minds of citizens in order to indoctrinate an imperialist attitude' (2006). A \$23.5 billion industry in the U.S. alone (Morris 2016) and estimated to rake in nearly \$100 billion worldwide in 2016 (Minotti 2016), gaming is not only big business, it is also political particularly as 'allusions to real events in videogames have grown in parallel with their technical complexity' (Courmont and Clément 2014, 31). As Bos' chapter in this volume attests, the use of foreign and domestic landscapes for first-person shooter (FPS) games is charged with power, and directly influenced by geographical imagination and highly resilient geopolitical codes. It should come as no surprise then that certain governments have taken offence to their countries being depicted in such gameplay, particularly when the player faces stereotypical antagonists or must confront hostile environments, jaundiced by decades or even centuries of Orientalist prejudices.

Once again, North Korea leads the charge in pushing back against pop culture. In 2005, Pyongyang issued a strident condemnation of Western videogames, representing the regime as a military aggressor in the Pacific Rim, singling out Tom Clancy's *Ghost Recon 2* (2004). A government newspaper issued the following ominous message directed at American game players: 'This may be just a game to them now, but it will not be a game for them later. In war, they will only face miserable defeat and gruesome deaths' (qtd. in Brooke 2005). Similarly, Venezuela, fearful of a Bay of Pigs-style incursion by the U.S., was equally alarmed at the premiere of the videogame *Mercenaries 2: World in Flames* (2008), stating that it was simply justifying 'imperialist aggression' on the part of Washington (Surette 2006).

Flipping the standard dynamic of the popular culture-world politics feedback loop (see Saunders and Strukov 2017), a number of videogames produced in the Arab world have led to handwringing and even official condemnation on the part of some Western governments. Hezbollah Central Internet Bureau's *Special Force* (2003) is a paradigm of this sort of controversial popular culture, given its 'explicit political agenda' and use of real-world geographies and events, namely Hezbollah's campaign against Israeli forces in occupied southern Lebanon in 2000 (Extreme Tech 2006). The second iteration of the game triggered the following response from Israeli Foreign Ministry spokesman Mark Regev: 'It should come as a surprise to no one that Hezbollah teaches children that hatred and violence are positive attributes' (CNN 2006). Tel Aviv has gone on the offensive against this and a host of other games, which it claims promote violence against the Jewish state, including Iranian-produced *Satan's Den 2* (2010), *Attack on Freedom Flotilla* (2010), and *Missile Strike* (2015). In several cases, the U.S. has supported its ally Israel, blocking the distribution of such games on popular distribution platforms, or calling on providers not to carry the content. Even the largest media companies have thus been embroiled in questions of free speech over such issues, mostly notably with Apple's controversial decision to allow *Liyla & The Shadows of War* (2016), a story-driven application focussed on a Palestinian fleeing Israeli attacks, in its popular app store (Kamen 2016).

The impact of videogames goes far beyond their role as a diversion for youth. As Hussain (2016) argues, years of gameplay have birthed a generation of filmmakers that employ the aesthetic sensibilities of FPS games in crafting propaganda videos,[13] most notably among the information brigades of the Islamic State (IS). As he points out, GoPro technology is employed to provide point-of-view shots that are not only intelligible to potential recruits, but which trigger affective responses due to somatic markers implanted over thousands of hours of gameplay. In other words, videogames are changing the very nature of psy-ops in the globalised media milieu, given that virtual environment-filmic representation like *Rage Wind* (2016), which Hussain argues looks just like a videogame, can be easily and surreptitiously distributed over the Internet to viewers around the world. In a recent interview with a convicted IS leader, Somali-American Abdirizak Warsame, U.S. television news magazine *60 Minutes* attempted to assess the importance of the virtual realm in terrorism-recruitment efforts, opining: 'The route to "action" was a link away in the recruitment videos of ISIS. Music videos, a language the boys could understand' (Pelley 2016).

Scott Pelley: YouTube became more real to you than your neighbourhood in Minnesota?

Abdirizak Warsame: Yes.

Scott Pelley: How could that be?

Abdirizak Warsame: It kind of takes control of you. And you think you're doing something for a greater cause. And you think you're doing it for good.

Scott Pelley: And what was that?

Abdirizak Warsame: Most of the videos would talk about how if you would engage in jihad you would be doing your family a favour. And that you would be saving their lives from eternal hell fire.

As a relatively unregulated digital realm, where the cost of entry is low and access to wide and distant audiences is comparatively easy, cyberspace has radically reduced the financial and technological hurdles associated with producing and distributing popular culture. By allowing almost anyone with access to a video camera, a personal computer, and Internet access the ability to deliver content to the 'world', cyberspace has rewritten the rules of broadcasting, privileging all sorts of new actors who are free to deliver high quality media content far and wide. Religious, ethnic, and social minorities have been direct beneficiaries of this trend, allowing marginalised groups to make their voice heard, engage in political organisation, and 'normalise' their position vis-à-vis the majority (see Brinkerhoff 2009; Saunders 2010; Chaplin 2014). However, the diminishing of the barriers to entry has come with other effects as well, especially the transmission of media products across national borders. This has resulted in a variety of state-based attempts as policing cyberspace and preventing 'foreign' content from making its way in the homes and onto the phones of its citizens (Goldsmith and Wu 2006); although in reality, such efforts have often proved to be limited in their impact, more powerful states have been able to bend the structure of cyberspace in ways that add to rather than detract from their influence in world affairs (Powers and Jablonski 2015). Regardless, the Internet is now a global media distribution platform connected to countless production sites where inexpensive and relatively simple software programmes allow for the creation of content, including output that is decidedly geopolitical in nature.

An obvious manifestation of this new world order is the series of events that surrounded the release of the 14-minute YouTube trailer for *The Innocence of Muslims*. Nakoula Basseley Nakoula, an Egyptian-born Coptic Christian immigrant to the U.S., undertook the project under highly questionable circumstances (namely, hiding the fact that it was about the Prophet Muhammad from the actors). Described as suffering from laughably poor production values, reminiscent of a low-budget pornographic film (Detweile 2012), the raw footage from *The Innocence of Muslims* trailer was uploaded to YouTube in July 2012 to little fanfare; however, as the anniversary of the 9/11 attacks approached, Islamist activists in the Middle East and Europe took notice resulting in widespread protests in Egypt. The unrest spread to other locales, and, while hotly disputed in foreign policy circles to this day, was initially connected to the infamous attacks on the U.S. consulate in Benghazi, Libya, which resulted in the deaths of Ambassador J. Christopher Stevens, U.S. Foreign Service Information Management Officer Sean Smith, and two CIA contractors. While the Obama Administration attempted to prove that the attack on the diplomatic compound was not linked to the wave of unrest associated with the YouTube video, the imbrication of the two remained real for many around the world and has even been affirmed by those on the American left (see, for instance, Drum 2014).

Anyone who closely followed the 2016 U.S presidential election and its aftermath will agree that 'fake news' is a medium whose time has come (a consummate example of phronesis). Departing from the satirical culture jamming (Warner 2007) and well-intentioned attempts at discursive integration (Baym 2005) associated with *The Daily Show with Jon Stewart* (1999–2015), and moving far afield from the faux 'reporting' of *The Onion* (Darling 2008; Saunders 2012), fake news is now deadly real. While there are countless examples of instances where 'fake news' has impacted world affairs, Lars Klevberg's staged 'Syria Hero Boy' (2014) video is perhaps the most emblematic of the issues that are created by the Internet when it comes to contemporary geopolitics. This elaborate hoax involved footage a boy braving sniper fire in Syria to save a girl. The Norwegian director, who used grant money from the Norwegian Film Institute and Arts Council Norway to finance the project, which was filmed in Malta on a set that had once served as a backdrop for Ridley Scott's 2000 sword-and-sandals epic *Gladiator* (Mackey 2014). After the video received massive media attention and was viewed over 5 million times, it was unveiled as a fake, prompting its creators to defend their actions as 'independent filmmakers' by arguing that they were doing the world a public service by critiquing the prevalence of false and misleading information and visuals involved in media coverage of world affairs. Klevberg later apologised for the stunt, recognising that it added to doubts about the overall veracity of news agencies' reporting of facts on the ground in the Syrian Civil War (2011–present) and served as potential fodder for the governing regime to deflect criticism of its actions in the future.

Returning to the 2016 U.S. election, it is becoming increasingly apparent that Russians (and mostly likely, paid agents of the Russian state), employed popular culture as a novel platform for geopolitical intervention. Originally, it appeared the highly targeted distribution of spurious news stories impugning the candidacy, campaign, and conduct of former U.S. Secretary of State Hillary Rodham Clinton were the result of actions taken by teenaged Macedonian entrepreneurs hoping to profit from click-happy users of social media (Woolf 2016). However, at the time of writing, it had become apparent that the use of social media networks by parties outside the U.S. was far more strategic. Recognising that two-thirds of U.S. adults rely on social media sites for part or most of their news (Shearer and Gottfried 2017), false stories about Clinton and her allies were promoted via advertisements, bot-driven posts, and other forms of weaponised disinformation. In late October 2017, Facebook reported that some 80,000 pieces of 'divisive content', originally generated by the Kremlin-linked Internet Research Agency, reached some 30 million Americans (while Twitter reported 131,000 tweets generated by the IRA); ultimately a combination of Facebook ads and shared stories were viewed by 126 million U.S. citizens in the lead-up to the lection (Isaac and Wakabayashi 2017). Unlike propaganda that characterised the Cold War, Russia's new popular culture based tactics are 'calibrated to confuse, befuddle and distract', rather than promote a specific agenda or influence the audience's ideological orientations (Lucas and Nimmo 2015, 4). Pioneered through Russia's state-owned media platforms like RT and Sputnik, such efforts were delinked from 'official' sites of Russian policy, instead being grafted on to pop culture production in the West, i.e. social media sites and their content, thus (ostensibly) removing the 'state' from the process. Forged in a conspiracy-driven realm of competing realities, the anti-Clinton 'news' campaign achieved a sort of life of its own, resulting—according to some commentators—in tangible results, including a lone-wolf terror attack on a pizzeria referenced as a front for Clinton's purported child sex-slave ring (Kang and Goldman 2016) and perhaps even the installation of a reality-TV star as the 45th president of U.S. (White 2017). While the U.S. regularly involves itself in elections oversees, and most dramatically the 1996 Russian presidential election (an intervention that was depicted in the motion picture *Spinning Boris* [2003]), this new form of pop culture-based manipulation represents a new form of foreign policy engagement, one which is likely to expand over time.

What does it all mean? Geopolitics, popular culture, and the codes we live by

Critical geopolitics began with a mission to excavate the reality of how power manifests in places and through spaces, attempting to shift from the 'masculinist gaze' (Koch 2014, 231) of the classical and/or academic geopolitical commentator towards a more holistic evaluation of the everyday forces that shape international relations, world affairs, and the projection of power around the globe. Taking a page from the Frankfurt School and influenced by Foucault, Gramsci, Habermas and other scholars, the first generation of critical geographers, led by Yves Lacoste who founded the journal *Hérodote* in 1976, sparked a revolution in the study of geopolitics (Kelly 2006). Later influenced by the works of Said (1978), Deleuze and Guattari (1987), Jay (1988), scholars like Sharp and Dodds pushed the field into new and exciting directions, one of which being popular geopolitics (see Dittmer's chapter in this volume), producing a practical wisdom (phronesis) about the popular culture-world politics continuum. As discussed in the introduction to this volume, there is a new set of shifts going on that demand increasing interdisciplinary attention to the imbrications of popular culture and geopolitics, or in the words of Rowley and Weldes: 'Popular culture *is* the "real world", providing us with meanings, including about world politics' (2012, 514). However, as Strukov points out in his chapter in the volume, everything is dependent on the vantage point of one's gaze.

In this chapter, I have endeavoured to point out just a few of the myriad ways in which (geopolitical) power and (popular) culture are intertwining, and more specifically how popular culture is 'interactive with other representations of political life' (Nexon and Neumann 2006). While this may not be a new phenomenon, the speed and impact of this twinning is undeniable. As I have pointed out, North Korea maintains a constant vigil over its representation in popular culture, bristling at comic books, videogames, and movies alike. Russia, China, Brazil, and India are all scrambling to get into the pop culture game in an effort to burnish their own images and blunt negative portrayals emanating from Hollywood (see

Dennison 2016). Filmmakers are now in a position where they literally hold the lives of people in their hands, given that a piece of pop culture can now spark riots across a region, lay low a multinational company, or even provoke a terror attack. Popular culture is no laughing matter in the world of IR. Like tanks, submarines, and long-range bombers, it can be deployed to attack or strategically stationed to defend the national interest. Movies, videogames, television series, and online videos shape our reality, and often punctuate it with their narratives, depictions, and representations. Through such praxis, geopolitical codes are generated in the mediated visions of cultural producers, whether or not they are paid by the state and regardless of where they produce their craft. Ignoring this truism puts us all at peril, and the sooner popular culture gains the respect it deserves in foreign policy circles, the safer the world will be.

Notes

1 This chapter is partially based on my short essay published by E-International Relations entitled '"The Interview" and the Popular Culture-World Politics Continuum' (2014).
2 Connelly defines a somatic marker as a 'culturally mobilised corporeal disposition, through which affect-imbued, preliminary orientations to perception and judgement scale down the material factored into cost- benefit analyses, principled judgement, and reflective experiments' (2002, 34).
3 In terms of data collection, this chapter draws my larger body of research which focusses on state reactions to Anglophone popular-culture representation; for more information, see Saunders 2017a.
4 For an interrogation of 'doing' popular geopolitics, see my essay on Max Brooks, colonialization, and *The Zombie Survival Guide* (Saunders 2013) and Dittmer's (2015) essay on *Captain America* and doing popular culture in social sciences.
5 Echoing a statement made by the French philosophe Voltaire in 1793, the first appearance of Spider-Man in Marvel Comics' *Amazing Fantasy* #15 (August 1962) included the timeless maxim 'WITH GREAT POWER THERE MUST ALSO COME – GREAT RESPONSIBILITY!' (later simplified to the version which appears above). Importantly, Obama, a self-avowed fan of comics (NPR 2008) and whose visage graced the cover of *The Amazing Spider-Man* #583 a week before the presidential inauguration in January 2009, employed a version of this phrase in a 2010 address to Indian policymakers on the issue of expanding the United Nations Security Council to include additional permanent members.
6 Intentionally reflecting the conceptual sloppiness of journalistic shorthand, I use the term 'Hollywood' in quotes to stand in for mainstream popular cultural production associated with the Anglophone west (see Saunders 2017a).
7 While the table above clearly reflects the power of 'Hollywood', it can also be viewed as an ecumenical framework in that is potentially applicable to any state-centric production ecosystem from North Korea to the Russian Federation to Australia.
8 Quasi-diplomacy here refers to forms of trans-border political interaction conducted outside the traditional structures and institutions of post-Westphalian interstate relations; however, like traditional definitions of diplomacy, I recognize that quasi-diplomacy is a multidimensional and relational concept dependent on identifying the 'self' and the 'other' (see Carta 2013).
9 Despite playing along in the West, Kazakhstani authorities did not permit the film to be shown in the country's movie theatres. Russia and a number of other sympathetic nations also de facto banned the motion picture.

10 Frank Miller unabashedly declares his antipathy towards Islam and emerged as a pop-culture firebrand in the wake of the 9/11 attacks on the U.S. His attempts to transform the iconic superhero Batman into an anti-Islamist vigilante were rejected by DC Comics as too controversial for the imprint. Miller ultimately created another character named The Fixer to achieve these ends; this un-caped crusader appeared in the hardcover graphic novel *Holy Terror* (2011) published by Legendary Comics (see Ackerman 2011).

11 The original film revolved around a group of Colorado teens taking up a guerrilla campaign against an occupying force of Cubans and their Soviets advisors, following an invasion of the U.S. by the Warsaw Pact allies.

12 Despite being lambasted as anti-Asian by some critics, *The Interview*'s representation of China is extremely positive, showing one of the main characters, Aaron Rapaport (Seth Rogen), touring the country, immersing himself in its culture, and partying with locals before things turn grim in North Korea.

13 As I have argued elsewhere (Saunders 2017b), videogames are also impacting the aesthetics of television production, most notably in the increase in geographic representation and the use of landscape cutscenes to advance the narrative of many series, particularly those with a geopolitical bent, e.g. *The Border* (2014–), *Tyrant* (2014–), *Deutschland 83* (2015–), and *Occupied* (2015–).

References

Ackerman, Spencer. 2011. 'Frank Miller's *Holy Terror* Is Fodder for Anti-Islam Set'. *Wired*, available at http://www.wired.com/2011/09/holy-terror-frank-miller/ [last accessed 18 December 2014].

Adorno, Theodor W. (ed.) 1991. *The Culture Industry: Selected Essay on Mass Culture.* London: Routledge.

Alford, Matthew. 2010. *Reel Power: Hollywood Cinema and American Supremacy.* New York: Pluto Press.

Arkin, Daniel. 2014. '"The Interview" Isn't Alone: 5 Movies That Riled World Leaders'. *NBC News*, available at http://www.nbcnews.com/pop-culture/movies/interview-isnt-alone-5-movies-riled-world-leaders-n270786 [last accessed 10 September 2016].

Baym, Geoffrey. 2005. 'The Daily Show: Discursive Integration and the Reinvention of Political Journalism'. *Political Communication* 22 (3):259–276.

Booker, Keith M. 2010. *Disney, Pixar, and the Hidden Messages of Children's Films.* Santa Barbara: ABC-CLIO.

Brinkerhoff, Jennifer M. 2009. *Digital Diasporas: Identity and Transnational Engagement.* Cambridge: Cambridge University Press.

Brooke, James. 2005. 'Pop Culture Takes Aim at North Korea's Kim'. *New York Times*, available at http://www.nytimes.com/2005/06/01/world/asia/01iht-north-5079850.html?pagewanted=all&_r=0 [last accessed 19 December 2014].

Carta, Caterina. 2013. *The European Union Diplomatic Service: Ideas, Preferences and Identities.* London: Routledge.

Chaplin, Tamara. 2014. 'Lesbians Online: Queer Identity and Community Formation on the French Minitel'. *Journal of the History of Sexuality* 23 (3):451–472.

CNN. 2006. 'Hezbollah Video Game: War with Israel'. *CNN*, available at http://www.cnn.com/2007/WORLD/meast/08/16/hezbollah.game.reut/index.html?iref=nextin [last accessed 21 November 2016].

Connolly, William E. 2002. *Neuropolitics: Thinking, Culture, Speed.* Minneapolis: University of Minnesota Press.

Courmont, Barthélémy, and Pierre-Alain Clément. 2014. 'When Geopolitics Meets the Game Industry: A Study of Arabic Video Games and What They Teach Us'. *Hemispheres* 29 (1):31–46.

Crampton, Andrew, and Marcus Power. 2005. 'Reel Geopolitics: Cinemato-graphing Political Space'. *Geopolitics* 10:193–203.

Darling, Cary. 2008. 'Onion's 'Our Dumb World' Could Pass for What It's Mocking'. *Chicago Tribune*, available at http://www.chicagotribune.com/chi-0103reader_jan03-story.html [last accessed 28 June 2013].

Deleuze, Gilles, and Félix Guattari. 1987. *A Thousand Plateaus: Capitalism and Schizophrenia*. Minneapolis: University of Minnesota Press.

Dennison, Stephanie. 2016. 'Soft Power, Cinema and the BRICS'. *Research Beyond Borders*, available at https://researchbeyondborders.wordpress.com/2016/08/12/soft-power-cinema-and-the-brics/ [last accessed 5 December 2016].

Der Derian, James. 2009. *Virtuous War: Mapping the Military-Industrial-Media-Entertainment Network*. London: Routledge.

Der Derian, James. 2010. 'Imagining Terror: Logos, Pathos, and Ethos'. In *Observant States: Geopolitics and Visual Culture*, edited by Fraser MacDonald, Rachel Hughes and Klaus Dodds, 23–40. London and New York: I. B. Tauris.

Detweile, Craig. 2012. 'The Innocence of Muslims: Blasphemy, Pornography, and Christianity'. *Patheos*, available at http://www.patheos.com/blogs/dochollywood/2012/09/the-innocence-of-muslims-blasphemy-pornography-and-christianity/ [last accessed 23 November 2016].

Ding, Sheng, and Robert A. Saunders. 2006. 'Talking Up China: An Analysis of China's Rising Cultural Power and Global Promotion of the Chinese Language'. *East Asia* 23 (2):3–33.

Dittmer, Jason. 2005. 'Captain America's Empire: Reflections on Identity, Popular Culture, and Post-9/11 Geopolitics'. *Annals of the Association of American Geographers* 95 (3):626–643.

Dittmer, Jason. 2010. *Popular Culture, Geopolitics, and Identity*. Lanham, MD: Rowman & Littlefield.

Dittmer, Jason. 2015. 'On *Captain America* and "Doing" Popular Culture in the Social Sciences'. In *Popular Culture and World Politics: Theories, Methods, Pedagogies*, edited by Federica Caso and Caitlin Hamilton, 45–50. Bristol, UK: E-International Relations.

Dittmer, Jason, and Klaus Dodds. 2008. 'Popular Geopolitics Past and Future: Fandom, Identities and Audiences'. *Geopolitics* 13 (3):437–445.

Dodds, Klaus. 2003. 'License to Stereotype: Popular Geopolitics, James Bond and the Spectre of Balkanism'. *Geopolitics* 8 (2):125–156.

Dodds, Klaus. 2005. 'Screening Geopolitics: James Bond and the Early Cold War Films (1962–1967)'. *Geopolitics* 10 (2):266–289.

Dodds, Klaus. 2008. 'Hollywood and the Popular Geopolitics of the War on Terror'. *Third World Quarterly* 29 (8):1621–1637.

Drezner, Daniel W. 2016. 'You Should Watch the Best Show about International Relations on Television Right Now'. *Washington Post*, available at https://www.washingtonpost.com/posteverything/wp/2016/02/19/you-should-watch-the-best-show-about-international-relations-on-television-right-now/?utm_term=.cf8e0109b57c [last accessed 20 August 2016].

Drum, Kevin. 2014. 'Yes, the 'Innocence of Muslims' Video Really Did Play a Role in the Benghazi Attacks'. *Mother Jones*, available at http://www.motherjones.com/

kevin-drum/2014/05/yes-innocence-muslims-video-really-did-play-role-benghazi-attacks [last accessed 26 November 2016].

Edensor, Tim. 2002. *National Identity, Popular Culture and Everyday Life*. Oxford: Berg.

Es, Murat. 2011. 'Frank Miller's *300*: Civilization Exclusivism and the Spatialized Politics of Spectatorship'. *Aether: The Journal of Media Geography* 8 (2):6–39.

Extreme Tech. 2006. 'Islamogaming: Looking for Video Games in the Muslim World'. *FOX News*, available at http://www.foxnews.com/story/2006/09/11/islamogaming-looking-for-video-games-in-muslim-world.html [last accessed 21 November 2016].

Glynn, Kevin, and Julie Cupples. 2015. 'Negotiating and Queering US Hegemony in TV Drama: Popular Geopolitics and Cultural Studies'. *Gender, Place & Culture: A Journal of Feminist Geography* 22 (2):271–287.

Goldsmith, Jack, and Tim Wu. 2006. *Who Controls the Internet?: Illusions of a Borderless World*. Oxford: Oxford University Press.

Grayson, Kyle, Matt Davies, and Simon Philpott. 2009. 'Pop Goes IR? Researching the Popular Culture-World Politics Continuum'. *Politics* 29 (3):155–163.

Guardian, The. 2014. 'North Korea Calls New Channel 4 Show a "Slanderous Farce"'. *The Guardian*, available at https://www.theguardian.com/world/2014/sep/01/north-korea-channel-4-opposite-number [last accessed 21 November 2016].

Hendershot, Cynthia. 2001. *I Was a Cold War Monster: Horror Films, Eroticism, and the Cold War Imagination*. Madison, WI: Popular Press.

Hussain, Murtaza. 2016. 'How Video Games Are Influencing War Propaganda in Syria'. *The Intercept*, available at https://theintercept.com/2016/09/21/in-syria-a-new-style-of-war-propaganda-emerges-influenced-by-video-games/ [last accessed 26 November 2016].

Isaac, Mike, and Daisuke Wakabayashi. 2017. 'Russian Influence Reached 126 Million Through Facebook Alone'. *New York Times*, available at https://www.nytimes.com/2017/10/30/technology/facebook-google-russia.html [last accessed 31 October 2017].

Jay, Martin. 1988. 'Scopic Regimes of Modernity' in *Vision and Visuality*, edited by Hal Foster, 3–23. Seattle: Bay Press.

Joneidi, Majid. 2007. 'Iranian Anger at Hollywood "Assault"'. *BBC News*, available at http://news.bbc.co.uk/2/hi/6455969.stm [last accessed 18 December 2014].

Kamen, Matt. 2016. 'Gaza Strip iOS Game Isn't a Game, Says Apple'. *Wired*, available at http://www.wired.co.uk/article/palestinian-game-liyla-apple-app-store-ban [last accessed 21 November 2016].

Kang, Cecilia, and Adam Goldman. 2016. 'In Washington Pizzeria Attack, Fake News Brought Real Guns'. *New York Times*, available at https://www.nytimes.com/2016/12/05/business/media/comet-ping-pong-pizza-shooting-fake-news-consequences.html [last accessed 30 October 2017].

Kelly, Phil. 2006. 'A Critique of Critical Geopolitics'. *Geopolitics* 11 (1):24–53.

Kirchik, James. 2016. 'Cold War Nostalgia: What German Television Tells Us about Today's Europe'. *Weekly Standard*, available at http://www.weeklystandard.com/print/cold-war-nostalgia/article/2002366# [last accessed 5 December 2016].

Koch, Natalie. 2014. 'Grounding Central Asian Geopolitics'. *Geopolitics* 19 (1):227–233.

Landreth, Jonathan. 2010. 'Chinese Press Rails against "Red Dawn"'. *Hollywood Reporter*, available at http://www.hollywoodreporter.com/news/chinese-press-rails-red-dawn-24152 [last accessed 19 December 2014].

Levine, Elena. 2011. 'Teaching the Politics of Television Culture in a 'Post-television' Era'. *Cinema Journal* 50 (4):177–182.

Lipschutz, Ronnie D. 2001. *Cold War Fantasies: Film, Fiction, and Foreign Policy*. Lanham, MD: Rowman & Littlefield.

Lucas, Edward, and Ben Nimmo. 2015. Information Warfare: What Is It and How Win It? In *CEPA Inforwar Papers*, edited by Centre for European Policy Analysis. Washington, DC.

Lynch, Colum. 2014. 'Russia Vetoes House of Cards'. *Foreign Policy*, available at http://foreignpolicy.com/2014/07/02/exclusive-russia-vetoes-house-of-cards/ [last accessed 21 November 2016].

McCurry, Justin. 2014. 'North Korea Threatens 'Merciless' Response over Seth Rogen Film'. *The Guardian*, available at http://www.theguardian.com/world/2014/jun/25/north-korea-merciless-response-us-kim-jong-un-film [last accessed 19 December 2014].

Mackey, Robert. 2014. 'Norwegian Filmmakers Apologize for Fake Syria Video'. *New York Times*, available at http://www.nytimes.com/2014/11/19/world/europe/norwegian-filmmakers-apologize-for-fake-syria-video.html?_r=3 [last accessed 26 November 2016].

Minotti, Mike. 2016. 'Video Games Will Become a $99.6B Industry This Year as Mobile Overtakes Consoles and PCs'. *VentureBeat*, available at http://venturebeat.com/2016/04/21/video-games-will-become-a-99-6b-industry-this-year-as-mobile-overtakes-consoles-and-pcs/ [last accessed 21 November 2016].

Moïsi, Dominique. 2016. *La géopolitique des séries ou le triomphe de la peur*. Paris: Stock.

Morley, David, and Kevin Robins. 1995. *Spaces of Identity: Global Media, Electronic Landscapes and Cultural Boundaries*. London and New York: Routledge.

Morris, Chris. 2016. 'Level up! Video Game Industry Revenues Soar in 2015'. *Fortune*, available at http://fortune.com/2016/02/16/video-game-industry-revenues-2015/ [last accessed 21 November 2016].

New York Times. 2016. 'What Donald Trump Said About Russian Hacking and Hillary Clinton's Emails'. *New York Times*, available at http://www.nytimes.com/2016/07/28/us/politics/trump-conference-highlights.html?_r=0 [last accessed 27 July 2016].

Nexon, Daniel H., and Iver B. Neumann. 2006. 'Introduction: Harry Potter and the Study of World Politics'. In *Harry Potter and International Relations*, edited by Daniel H. Nexon and Iver B. Neumann, 1–26. Lanham, MD: Rowman & Littlefield.

NPR. 2008. 'Analyzing Obama's Love of Comic Books'. *Day to Day*, available at http://www.npr.org/templates/story/story.php?storyId=97260170 [last accessed 10 September 2016].

Ó Tuathail, Gearóid. 1996. *Critical Geopolitics: The Politics of Writing Global Space*. Minneapolis: University of Minnesota Press.

Ó Tuathail, Gearóid. 2003. '"Just Out Looking for a Fight": American Affect and the Invasion of Iraq'. *Antipode* 35 (5):856–870.

Peerson, Joe. 2015. 'Kazakhstan Named One of World's Most Valuable Brands'. *Silk Road Reporters*, available at http://www.silkroadreporters.com/2015/10/14/kazakhstan-named-one-of-worlds-most-valuable-national-brands/ [last accessed 5 December 2016].

Pelley, Scott. 2016. 'In God's Name'. *60 Minutes*, available at http://www.cbsnews.com/news/60-minutes-american-teen-isis-cell-leader-scott-pelley/ [last accessed 26 November 2016].

Philpott, Simon. 2010. 'Is Anyone Watching? War, Cinema and Bearing Witness'. *Cambridge Review of International Affairs* 23 (2):325–348.

Powers, Shawn M., and Michael Jablonski. 2015. *The Real Cyber War: The Political Economy of Internet Freedom*. Urbana, Chicago and Springfield: University of Illinois Press.

Rai, Shailza. 2011. 'Identity and its Monsters: Borders Within and Without' in *Cartographies of Affect: Across Borders in South Asia and the Americas*, edited by Debra A. Castillo and Kavita Panjabi, 325–345. Delhi: Worldview Publications.

Rowley, Christina. 2010. 'Popular Culture and the Politics of the Visual'. In *Gender Matters in Global Politics: A Feminist Introduction to International Relations*, edited by Laura J. Shepherd, 309–325. London: Routledge.

Rowley, Christina, and Jutta Weldes. 2012. 'The Evolution of International Security Studies and the Everyday: Suggestions from the Buffyverse'. *Security Dialogue* 43 (6):513–530.

Sacks, Ethan. 2014. ''The Interview' is Latest Film to Spark International Uproar'. *New York Daily News*, available at http://m.nydailynews.com/entertainment/movies/interview-films-sparking-international-uproar-article-1.2047598 [last accessed 10 September 2016].

Said, Edward. 1978. *Orientalism*. New York: Vintage Books.

Saunders, Robert A. 2007. 'In Defence of *Kazakshilik:* Kazakhstan's War on Sacha Baron Cohen'. *Identities: Global Studies in Culture and Power* 14 (3):225–255.

Saunders, Robert A. 2008a. 'Buying into Brand Borat: Kazakhstan's Cautious Embrace of its Unwanted 'Son''. *Slavic Review* 67 (1):63–80.

Saunders, Robert A. 2008b. 'The Ummah as Nation: A Reappraisal in the Wake of the 'Cartoons Affair''. *Nations and Nationalism* 14 (2):303–321.

Saunders, Robert A. 2010. *Ethnopolitics in Cyberspace: The Internet, Minority Nationalism, and the Web of Identity*. Lanham, MD: Lexington Books.

Saunders, Robert A. 2012. 'Brand Interrupted: The Impact of Alternative Narrators on Nation Branding in the Former Second World'. In *Branding Post-Communist Nations: Marketizing National Identities in the 'New' Europe*, edited by Nadia Kaneva, 49–78. New York and London: Routledge.

Saunders, Robert A. 2013. 'Zombies in the Colonies: Imperialism and Contestation of Ethno-Political Space in Max Brooks' *The Zombie Survival Guide*'. In *Monstrous Geographies: Places and Spaces of the Monstrous*, edited by Sarah Montin and Evelyn Tsitas, 19–46. Oxford: Inter-Disciplinary Press.

Saunders, Robert A. 2016. 'Geopolitical Television: A Brief History and Tentative Taxonomy'. Paper read at Popular Culture and World Politics 9, at Balsillie School of International Affairs, Waterloo, Canada (13 November).

Saunders, Robert A. 2017a. *Popular Geopolitics and Nation Branding in the Post-Soviet Realm*. London and New York: Routledge.

Saunders, Robert A. 2017b. 'Small Screen IR: A Tentative Typology of Geopolitical Television'. *Geopolitics*, DOI: https://doi.org/10.1080/14650045.2017.1389719.

Saunders, Robert A., and Vlad Strukov. 2017. 'The Popular Geopolitics Feedback Loop: Thinking beyond the "Russia versus the West" Paradigm'. *Europe-Asia Studies* 69 (2):303–324.

Schopp, Andrew, and Matthew B. Hill. 2009. *The War on Terror and American Popular Culture: September 11 and Beyond*. Madison and Teaneck, NJ: Fairleigh Dickinson Press.

Shaheen, Jack. 2001. *Reel Bad Arabs: How Hollywood Vilifies a People*. New York: Olive Branch Press.

Shane, Scott. 2017. 'The Fake Americans Russia Created to Influence the Election'. *New York Times*, available at https://www.nytimes.com/2017/09/07/us/politics/russia-facebook-twitter-election.html [last accessed 30 October 2017].

Shapiro, Michael J. 2008. *Cinematic Geopolitics*. Abingdon, UK: Taylor & Francis.

Sharp, Joanne P. 1993. 'Publishing American Identity: Popular Geopolitics, Myth and *The Reader's Digest*'. *Political Geography* 12 (6):491–503.

Shearer, Elisa, and Jeffrey Gottfried. 2017. 'News Use Across Social Media Platforms 2017'. *Pew Research Center*, available at http://www.journalism.org/2017/09/07/news-use-across-social-media-platforms-2017/ [last accessed 30 October 2017].

Shepherd, Laura J. 2013. *Gender, Violence and Popular Culture: Telling Stories*. London and New York: Routledge.

Spigel, Lynn, and Jan Olsson. 2004. *Television after TV: Essays on a Medium in Transition*. Durham and London: Duke University Press.

Stewart, Will. 2013. 'New NBC Show *Siberia* is Scorned by Russians for Stereotyping and Being 'Frozen in a Cold War Time Warp''. *Daily Mail*, available at http://www.dailymail.co.uk/news/article-2353073/NBC-Siberia-scorned-Russians-frozen-Cold-War-time-warp.html#ixzz3IDG5HX3I [last accessed 3 November 2014].

Strukov, Vlad. 2016. 'Digital Conservatism: Framing Patriotism in the Era of Global Journalism'. In *Eurasia 2.0: Russian Geopolitics in the Age of New Media*, edited by Mikhail Suslov and Mark Bassin, 185–208. Lanham, MD: Lexington Books.

Suebsaeng, Asawin. 2014. 'When Countries Lose Their Shit Over American Movies' *The Daily Beast*, available at http://www.thedailybeast.com/articles/2014/12/17/when-countries-lose-their-shit-over-american-movies.html [last accessed 10 September 2016].

Surette, Tim. 2006. 'Venezuelan Government Takes Issue with *Mercenaries 2*'. *Gamespot*, available at http://www.gamespot.com/articles/venezuelan-government-takes-issue-with-mercenaries-2/1100-6151849/ [last accessed 21 November 2016].

Tucker, Reed. 2014. '5 Times Hollywood has Angered Foreign Governments'. *New York Times*, available at http://nypost.com/2014/12/17/5-times-hollywood-has-angered-foreign-governments/ [last accessed 10 September 2016].

Van Jelgerhuis, Daniel. 2015. *Resurgence of Cold War Imagery in Western Popular Culture*. Tallahasse, FL: Florida State University.

Warner, Jamie. 2007. 'Political Culture Jamming: The Dissident Humor of 'The Daily Show with Jon Stewart''. *Popular Communication* 5 (1):17–36.

Weber, Cynthia. 2005. *Imagining America at War: Morality, Politics and Film*. London and New York: Routledge.

Weldes, Jutta. 2003. *To Seek Out New Worlds: Exploring Links between Science Fiction and World Politics*. Houndsmills, UK: Palgrave Macmillan.

White, Jeremy B. 2017. 'How Russia's Election Meddling Could Spell the End of an Era for Facebook, Google, and Twitter'. *The Independent*, available at http://www.independent.co.uk/life-style/gadgets-and-tech/russia-facebook-google-twitter-election-meddling-end-of-era-analysis-a8009976.html [last accessed 30 October 2017].

Woolf, Christopher. 2016. 'Kids in Macedonia Made Up and Circulated Many False News Stories in the US Election'. *PRI's The World*, available at http://www.pri.org/stories/2016-11-16/kids-macedonia-made-and-circulated-many-false-news-stories-us-election [last accessed 26 November 2016].

Part II

Popular geopolitics goes global and looks into the future

6 The convenient fiction of geopolitics

Rethinking 'America' in the geopolitical imagination of Yugoslav culture[1]

Maša Kolanović

Introduction

The advent of the Cold War overlaps with the formative beginnings and emergence of global popular culture in its contemporary meaning both symbolically and in practice. According to one of the most extensive historical overview of popular culture by Cantor and Werthman (1968), during the second half of the twentieth century, a global market of popular culture emerged with its roots dominantly situated in the U.S. Similar formulations of contemporary popular culture, condensed (or reduced) to its American twentieth-century image and, even more narrowly, to its post-World War II appearance, can be found in other histories of popular culture (see Hebdige 1980; Hoggart 1992). Maltby (1989), for example, emphasises the crucial role of U.S. popular culture in the country's post-war domination over the rest of the world, while Betts (2004) shows a connection between the emergence of global popular culture and the post-World War II development of rapid communications, widespread consumption, the market economy, and the virtualisation of reality through which the U.S. established its role as world leader.

This historically verifiable congruence of the post-World War II period, both symbolically and taking place in practice, and the emergence of global popular culture in a geopolitically divided world point to the historical significance of the Cold War during which the discipline of popular geopolitics inscribed a prehistory of its own. Namely, the Cold War was a period of intense world-making, world-writing, and world-picturing of the mental and moral geographies of the East and West, often identified with the ideologies of communism and capitalism. Those perceptions were not just conducted through international politics but—even more critically—through the spheres of popular culture and everyday life.[2] The moral geography of the Cold War was thus 'a perfect example of the naturalization of certain metageographical concepts... within the realms of popular culture', so that 'these political geographies are rendered politically inoffensive, as common sense' (Sharp 2000, 168). Though the period of geopolitical crisis, in general, necessarily evokes the analytical tools of critical geography and popular geopolitics,[3] the Cold War holds a prominent place in terms of geopolitical discourses since it was the *longue durée* structure of a specific mental and moral geography that, to a certain extent, has not vanished after the end of the Cold

War.[4] By considering popular geopolitics as an evolving interdiscipline, it is thus important to repeat the lessons from the Cold War in analysing the making (and breaking) of the popular images of nations and places from rearticulated theoretical and methodological positions.

The Cold War was also the period during which the disciplines and approaches that critically deal with popular culture, such as cultural studies,[5] were formed and developed, especially due to its interest in the political dimensions of popular culture and imagology, a specific branch of comparative literature that deals with the images and perceptions of foreign countries in national literatures (Beller and Leerssen 2007; Dukić 2012). Both analytical approaches, cultural studies and imagology, share a common basis, and are critical components in developing and understanding the discipline of popular geopolitics. Particularly, the methodological focusses of imagology and popular geopolitics, though diversified in their interests in materials and sources, meet at the analytical point of construction and reproduction of national characters within a specific contextual frame. According to their basic, premise, cultural texts and practices are permeable to their social and political contexts and, at the particular crossroads between 'the text as verbal issue and the text as social act' (Leerssen 1991, 126), lies the challenge of analysis that takes into account contextual geopolitical complexity. In a certain way, the analytical focus on the study of either national or geopolitical misperceptions encouraged both disciplines to pay more attention to the national stereotypes that could be considered useful, while keeping in mind their reduced and simplified premises. Stereotypes, of course, take in a large part of geopolitical and cultural representations, but cultural practices can also question or oppose some entrenched stereotypes of a foreign country in the home culture, making a field of tension where unexpected and new meanings can occur.

In an attempt to be aware of these methodological challenges and disciplinary linkages, this chapter will therefore try to combine the methodological approaches of imagology, popular geopolitics and cultural studies. According to Sharp's lucid insight:

> ironically, geopolitical arguments are, in one significant sense, profoundly ageographical. Rather than being concerned with understanding geographical process, geopolitics reduces spaces and places to concepts or ideology. Space is reified into units that tautologically display evidence of the characteristics that are used to define the spaces in the first place: the "Orient" is exoticism, the USSR is communism, Iran is fundamentalism, and so on.
>
> (Sharp 2000, 27–28)

This observation leads us to another important theoretical notion, which is that of the great importance of language, discourse and the metaphorical character of geopolitics. Namely, what are 'the Orient', 'the East', 'the West' or 'America', 'USSR', 'Yugoslavia', 'Kazakhstan' or 'Siberian' other than geopolitical metaphors that not just point to a certain place but hold specific cultural and ideological meanings? These and similar 'conceptual metaphors' (Lakoff and Johnson

2003) could be perceived as the privileged agents of the geopolitical imagination.[6] In this sense, Ó Tuathail's critical (and a bit ironical) notion of geopolitics as 'convenient fiction'[7] could be further interpreted as geopolitical constructions that are in a large part indeed fictional constructions (such as the 'Evil Empire', 'the Third World', and 'the Iron Curtain') and that they are also constructed, disseminated, and challenged through fiction, such as textual and visual cultural representations.

From language to geopolitics (and back again)

Existing literature on Cold War geopolitical imaginations has so far focussed on Western constructions and representations of the East as the communist Other (see Sayre 1979; MacDonald 1985; Barson 1992; Hendershot 2001; Shaw 2006). In this chapter, I focus on one particularly significant geopolitical narrative in socialist Yugoslavia, a country that challenges the political, economic, and cultural uniqueness of the Cold War binaries of the East and the West. My case study focusses on the narrative of 'America' in the geopolitical imagination of Yugoslav culture. This narrative had a prominent place in Yugoslav geopolitical and cultural discourses, especially after 1948 when the Yugoslav narrative, not completely unambiguously, prevailed over the Soviet version.[8] I use the term 'America', instead of the 'United States of America' or 'USA', to highlight the contradictions in that naming. Sharp notes on the relationship between geopolitics and language:

> language is not unproblematic, somehow simply describing what is there. Language is metaphorical, explaining through reference to other, known concepts. There is always a choice of words and metaphors. The type of terms used – the conceptual links made – affects meaning. There is, as a consequence, a politics of language.
>
> (Sharp 2000, 25)

Although the term 'America', in its most common official meaning, indicates the geographical space of the two continents, it is often reduced as a signifier of the United States of America. As discussed by Vladimir Mayakovsky, the Soviet avant-garde poet and artist, the U.S. clearly annexed the word 'America' (2005, 74). Though use of 'America' in this chapter, at first glance, does not seem to challenge this customary usage, thus continuing to produce its conceptual inequalities (Sardar and Davies 2003, 10), I want to stress that my use of the word 'America' primarily implies not the real or lived 'United States of America', but the geopolitical cultural image of the U.S. that stands between the real and the imagined nation, which is why I put the word in quotation marks. The cultural creation and dissemination of its meanings is part of the problem that I discuss in this chapter.

Arguably, the idea of 'America' has become one of the 'metaphors we live by' (Lakoff and Johnson 2003). Several of these metaphorical meanings are also rooted in the Croatian language and with the words, phrases, and expressions semantically connected with 'America'. Here, I will enumerate just a few of them taken from a variety of dictionaries:

AMERICA (noun): in slang, 'a rich man, a place with a lot of money, a wealthy country in general; wealth; a great opportunity to steal something; a paradise'; when used as an exclamation, 'America!' it usually means 'a great, wonderful, excellent thing'.

AMERICANS (noun): used in slang, 'usually in situations in which the differences between Americans and Europeans are being considered or shown (designating strong military power and particular warfare methods or a feat by an individual or his/her wealth)'; 'Croatian migrant to America'; 'an American product'.

AMERICAN (adjective): in figurative meaning, 'one of excellent and of high quality'.

AMERICAN DREAM (noun): 'each American's mythical dream, variously defined in the terms of material goods'; 'in general every generation's wish to live better than the one that came before it'; 'rapid success at any cost'.

AMERICANIDE (noun): 'stereotypical movie or TV show of low quality, usually with a melodramatic plot, a happy ending and typical elements of mass Hollywood production'; 'aggressive propaganda or advertising for some product, also used in political election campaigns'.

AMERICANMANIA (noun): 'uncritical worship of the American way of life'.[9]

Significantly, we can tease out how meanings of consumerist and material well-being dominate the semantics of the aforementioned meanings in dictionaries published during and after the Cold War[10] and in this brief *précis* of the aforementioned words and phrases in the Croatian language connected with the idea of 'America', alternatively providing positive and negative connotations. These meanings were dominantly created in the historical context in which the U.S. was strengthening its force as an 'irresistible empire' that was, according to de Grazia (2005, 6), more informal and different than a classic totalitarian empire. It was, first of all, the 'market empire', which de Grazia variably defines as an 'empire by invitation', 'empire by consensus', and 'empire of fun'. Especially when it comes to 'empire of fun', one of the U.S.'s recipes for success was recognising the importance of trade in symbolic values through cultural products, whereby popular and consumerist culture became one of its most attractive exports. Through popular culture, the U.S. established almost archetypal utopian fantasies about the land of prosperity and material utopia. Thus articulated, the narrative of 'America' experienced its peak during the Cold War, and had its specific reflections in socialist countries. These issues become even more complex when we speak about Yugoslavia, 'a somewhat eccentric member of the family of socialist systems' (Hobsbawm 2004, 302). In rethinking the narratives of geopolitical imagery of 'America' in the different decades of Yugoslav socialism, the accompanying analytical scope will be a narrative of Yugoslav socialism in and of itself since the imagological approach and the heteroimage (i.e. the United States) are, at the same time, always an autoimage (i.e. Yugoslavia). In this particular case study of Yugoslavia, we are dealing with a very dynamic and complex geopolitical imagination, one that was quite heterogeneous during the Cold War period. The challenge of reflecting on these images also lies in the fact that the 'orthodox' ideological perceptions of 'America' in Yugoslav literature and popular culture often met resistance from various positions, a resistance that often became part of the cultural mainstream narratives.

In the rest of this chapter, I will present an analytical overview of the geopolitical imagination through the representative 'cultural patterns' (Williams 2006, 39–40) of Yugoslav culture by considering culture as 'a whole way of life' (Williams 1997). Thus, in analysing various and stratified cultural patterns of the Yugoslav geopolitical image of 'America', I will consider different cultural texts, artefacts and practices: from a text by the Yugoslav Nobel Prize laureate to popular novels, from political speeches to popular songs and poems, from the travelogue of a Yugoslav diplomat to geography textbooks, literary theory, translations, movies and commercials, which all show structural concordance within the creation of metageographical meanings.[11] All these sources show interlinkage and occasionally also form a field of cultural struggle that can also be perceived as a field of geopolitical struggle.

My methodological approach will combine the disciplinary strategies of imagology, popular geopolitics, and cultural analysis in order to show they are necessarily interweaving as well as to point to the transdisciplinary character of popular geopolitics in revealing the geopolitical assumptions of different cultural texts and practices, though, at first glance, geopolitical claims do not seem to be their ambition. With the help of these three fields of analysis, I will show how the geopolitical imaginations of foreign countries are not just 'arbitrarily accessible and available documentary material' (Fischer 2009, 54), or 'mimetic representation of empirical reality but particular *objets discursifs'* (Leerssen 2009, 101) with specific social performances that engage in everyday perceptions. It is wrong to measure those images against reality, just as it is similarly impossible to ignore the impact of the context on their design and their reciprocal influence on the imaginary and its everyday geopolitical repercussions. Thus, the relation between the 'real' and the 'imagined' is at the crux of this analytical problem.

The issue becomes even more complicated when we think about how a society is imagined beyond its boundaries. The very idea of 'America' cannot be thought about without bringing to mind the political, social and economic aspects of 'real' U.S. society, nor can the so-called 'real' U.S. be understood without keeping in mind all the cultural narratives that have articulated and even predestined its existence. Moreover, the 'real' nation is always a kind of 'imagined community' (Anderson 1990), both to its own members and those outside the national community. Further, if all nations are imagined, then the 'United States is the imagined community *par excellence*' (Campbell in Sharp 2000, 28). Nations do not necessarily have to border with each other to accomplish geopolitical and cultural exchange. Particularly, during the Cold War,

> America was projected beyond its state territory to become a transcendental force without necessary borders, an embodiment of a certain morality that, merged with the rhetoric of Manifest Destiny, could make necessary the extension of its influence beyond American borders.
>
> (Ó Tuathail and Agnew in Sharp 2000, 75)

The analytical overview that follows will point to the symbolic and political impact of 'America' in the Yugoslav geopolitical imagination as read through cultural text practices, revealing the notion that 'our view of America is largely shaped by the images of America rather than its reality' (Fischer 2009, 45).

'Bandit boogie-woogie civilisation'

During Yugoslav's early period of socialism (1945–1948), the country was briefly under Soviet political dominance, which also meant the dominance of the socialist realist cultural paradigm.[12] Consequently, the image of 'America' and the West in Yugoslavia was predominantly related to an opposing ideology and, as such, was marked by negative qualifications. Imagining 'America' basically meant a geopolitical imagining of the hostile ideology of capitalism. In the spectrum of those negative notions, one of the most frequently used metaphors was that of the 'decadent West', an umbrella concept for the Western, capitalist world (see, for instance, Kašić 1991). Specifically, in the cultural and literary criticism of the period, negative notions were often derived from discursive ingredients of American popular and consumer culture. In this regard, we can look at several conceptual metaphors that functioned as representative of U.S. society, such as: 'jazz', 'boogie-woogie', 'chewing gum', and others.[13] Thus, for example, in 1947 the propaganda director of the Yugoslav Communist Party, Milovan Đilas, claimed that 'America is our sworn enemy, and jazz, likewise, as its product' (Đilas in Vučetić 2012, 125).[14] Another example of a similar creation of metaphors with negative connotations connected with the sphere of American economy and consumer goods is the text *All Quiet on the Western Front* [*Na zapadu ništa novo*] by film critic Fedor Hanžeković.[15] As Hanžeković claims: 'Fur sellers became weapon industrialists and newspaper trust shareholders; suit cleaners became bankers and chewing gum producers and perfume merchants are shaking up today's American (and perhaps not only American) radio station networks. Additionally, they are sponsoring the movie industry and MacArthur' (Hanžeković 1947, 131). The cumulative use of the signifiers of American industry in the aforementioned example highlights the negative effect of the interference of capitalist economy and culture, especially films. Within Hažенković's special focus on the movie industry as a representative feature of U.S. society, we can notice the same issue as Davies and Wells argue: 'the fact remains, though, that the politics of the most powerful nation in the world [the USA] cannot be divorced from the most far-reaching entertainment medium in the world' (Davies and Wells in Dodds 2005, 77).

Other, similarly negative metaphors of the time were also constituted via American pop culture in the form of music and rhythms. In a text about Yugoslav magazines, the prominent literary critic Marin Franičević writes that: 'Our [Yugoslav] youth, who was at the centre of all the historical events and who is, through its heroism, sacrifice, enthusiasm, and awareness, the first youth behind only the Soviet Union youth, cannot look at art through the dehumanized eyes of the bandit boogie-woogie civilization' (Franičević 1948, 10). Of course, 'bandit boogie-woogie civilisation' refers here to the U.S. Though severely criticised, the examples mentioned here also point to a hidden desire for 'America', especially by the Yugoslav youth who needed to be morally taught to worship genuine socialist values and not to be seduced by the glow of American consumer, popular, and material culture.

Official condemnation of American consumer culture was also reflected in the critical discourses of Yugoslavian cultural production of the era. One of the earliest literary examples in which we can trace the geopolitical image of 'America' is the short story *Seed from California* [Seme iz Kalifornije] by the Yugoslav Nobel Prize laureate Ivo Andrić. The story develops an image of 'America' at the level of allegory and myth. The plot unfolds in an unnamed Bosnian town immediately after World War II. The narrative focusses on the sale of bean seeds from California that provoke special interest among the people who talk about their extraordinary features. Consumeristic mystifications of the seed build on the discursive tradition of fairy tales such as *Jack and the Beanstalk* and are accompanied with mythical stories about California and 'America' itself. The characters talk about 'America' as the land of material utopia where everything is big, bright and more advanced in the material aspect in comparison with the local context. But, of course, there's a very simple consumeristic catch to this exaggerated bean talk. The Californian seed is ten times more expensive than the domestic one, but people do not mind. Only the unnamed third-person narrator (who possesses a clear ideological distance) is critical of this kind of consumeristic craze, claiming: 'Bewitched by the vision of the American fruits of stunning size, the people are buying the seeds and not asking how much they cost. The seed's high price just makes it more appealing and people are encouraged to compete for foreign seed' (Andrić 2008, 368). After the time necessary for the seed to sprout, the myth of the American bean is unmasked and shown to be illusory and that it is no better than the domestic bean. After this turning point in the story, everybody starts to mock the so-called amazing California seed. Andrić's short story is built on the semantic destruction of narratives about 'America' as a land of material utopia, especially in deconstructing the idea of material prosperity connected with post-World War II USA.[16] 'America' in this story functions as the specific imagotype that must be understood according to its geopolitical but also literary context, which is the socialist realism poetic in the early stage of Yugoslav socialism. The image of 'America' is here a 'structural element in the shape of allegory, myth, parable etc., which could be interpreted only in the whole of literary art, therefore, within the meaning of the literary and modifying context' (Fischer 2009, 43). The critique within this narrative was performed through a socialist realist style with a clear moral and ideological lesson in condemning blind worship of 'America' and its products, which undoubtedly existed but needed to be ideologically put under control.

The aforementioned geopolitical imagining of 'America' coalesces with some of what were then rare accounts of contact with the real U.S., such as the travelogue of Yugoslav partisan and politically active journalist Vladimir Dedijer, who went as Yugoslavia's representative to the founding meeting of the United Nations. After which he documented his journey around the country in *Notes from America* [Beleške iz Amerike], published in 1945. In his travelogue Dedijer depicted the U.S. as a land of brutal capitalism and failed democracy. Financial capital, according to Dedijer (1945), ruled all aspects of cultural life in America, especially the media. Dedijer's severe criticism of U.S. media illustrates a view of the Yugoslav socialist's ideologically sharpened Other, which was at that time 'America' under capitalism. His notes, among others also found in *Reader's Digest* (Dedijer 1945, 96), are a significant source for understanding how the Yugoslav 'communist eye' (Drakulić 1993, 119) was sensitive to class and race (though not gender!) inequalities in capitalist America. Dedijer notices the imperialistic, even fascist characteristics of modern American culture, but interestingly, he also points out, though rarely found, the revolutionary aspects of American popular culture within leftist orientated artists such as Charlie Chaplin, who even sent a letter to the Yugoslav people via Dedijer, congratulating them on their brave and genuine partisan struggle.[17] Such ideological recourse is symptomatic of the early Yugoslav period when 'America', as a sign within the context of official ideology, semantically covered most of the negative meanings in the metaphor of the 'decadent West'. These cultural practices are indubitably carriers of geopolitical claims and their concordance illustrates the ideologically homogenous representation of 'America' by the dominant ideology in the period when Yugoslavia was still under Soviet cultural dominance, and ideas of American prosperity and democracy were strongly attacked in the cultural representation of Yugoslav socialist envisions.

'Socialism on American wheat'

Despite the strong negative connotations that arise from the aforementioned representations, the period that was unambiguously unfavourable to 'America' possessed a rather short expiration date. In the history of socialist Yugoslavia, the year 1948 marked the beginning of dynamic changes in the country's international geopolitical position. Three years after the end of World War II, Yugoslav leader Tito gave his legendary 'No' to Joseph Stalin, the Soviet premier and leader of the socialist world, thus ending the unquestioned dominance of the Soviet Union over socialist states on the Balkan Peninsula. From that period on, Yugoslavia pursued its own unique path to socialism or, as stated in a historical document by the U.S. Central Intelligence Agency, entitled *The Yugoslav Experiment* 20 years after this crucial event:

> Yugoslavia is a Communist state in name and theory, but in practice it is a fully independent state, which has rejected most of the 'socialist' experiences of other Communist states, including the USSR. It has deliberately removed a large portion of its economy from direct centralized controls, and despite its retention of a one party political system, it has largely freed its people from arbitrary authority. The Yugoslav Experiment appears to be

> progressing satisfactorily. Since the breakup with Stalin nearly 20 years ago, the Yugoslavs have managed to build a viable and independent economy, to go a good distance down the road to decentralization and the democratization of public institutions, and to establish a position in world affairs considerably more significant than their power and resources would seem to warrant.
>
> (Yugoslav Experiment National Intelligence Estimate 1967, 1)[18]

Geopolitical distance from the Soviet Union at that time allowed for the creation of a way of life that was typically Yugoslav, characterised by an enthusiasm for Westernisation combined with a commitment to Western influences and shifts in its socialist economy. Once admitted into the Yugoslav symbolic space, Western influences proved that they were there to stay. On the level of everyday life, this is the period in which a unique form of Yugoslav hybridity between the East and West was created, with pro-Western tendencies on the one hand and the basic aspirations of the dominant ideology leaning towards the East on the other. In the contemporary post-Yugoslav humanities, this hybridity has presented itself in a variety of historiographical metaphors. For the title of his book on Yugoslav foreign policy between 1948 and 1963, Croatian historian Tvrtko Jakovina (2002) chooses the phrase *Socialism on American Wheat* [*Socijalizam na američkoj pšenici*], taken from a speech by Tito in 1958.[19] In a relatively short period of time, Yugoslavia passed down the road from the stigmatised American seed to American wheat as its economic base, which meant, of course, not only economic but also cultural permeability. Soon after the break with Stalin, the image of the ideological enemy abruptly changed. The doors to Western influence opened, while influence from the East underwent intense scrutiny. By way of example, Yugoslavia's intellectuals, now led by the same Ivo Andrić, condemned Soviet writers and artists who supported Cominform (Andrić 1949). Moreover, the dominance of the Russian language in Yugoslav schools was gradually decreased after 1949 in favour of other foreign languages (Bekić 1988, 134). Furthermore, the import of Western, and especially American, films increased just as the number of Russian plays in Yugoslav theatres decreased (see Dimić 1988; Batušić 2004).[20]

Tito's 'No' to Stalin also meant a 'Yes' to the West, particularly American popular culture. In this political climate, literature and popular culture became two of the most important indicators of, but also protagonists in, the specified changes. For example, literary criticism and theory as well as translation choices at the time indicated an increased interest for Western—primarily American—cultural values, as illustrated by the Zagreb-based magazine *Circles* [*Krugovi*, 1952–1958], which focussed on translations of Anglo-American modernist literature. Also, literary and cultural critics showed a tendency of using the signifiers of consumer origin in a different, more positive manner, such is in the example of the metaphors connected with chewing gum as one of the most widespread symbols of American popular culture. In contrast to previously mentioned examples from the earlier period of socialist realism that approached Western popular culture with clear ideological prejudices, when entering the 1950s we can notice a kind of discursive flirt with Western popular cultural aesthetics in texts by several authors.

This attitude is obvious in writer and critic Vlatko Pavletić's (1952) text *Let There Be Liveliness* [Neka bude život], which opened the first issue of *Circles*. Pavletić's text is bursting with the author's enthusiasm for the 'American spirit', which he compares to chewing gum. This is possibly the first time that someone associated the symbolic meaning of chewing gum with the positive connotations of flexibility, optimism, changeability, liveliness, etc., albeit not without a touch of irony.[21] Still, a shift in value towards a positive perception of chewing gum as a symbol of American popular culture was undeniable. Besides the chewing gum motif, Pavletić's text teems with other signifiers of popular culture, such as 'cinema screen', 'jazz trumpet', etc. In addition, the text promotes pro-Western attitudes and freedom of the arts as opposed to the 'blind alley' of socialist realism that, according to Pavletić, stifled artistic freedoms. Even in this theoretical sense, the image of 'America' is a mixture of affections and geopolitical ideas whose resonances are important to understand (see, for instance, Pageaux 2009, 128).

To the aforementioned gest of cultural opening, we can also add events such as the publication of the anthology *American Poetry* [Američka lirika] in 1952, which brought a selection of poems by T.S. Eliot, Ezra Pound, Robert Frost, Emily Dickinson, Walt Whitman, Herman Melville, Edgar Allan Poe, and others to a Yugoslav audience. At the House of Visual Arts in Zagreb in 1953 there was also an exhibition of selected American books. Similarly, the pivotal year 1956 brought an exhibition of Contemporary Art from the U.S. Consumerist 'America' also entered the official culture at the International Zagreb Fair of 1957, with the U.S. exhibition entitled *Supermarket 1957*. Different American products and symbols, such as cigarette machines, machines for sweets, washing machines, dryers, electric grills, air-conditioning devices, and televisions, were exhibited next to a large picture of U.S. president Dwight Eisenhower who was a World War II general and occupying commander before he was president. The central part of the pavilion was reserved for the American supermarket, which demonstrated a pleasant shopping experience (Jakovina 2003, 475). Yugoslav popular culture by that time had also been touched by the processes of Americanisation and Westernisation. In the form of film, one of the earlier example of the process is the iconic Yugoslav *Love and Fashion* [*Ljubav i moda*] by Ljubomir Radičević (1960). The topic of the movie is very consumeristic: the basic plot is based on the market competition between two fashion houses, as in Western style called Jugomoda and Jugochic, intertwined with a love story between a young fashion designer and a student model. The film shows the desirable luxury of Western fashion and prominent Yugoslav celebrities from the movie and music industries. Belgrade, the capital of Yugoslavia, is shown as a cosmopolitan city with modern technology and trends: there are many cars on the streets and many young, good-looking, and modern-dressed people; there are scenes with motorcycles, scenes with airplanes and the small talk is filled with the names of Western celebrities such as Brigitte Bardot, Gregory Peck, and others. Yugoslavia is, in short, represented as a country in which popular and consumer culture in the Western mode reigns.

Thus, Yugoslav culture in the 1950s became the terrain of ideological change and an agent of geopolitical exchange.

'Coca-Cola socialism'

At the beginning of the 1960s, Radičević's *Love and Fashion* represented a desire for material prosperity and Western fashion that was still not present in reality.[22] From the perspective of economic changes, the material prosperity under Western fashion would burst in the upcoming decades of Yugoslav socialism (see, for instance, Dimitrijević 2005; Polimac 2010). As moving further into the sixties, the film's scenery became much more realistic.[23] As the 1960s progresses, the metaphors of wheat and seed were replaced by the consumerist metaphor of *Coca-Cola Socialism* [*Koka-kola socijalizam*], as articulated by the Serbian historian Radina Vučetić (2012) in her book about the Americanisation of Yugoslav culture in the 1960s. In that period, different Yugoslav media[24] dedicated a lot of space to American popular culture that was seductive even to political elites.[25] In this early period of the Americanisation of Yugoslav culture, it is evident that the critical problematisation of 'America' in art and popular cultural forms was somewhat absent. Of course, Marxist criticism of the capitalist West and 'America' did not completely disappear from the stage of public discourse after 1948.[26] An interesting example is the text *Team Šoljan-Slamnig* [*Međuteam Šoljan – Slamnig*] by the literary critic Nikola Milićević (1953), about the writers of that young generation. Milićević writes with indignation about their literary experiments and compares them with the experiments of 'American jazz musicians' (1953, 717). Milićević's work was published in *Circles*, which shows that cultural openings were not without existing counter-narratives. Also, even in the 1960s an ideological comment about the U.S. was clear even in a geography textbook:

> In the United States, the biggest capitalist country, there is great inequality among the people. In contrast to the millions of workers there are hundreds of billionaires with fantastic incomes. United capitalist associations and banks have a significant impact on the government and its domestic and foreign policy. Because of its economic power, the United States strongly affects the economy and politics of many capitalist countries.
>
> (Jurin and Gliederer-Vivoda 1963, 138)

However, this and similar examples of defamation of the U.S. never had the same degree of homogeneity and acuity as occurred in the period from 1945 to 1948. But what was also missing at the time was a more complex problematisation and interpretation of 'America' in art and other cultural forms.

One possible reason for this lack may be that post-World War II American culture first had to be adopted before it could be problematised in a more complex and creative manner as was the case in Western Europe. Also, during the period of the late 1950s and early 1960s, the U.S. experienced the culmination of post-war prosperity as explained in *Consumers' Republic* (Cohen 2003), and it was only after this period that a serious erosion of the symbolic values of American society came into the picture. As has been well documented, the assassinations of John F. Kennedy[27] and Martin Luther King Jr. and the Vietnam War would have a sobering effect on U.S. worship and evidence can be found in the cultural texts of that period. Parallel with these turning points in American history, more and more complex criticism began to appear in Yugoslav cultural practices. It should be stressed, however, that these critics did not necessarily give up on American popular culture. In fact, disagreement with U.S. policies did not necessarily exclude sympathizing with American popular culture. One such example is the youth magazine *Verve* [*Polet*], which was published by the Socialist Youth of Croatia (see Krušelj, 2015). Especially interesting for this topic is the short article 'Pro Jeans' [*Za jeans*] by journalist Zoran Franičević. In his article, Franičević (1977) polemically replies to 'backward' ideological attacks on American fashion from a certain comrade named Stojić, and he says that being dressed in American fashion does not undermine his socialist habitus. As Franičević argues:

> Comrade Stojić, our youth is not as politically shallow and stupid as you see them; they don't buy foreign jeans because they love the USA or their army. As I believe you know, we live in an open society, and the more open we are, the more aware we are of the potential hidden agenda behind Western makeup, but some influences intertwine nevertheless, and this happens most easily in the realm of fashion. It is backwards to think that jeans with Western labels are incompatible with the Socialist Youth Union. I just fail to comprehend the idea that jeans are becoming a bone of contention among young people. [...] Yesterday I went to the Party meeting wearing my new pair of Levi's jeans, and then straight to the meeting of my youth organization.
>
> (1977, 3)

According to Franičević, the socialist youth is not made up of 'cultural dopes' (Hall 2006a, 302) that will, without question, accept anything that the capitalist market has to offer. Thus, the worship of American popular culture has not been significantly undermined, but from this period on the utopian imaginary of 'America' started to be reflected in a more and more complex manner. Franičević's defence of jeans could be interpreted as a defence of the 'imaginary West', a concept developed by the cultural anthropologist Alexei Yurchak (2006) to signify the redefinition and transmission of Western symbols into Soviet culture that is in the process of gaining meaning that is neither Western nor Soviet. That concept is useful for Yugoslavia but with a slightly different meaning that takes into account the specificity of Yugoslavia's geopolitical position as a country politically and culturally more open to the West than the USSR. In this sense, Franičević's negotiation with American popular culture could be interpreted as a form of deterritorialising[28] the system, or the internal displacement of the ideological discourse of the ruling system, which does not necessarily equate with dissidence since the young Yugoslav socialist is actually going to the Communist Party meetings while wearing denim.[29]

From this period, increasing criticism arose from the art works of the, at the time, youngest Yugoslav generation that had adopted American pop culture while growing up under 'Coca-Cola Socialism' to which, as a part of that process, they became critical. Of course, the reception of the American countercultures of the 1960s and 1970s and their symbolic politics of resistance to the U.S. establishment certainly influenced this generation. In addition to the aforementioned example of the jeans, there was the specific use of comic art, such as a wordless comic strip critical of Coca-Cola (1977) as a synecdoche for capitalism. Similarly we see the song 'Goodbye America' from 1976 by one of the most popular Yugoslav pop rock bands, Bijelo Dugme, which explicitly criticised American materialism with the following lyrics:

> Here I can buy everything / friends, glory, splendour and dreams, / everything's on sale for a dollar. / Hey, but do I know if anyone here really cares / about my sorrow? / That's not for sale for a dollar. / Goodbye, America! / Hey, America, Goodbye, America! / It's time to go home, / where my heart is / where my love is.[30]

The aforementioned practices are thus creatively appropriating American popular cultural forms and transforming them with a meaning that is neither socialist nor American, but carries a specificity of Yugoslav geopolitical hybridity.

Political crisis and a requiem for American (and Yugoslav) dreams

Bijelo Dugme's song can also be perceived as the beginning of the requiem for the American and Yugoslav dreams for a young Yugoslav generation, which would reach its peak in the last decade of Yugoslav socialism. In the late 1970s and throughout the 1980s, utopian images of both countries were politically challenged: the USA as an oasis of Western liberalism and democracy on the one hand, and Yugoslavia as a country where socialism combined the best of both worlds on the other, or, in short, the capitalist West and the communist East. As is well known, at the end of the 1970s, U.S. President Jimmy Carter, in his televised address known as the 'Malaise Speech', emphasised the crisis of American confidence and criticised American citizens' decaying set of values on which materialism and consumerism were founded in the first place.[31] The Vietnam War, political assassinations, the Watergate Scandal, and, on the threshold of the 1980s, the major economic crisis in the U.S., all culminated to question the image of prosperity and democracy as the main ingredients of American utopian appeal as in earlier decades. At the same time, the utopian image of Yugoslavia as a country between the Eastern and Western blocs was also put into question. Those were the years of, popularly speaking, decadent socialism marked by the death of President Tito, who had been an important factor of the cohesiveness of

Yugoslav multiculturalism, as well as economic crisis and the growth of national particularities that threatened the Yugoslav ideal of 'brotherhood and unity'. Both of these crises can be indirectly seen in Yugoslav cultural images of 'America' in the 1980s. Also, it needs to be stressed that in those years, American popular culture had already become a part of the Yugoslav cultural mainstream.[32] As imagologist Joep Leerssen claims: '[…] through time, images of different nations are often subject to changes [but] the old ideas are not replaced with the new phenomena, they are only temporarily removed and remain subconsciously present in social discourse that can always be reactivated' (2009, 109–110). Thus, images of 'America' in the last decade of Yugoslav socialism possessed a higher degree of historical memory than they had in previous decades, which also means that they had a higher degree of complexity. In terms of cultural memory, the last decade of Yugoslav socialism had a wide range of geopolitical images, from stigmatisation to mainstreaming and fascination, and it is this quality of almost schizophrenic ambivalence that predominantly characterised the texts and popular cultural forms during the decadent socialism era, which I will discuss through a few characteristic 'cultural patterns' that emerge from Yugoslav literature that are illustrative of this period.

Srđan Karanović's film *Something In-between* [*Nešto između*] (1983) is particularly significant for illustrating the tendency of problematising the utopian image of 'America'. The movie's main conflict centres on a comparison between 'America' and 'Yugoslavia' as embodied in characters of U.S. and Yugoslav origins. The young American journalist Eve's (Caris Crofman) first encounter with Yugoslavia is represented through a series of prejudices that she holds against socialist regimes. But soon after this initial misstep, she is captivated and seduced by Yugoslavia's particular geopolitical distinctiveness and cultural hybridity, which, as she finds out through first-hand experience with the young Westernised Yugoslav gigolo Marko (Dragan Nikolić), shows that it is not a typical communist country. It is, as Marko explains, something in-between. Through her love affair with the young Belgrade surgeon Janko (Predrag Manojlović), Yugoslavia begins to acquire more and more romantic connotations from Eve's point of view: it becomes an achieved utopia where people are more open and sincere, intellectual culture is of a high standard, and the land is beautiful. But, as her romantic relationship with Janko enters a crisis, triggered by differences in mentality that inevitably surface, Eve suddenly begins to see another, not-so-sunny side of Yugoslavia. She uncovers a country which was also in a crisis in the 1980s marked by power cuts, food shortages and drills in preparation for war. The culmination of the film represents a fight between Eve and Janko where the arguments and insults have more to do with national and ideological matters than personal qualities. The film's outcome does not overcome these prejudices and culminates in an unhappy ending. The movie plays with the Yugoslav geopolitical identity of 'in-betweenness', by first developing it as an exciting and attractive feature of Yugoslav geopolitical hybridity that, at the end of the movie, stands for a chaotic and politically insecure country as seen by this American character.[33]

Together with Karanović's movie, Goran Tribuson's novel *Made in USA* (1986) is also shaped through a tension between the critical and utopian, mythical and socio-political images of the USA that pervade the relation between the two main characters of this crime novel (the Yugoslav Detective Nikola Politeo and his American client Leo Wolf). Wolf hires Politeo to investigate at a moment when he is debating emigrating to the U.S., a country about which he has very detailed utopian presumptions. Politeo is a depressed Yugoslav obsessed with America, being familiar with American popular culture, especially the movies. His desire to quit Yugoslavia for 'America' has its roots in the Yugoslav crisis of the 1980s and perceptions of 'America' in the novel are regularly accompanied by reflections of the Yugoslav environment. Yugoslavia is thus marked by the economic and political crisis, which is reflected through the lack of consumer goods and political corruption, apathy and a general greyness, while 'America' is, in contrast, perceived as a promised land where everything is shiny and spectacular. As a counterpoint to Politeo's perceptions of 'America' stands Wolf, whose perception of Yugoslavia is a source of negative comments, mostly on economic and consumeristic backwardness. Yugoslavia is, from Wolf's perspective, 'a terrible country', a phrase that acts as a leitmotif for the novel. But, as a qualifying factor, Wolf is also very cynical about Politeo's desire for 'America', which he presents as a land of brutal capitalism.[34] A 'terrible country', a phrase that is used for Yugoslavia, thus becomes metaphorically applicable to 'America'.[35] At the end of the novel, Politeo gives up on his dream of going to 'America', allowing himself to be overcome by doubts about his mythical image of the country, which had been destroyed by Wolf.

Political crisis in both Yugoslavia and the U.S. shaped the geopolitical imaginations of the American and Yugoslav characters in the aforementioned works and could be interpreted as the anthropomorphisation of the geopolitical notions of 'America', as well as 'Yugoslavia'. Both 'America' and 'Yugoslavia' in the aforementioned examples from the 1980s function as a geopolitical mirror of the Other, in which the desire often carries with it traces of repulsion.

Lessons from the Cold War

To go or not to go to 'America' became a significant motif and also a common dilemma in the cultural texts of decadent socialism. In the contextual reading of both Karanović's movie and Tribuson's novel, this motive can be interpreted as a side effect of the evident economic and political crisis in Yugoslavia, which provoked an escapist longing for 'America' in charters such as Marko and Nikola. The motive for leaving for 'America' can be found in various other cultural patterns of the period. One such example can be found in the movie *In the Jaws of Life* [*U raljama života*] by Rajko Grlić (1984), in which the two main characters, Dunja (Gorica Popović) and Pipo (Bogdan Diklić), have the following conversation:

> So, you're going to America?

> Do you know how many people have already left? It will be nasty here, very nasty. For real. [36]

Besides confirming the idea of utopian American imagery of the time, this dialogue sounds prophetic and disturbing knowing the historical facts about the disintegration of Yugoslavia that would follow in a few years. In 1989, when the political aspects of socialism's decay were evident all over Eastern Europe, the Yugoslav rock band Ekatarina Velika released the single 'Amerika' with the lyrics:

> People come / And go, and go / It's boring here, everyone escaped / To some more entertaining place / It's far away / It's far away – America / And if I went somewhere it was only inside me / Inside me, inside me and not outward / Up, and down, and left, and right inside me / Inside me, inside me, inside me, and not outward / There is the purpose, there is a way / There is a sign / There is America.[37]

In the contextual reading of this popular song, which as a genre and source is still not appreciated enough within the discipline,[38] we notice that the atmosphere of leaving is connected to the evident political and economic crisis in Yugoslavia, as it was in Grlić's movie. The song's subject is aware that it is impossible to find a utopia in 'America', and that utopia exists only as an internal fictional construct, which is, paradoxically, called 'America'. Thus, in the song, 'America' works in a deconstructivist sense as a 'crossed term': the idea that you cannot think in the old way, without which, however, certain key questions cannot be thought (Hall 2006b, 358). This song from the coda of Yugoslav socialism points to inevitable political crisis but also to a crisis in the Yugoslav Cold War geopolitical imagination. Playing with the meaning of what is inside and outside, what is 'here' and what is 'there', what is 'America', etc., the aforementioned song articulates the geopolitical terms as purely subjective, arbitrary and intimate in the wake of seismic convulsions that would result in the bloody disintegration of Yugoslavia as a country balanced between the socialist East and capitalist West 'until it was no more'.

The geopolitical image of 'America' in Yugoslav culture indubitably shows its 'Janusian ambivalence and contradictory nature' (Leerssen 2009, 110). These cultural images activate different meanings in different segments of Yugoslav history in the Cold War period since representations of global political space do not exist in a geographical and historical vacuum (Dodds 2005, 13). Sometimes seen as repulsive, sometimes as desired, sometimes as the negotiated Other, the cultural image of 'America' is tightly interconnected with geopolitical events and processes, showing how 'the 'moment of geopolitics' is not fixed, but is diffused through society in the constant re-creation of images and understandings of the world and how it works' (Sharp 2000, 29). According to Ó Tuathail:

> The struggle over geography is also a conflict between competing images and imaginings, a contest of power and resistance that involves not only struggles

> to represent the materiality of physical geographic objects and boundaries but also the equally powerful and, in a different manner, the equally material force of discursive borders between an idealized Self and a demonized Other, between 'us' and 'them'. Viewed from the colonial frontier, geography is not just a battle of cartographic technologies and regimes of truth; it is also a contest between different ways of envisioning the world.
>
> (2000, 11–12)

In the patchwork of different sources utilied here, I have wanted to emphasise the plurality and historicity of Yugoslav cultural texts and practices in their geopolitical concordance, struggle and negotiation. Though I have looked at culture as 'a whole way of life' in Williams' terms, a larger part of the examples belonged to popular culture, which represents its particular importance to the analysis (see also Kolanović 2013, 2016). The importance of popular culture in the geopolitical imaginations of other nations also pointed to one of imagology's weak spots, which is showing how 'modern mass communication of the technical and media world would certainly be incomparably more effective than it has been in previous centuries, during which the role belonged to literature, travel books or personal meetings in foreign countries with its citizens and so on' (Fischer 2009, 45). In this sense, popular geopolitics demonstrates its significantly transdisciplinary character, unavoidable in discussing any aspect of the Cold War, which 'provided an important global map of moral geography from which American purpose and destiny could be reflected' (Sharp 2000, 167). Yet I hasten to add, not just American. The Cold War produced countless effects around the globe, particularly via its shaping of popular culture as a tool of geopolitics.

In this overview of 'America' in the geopolitical image of Yugoslav culture, I have tried to show how cultural texts and practices are deeply interconnected with geopolitical imagination. The cultural examples taken here as representative are not merely perceived as reflections of the turbulent events on the geopolitical stage, since they also anticipate, prepare and even contest geopolitical events, thus showing how 'the diplomatic conference room and the battlefield are not considered disconnected and/or divorced from public culture' (Dodds 2005, 75). Various cultural examples showed how the cultural imagination is inseparable from the geopolitical imagination even when that connection does not seem obvious at first glance. The cultural is always, to a greater or lesser degree, geopolitical, which is one of the most important lessons for the discipline acquired from the Cold War.

After all, the Cold War did not cease to organise geopolitical meanings after the fall of the Berlin Wall. The turbulent processes of the post-socialist transitions are occurring with a specific affection attached to capitalism, very often connected to 'America'. Social, political and economic change in post-socialism certainly brought a new quality to the way 'America' is imagined and, in doing so, the process has not bypassed the legacy of the Yugoslav cultural imagery and its negotiation with 'America', through which, to invoke Daniel-Henri Pageaux (2009, 127), one society was seen, defined, and dreamed of in a specific political, social, and historical context.

Notes

1 This research was supported by Croatian Science Foundation under the project IP-2016-06-2013.
2 From this angle the book *Condensing the Cold War: Reader's Digest and American Identity* by the British geographer Joanne Sharp (2000) is particularly interesting. She shows the importance of the Cold War period in constructing American national identity within the geopolitical construction of the communist Other. The Cold War has been, in institutional terms, widely recognized as fruitful ground for research in critical geopolitics. See, for example, Ó Tuathail 2000, Dodds 2005, and Dittmer 2010.
3 See, for example, Dittmer's (2005) analysis of the post-9/11 era in *Captain America.*
4 Following Frederick Dolan, Sharp speaks of a 'continuation of "Cold War metaphysics" even in the absence of Cold War geopolitics' (2000, 84). According to Ó Tuathail, 'The Cold War may be dead, but Cold War strategic culture and its society of security is undead. A vampirish culture, it lives on by sucking meaning out of the new global indeterminacies and contingencies as well as material resources for itself from the depleted reserves of society as a whole' (2000, 182).
5 By cultural studies I refer here to the critical post-Marxist approaches to culture that developed at the Birmingham Centre for Contemporary Cultural Studies and scholars such as Stuart Hall, Richard Hoggart, Raymond Williams, Angela McRobbie and others. See During (2000); Ferguson and Golding (1997); Duda (2002). See also Strukov's chapter in this volume.
6 Here I consider the term 'geopolitical imagination' as the representation, creation and subversion of geopolitical issues through cultural texts and practices. The term was used in the title of the text *The Geopolitical Imagination and the Enframing of Development Theory* (Slater 1993), though the author does not develop the concept in this direction. The importance of the so-called geopolitical fantasy or fiction is, however, recognised within the discipline but is termed in various ways. Dodds, for example, following G. Dijkink, writes about 'geopolitical visions' in the sense that ideas about places and populations are mobilized to construct 'geopolitical visions' (2005, 1).
7 'The term "geopolitics" is a convenient fiction, an imperfect name for a set of practices within the civil societies of the Great Powers that sought to explain the meaning of the new global conditions of space, power, and technology' (Ó Tuathail 2000, 12).
8 For more on Soviet-Yugoslav geopolitical and cultural relations, see Bekić 1988, Banac 1990, and Peruško 2015.
9 The dictionary entries are originally as follows: [*Amerika*], [*Amerikanci*], [*američki*], [*američki san*], [*amerikanijada*], and [*amerikomanija*]. The full names of the dictionary sources are listed in the bibliography. All translations from the original Croatian are by the author unless otherwise indicated. The meanings extracted here are also comparable with other Slavic languages (spoken in Yugoslavia) in various dictionaries, such as Saračević 2004, Čedić 2007, and Keber 2011.
10 Some specific words and phrases are even more geopolitically connected with the U.S. of that time, such as 'Truman's eggs' [*Trumanova jaja*], an expression coined in Yugoslavia after World War II for the powdered eggs that arrived as humanitarian aid from the United Nation's Relief and Rehabilitation Organization, which is nowadays used figuratively to mean American help to poorer or war-endangered countries in general.
11 According to Sharp, 'through the descriptions of the world and global politics in the speeches of political elites, in media coverage and school textbooks, orders are made and remade in the aim of producing the meaning of events' (2000, 23).
12 Some of the materials used in this analysis I already discussed in Kolanović (2011, 2013).
13 Chewing gum was especially connected with 'America' and has a long tradition in the stigmatisation of works of American popular culture, which is not specific to socialism.

In his text *The Hour of Chewing Gum* [*Die Stunde des Kaugummis*] from 1926, German art historian Ernst Lorsy severely criticizes young Germans' frenzy for chewing gum and all things American, suggesting serious political consequences resulting from this excitement and blaming the 'lack of revolutionary ardour to the fact that their jaws were too busy to allow their minds to work' (Lorsy in Confino and Koshar 2001).

14 It is worth mentioning that Đilas later became a political dissident and was sent to prison by the Yugoslav leader Josip Broz Tito. After serving his sentence, he emigrated to the U.S.

15 The title of text alludes on the novel by Erich Maria Remarque that, in Hanžeković's text, marks the stagnation, lack of change and cultural conservatism of U.S. society.

16 The idea of 'America' as a 'promised land' in the aspect of material prosperity has its mythical roots in pre-Columbian European imagination (see Ruland 1976).

17 The letter was published at the end of Dedijer's book along with Yugoslav writers' responses to Charlie Chaplin and their letter to John Garfield and mutual letters of Hollywood writers in War society and Belgrade Writer's Society (Dedijer 1945, 235–237).

18 In fact, the CIA's notions of Yugoslavia correspond with a representation of the country in *Reader's Digest* magazine from the same year in which, as shown by Sharp (2000), Tito was the only leader of a communist country visually represented as smiling.

19 The speech was directed toward Soviet president Nikita Khrushchev's remark that socialism could not be built using American credits, to whom Tito replied: 'Comrade Khrushchev often repeats that socialism cannot be built on American wheat. I think it can be done by those who know how, and those who don't know how cannot build socialism even with their own wheat. U.S. wheat, after all, is no worse than Soviet wheat, which, by the way, we are not receiving while we get it from the U.S., and this also works for other goods' (Broz in Jakovina 2002, 128). Similarly, a U.S. historian Dennison Rusinow (1977) uses the CIA's phrase *The Yugoslav Experiment* for the title of his book on, while art historian Branislav Dimitrijević (2005) uses Srđan Karanović's metaphor from the movie *Something In-between* [*Nešto između*], which will be discussed later in this chapter, and historian Predrag Marković uses the metaphor of a Yugoslav tower situated in the middle of the world divided into blocs. The tower itself was eclectic, with an Eastern structure at its base, but people were allowed to have windows, furniture, wallpaper in bright colors and luxurious Western fashion (see Marković 1995).

20 I here use the term 'Russian' rather than 'Soviet' since I am mostly referring to Russian culture and playwrights from the late nineteenth and early twentieth centuries.

21 As claimed by Pavletić (1952, 1): 'Semantics is the science of the meanings of words. But it is more interesting as a view of the world. Such conciliatory philosophical doctrine could only create the American spirit, witty and shallow, compromising in its own way and as resilient as "chuingumm". Derived from the specifically American feeling of changeability and the impermanence of all life values, it does not tend to scepticism, pessimism and resignation as, for the example, European or Eastern relativism does, but rather it is a clear wellspring of optimism and the driving force of activities and focussed on materiality, which is more stable than all the empty concepts'.

22 According to IMDB movie site, 'one scene required an urban traffic jam. In 1960, there were few cars in Belgrade, Yugoslavia. In order to shoot this scene, they had to divert traffic in the city into the main street to generate the impression of a jam. By today's standards, the traffic in the scene still moves very well. In real life this street has been jammed with cars most of the time since the seventies'. See http://www.imdb.com/title/tt0179945/.

23 For a more detailed movie interpretation, also see an online lecture by the art historian Branislav Dimitrijević (2014), entitled *Love and Fashion: Consumer Culture in Yugoslav Socialism*. The lecture is available at https://www.mixcloud.com/CSEES/love-and-fashion-consumer-culture-in-yugoslav-socialism-branislav-dimitrijevic/ [last accessed 15 September 2015].

24 Such as pop magazines: *Blue Novuelles* [*Plavi vjesnik*], *Jukebox* [*Džuboks*], *Pop Express*, *The World* [*Svijet*], etc.
25 In this sense, one cannot escape the iconic pictures of Tito together with movie celebrities such as Elizabeth Taylor, Richard Burton, Sophia Loren, and others.
26 For more about this discursive hybridity in the discourse of Yugoslav intellectuals, see Kolanović 2011.
27 As a curiosity, it is interesting to mention that Jozo Karamatić, a Yugoslav folk singer and fiddler, dedicated a song to the assassination of J. F. Kennedy in 1963, which was released as the record *Death in Dallas* [*Smrt u Dalasu*] in 1965.
28 I am here referring to the concept of 'deterritorialisation' as developed in the works of Gilles Deleuze and Félix Guattari (1990; 2013), which was used by Yurchak (2006) in developing his concept of the 'imaginary West'.
29 This tendency could also be interpreted in the larger cultural context of global popular culture that spread across the West and the Eastern Bloc, wherein young activists opposed the system in whatever form it took while wearing denim. It was more precisely this kind of global counterculture that symbolically connected the progressive youth of the East and West.
30 Bijelo Dugme: *Single Records 1976–1980* [*Singl ploče* 1976–1980.], Jugoton, 1982. Lyrics taken from http://tekstovi-pesama.com/bijelo-dugme/goodbye-america/291/1/, 12.9.2015.
31 The transcript of Carter's speech is available at http://www.pbs.org/wgbh/americanexperience/features/primary-resources/carter-crisis/, 15.9.2015.
32 Just to mention the fact that in 1983 Levi's opened a factory in Yugoslavia; in 1987 Coca-Cola produced a bottle with the sign of the student Olympics in Zagreb; the American TV show *Dynasty* was on Yugoslav TV in the mid-1980s and extremely popular.
33 For an interpretation of the film and the specific metaphor of 'in-betweenness' that it develops, also see Dimitrijević 2005.
34 'You'll work a boring and stupefying job, the same you can actually find here and you'll be dying for a lunch break. First you'll take a menu on which the food is described with words; then you'll order from a menu with numbers; then you'll just swallow a hot dog and at the end you'll only have time for a pill. You'll swallow pills for hunger and then pills for thirst, pills for hunger, pills for thirst, pills for sleep and waking up, and then a pill which replaces two to three pills and you'll end up taking on a pill to release you from the addiction to the pills' (Tribuson 1986, 210).
35 The overlapping of the utopian and ideological representations of both the Yugoslavia and 'America' of that time had its particularly successful parodic version in the Yugoslav humorous TV show *Surrealist Charts* [*Top lista nadrealista*] where the signs of the Yugoslav economic crisis were attached in the dialogues to clips from the famous American TV show *Dynasty*. I am thankful to Danijela Lugarić for this information.
36 The movie is an adaptation of the novel *Steffie Speck in the Jaws of Life* [*Štefica Cvek u raljama života*, 1981] by Dubravka Ugrešić, who was also co-author of the movie script together with the director Rajko Grlić.
37 Ekatarina Velika, 1989. The lyrics are taken from http://www.lyricsmode.com/lyrics/e/ekv/amerika_lyrics.html [last accessed 9 September 2016].
38 According to Dodds. 'geopolitical writers have not fully explored the geographical soundscapes and political worlds created by music' (2005, 99).

References

Anderson, Benedict. 1990. *Nacija: zamišljena zajednica. Razmatranja o porijeklu i širenju nacionalizma*. Zagreb: Školska knjiga.

Andrić, Ivo. 1949. 'Odgovor jugoslavenskih književnika sovjetskim književnicima F. Glatkovu, N. Tihonovu i drugima'. *Republika* 2–3:182–187.

Andrić, Ivo. 2008. *Sabrane pripovetke*. Beograd: Zavod za udžbenike.

Banac, Ivo. 1990. *Sa Staljinom protiv Tita. Informbiroovski rascjepi u jugoslavenskom komunističkom pokretu*. Zagreb: Globus.

Barson, Michael. 1992. *'Better Dead than Red!' A Nostalgic Look at the Golden Years of Russiaphobia, Red-Baiting, and Other Commie Madness*. New York: Hyperion.

Batušić, Nikola. 2004. 'Kazalište pedesetih'. In *Pedesete godine u hrvatskoj umjetnosti*, edited by Zvonko Maković, 158–177. Zagreb: Hrvatsko društvo likovnih umjetnika.

Bekić, Darko. 1988. *Jugoslavija u hladnom ratu: odnosi s velikim silama 1949–1955*. Zagreb: Globus.

Beller, Manfred, and Joep Leerssen. 2007. *The Construction and Literary Representation of National Characters: A Critical Survey*. Amsterdam and New York: Rodopi.

Betts, Raymond. 2004. *A History of Popular Culture: More of Everything, Faster and Brighter*. London and New York: Routledge.

Cantor, Norman F., and Michael S. Werthman. 1968. *The History of Popular Culture*. New York and London: The Macmillan Company.

Čedić, Ibrahim. 2007. *Rječnik bosanskog jezika*. Sarajevo: Institut za jezik.

Cohen, Lizabeth. 2003. *A Consumers' Republic: The Politics of Mass Consumption in Post War America*. New York: Vintage Books.

Confino, Alon, and Rudy Koshar. 2001. 'Régimes of Consumer Culture: New Narratives in Twentieth-Century German History'. *German History* 19 (2):131–161.

de Grazia, Victoria. 2005. *Irresistible Empire: America's Advance through Twentieth Century Europe*. Cambridge, MA and London: Harvard University Press.

Dedijer, Vladimir. 1945. *Beleške iz Amerike*. Beograd: Kultura.

Dimić, Ljubodrag. 1988. *Agitprop kultura: agitpropovska faza kulturne politike u Srbiji: 1945–1952*. Beograd: Rad.

Dimitrijević, Branislav. 2005. 'Socialist Consumerism, Westernization and Cultural Reproduction: 'Post-communist' Transition in Tito's Yugoslavia'.unpublished manuscript.

Dittmer, Jason. 2005. 'Captain America's Empire: Reflections on Identity, Popular Culture, and Post-9/11 Geopolitics'. *Annals of the Association of American Geographers* 95 (3):626–643.

Dittmer, Jason. 2010. *Popular Culture, Geopolitics, and Identity*. Lanham, MD: Rowman & Littlefield.

Dodds, Klaus. 2005. *Global Geopolitics: A Critical Introduction*. Harlow, UK: Prentice Hall.

Drakulić, Slavenka. 1993. *How We Survived Communism and Even Laughed*. London: Vintage.

Duda, Dean. 2002. *Kulturalni studiji: ishodišta i problemi*. Zagreb: AGM.

Dukić, Davor. 2012. *Imagology Today: Achievements, Challenges, Perspectives*. Bonn: Bouvier Verlag.

During, Simon. 2000. *Cultural Studies: A Critical Introduction*. New York and London: Routledge.

Ferguson, Marjorie, and Peter Golding. 1997. *Cultural Studies in Question*. London: Sage Publications.

Fischer, Manfred S. 2009. 'Komparatistička imagologija: za interdisciplinarno istraživanje nacionalno-imagotipskih sustava'. In *Kako vidimo strane zemlje. Uvod u imagologiju*, edited by Davor Dukić, Zrinka Blažević, Lahorka Plejić Poje and Ivana Brković, 37–56. Zagreb: Srednja Europa.

Franičević, Marin. 1948. 'Uvod u diskusiju o našim časopisima'. *Republika* 1:8–17.

Franičević, Zoran. 1977. 'Za jeans'. *Polet*, 23 September.

Grlić, Rajko. 1984. U raljama života. Beograd: Art Film Beograd, Croatia Film, Jadran Film, Kinematografi Zagreb, Union Film Beograd.

Hall, Stuart. 2006a. 'Bilješke uz dekonstruiranje popularnog'. In *Politika teorije, zbornik rasprava iz kulturalnih studija*, edited by Dean Duda, 279–311. Zagreb: Disput.

Hall, Stuart. 2006b. 'Kome treba 'identitet'?' In *Politika teorije, zbornik rasprava iz kulturalnih studija*, edited by Dean Duda, 357–374. Zagreb: Disput.

Hanžeković, Fedor. 1947. 'Na Zapadu ništa novo. Razmatranja o filmu'. *Republika* 2:130–139.

Hebdige, Dick. 1980. *Potkultura: značenje stila*. Beograd: Rad.

Hendershot, Cynthia. 2001. *I Was a Cold War Monster: Horror Films, Eroticism, and the Cold War Imagination*. Madison, WI: Popular Press.

Hobsbawm, Eric J. 2004. *Doba ekstrema: istorija Kratkog dvadesetog veka, 1914–1991*. Beograd: Dereta.

Hoggart, Richard. 1992. *The Uses of Literacy: Aspects of Working-Class Life with Special Reference to Publications and Entertainments*. Harmondsworth: Penguin Books.

Jakovina, Tvrtko. 2002. *Socijalizam na američkoj pšenici*. Zagreb: Matica hrvatska.

Jakovina, Tvrtko. 2003. 'Narodni kapitalizam protiv narodnih demokracija. Američki super-market na Zagrebačkom velesajmu 1957. godine'. In *Zbornik Mire Kolar Dimitrijević*, edited by Damir Agičić, 469–496. Zagreb: FF press.

Jurin, Vinko, and Ivančica Gliederer-Vivoda. 1963. *Naš svijet 2. Geografija za VII razred osnovne škole*. Zagreb: Školska knjiga.

Karanović, Srđan. 1983. *Nešto između*. Beograd: Centar film.

Kašić, Biljana. 1991. 'Značajke partijske ideologije u Hrvatskoj (1945–1948)'. *Časopis za suvremenu povijest* 1–3:243–263.

Keber, Janez. 2011. *Slovar slovenskih frazemov*. Ljubljana: Založba ZRC and ZRC SAZU.

Kolanović, Maša. 2011. *Udarnik! Buntovnik? Potrošač..Popularna kultura i hrvatski roman od socijalizma do tranzicije*. Zagreb: Naklada Ljevak.

Kolanović, Maša. 2013. 'Utopija pod upitnikom. Predodžba "Amerike" u stihovima dekadentnog socijalizma'. In *Socijalizam na klupi. Jugoslavensko društvo očima nove postjugoslavenske humanistike*, edited by Lada Duraković and Andrea Matošević, 177–216. Pula-Zagreb: Srednja Europa.

Kolanović, Maša. 2016. ''Amerika' i drugovi: prvost Sjedinjenih Američkih Država i trećost Jugoslavije'. In *Drugi. Alteritet, identitet, kontakt u hrvatskome jeziku, književnosti i kulturi. Zbornik radova 44. Seminara Zagrebačke slavističke škole*, edited by Tatjana Pišković and Tvrtko Vuković, 107–128. Zagreb: FF press and Zagrebačka slavistička škola.

Lakoff, George, and Mark Johnson. 2003. *Metaphors We Live By*. Chicago and London: University of Chicago Press.

Leerssen, Joep. 1991. 'Ecohes and Images: Reflections upon Foreign Space'. In *Alterity, Identity, Image: Selves and Others in Society and Scholarship*, edited by Raymond Corbey and Joep Leerssen, 123–128. Amsterdam and Atlanta: Rodopi.

Leerssen, Joep. 2009. 'Retorika nacionalnog karaktera: programatski pregled'. In *Kako vidimo strane zemlje. Uvod u imagologiju*, edited by Davor Dukić, Zrinka Blažević, Lahorka Plejić Poje and Ivana Brković, 99–124. Zagreb: Srednja Europa.

MacDonald, J. Fred. 1985. *Television and the Red Menace. The Video Road to Vietnam*. New York: Praeger.

Maltby, Richard. 1989. *Dreams for Sale: Popular Culture in the 20th Century*. London: Harrap.

Marković, Predrag. 1996. *Beograd između istoka i zapada, 1948–1965*. Beograd: Službeni list.

Mayakovsky, Vladimir. 2005. *My Discovery of America*. London: Hesperus Press.
Milićević, Nikola. 1953. 'Među-team Šoljan – Slamnig'. *Krugovi* 8:714–721.
Ó Tuathail, Gearóid. 2000. *Critical Geopolitics: The Politics of Writing Global Space*. London: Taylor & Francis.
Pageaux, Daniel-Henri. 2009. 'Od kulturnog imaginarija do imaginarnog'. In *Kako vidimo strane zemlje. Uvod u imagologiju*, edited by Davor Dukić, Zrinka Blažević, Lahorka Plejić Poje and Ivana Brković, 125–150. Zagreb: Srednja Europa.
Pavletić, Vlatko. 1952. 'Neka bude živost'. *Krugovi* 1:1–8.
Peruško, Ivana. 2015. 'Čudovišni SSSR i mitska zemlja Jugoslavija'. *Philological Studies* 19 (9): available at http://philologicalstudies.org/index.php?option=com_content&task=view&id=479&Itemid=187&lang=hr
Polet. 1977. 'Coca-Cola'. *Polet* 28:no pp.
Polimac, Nenad. 2010. 'Ljubav i moda: kako se 60-ih slavilo potrošačko društvo'. *Jutarnji list*, 3 July.
Radičević, Ljubomir. 1960. *Ljubav i moda*. Beograd: Avala film.
Ruland, Richard. 1976. *America in Modern European Literature: From Image to Metaphor*. New York: New York University Press.
Rusinow, Dennison. 1977. *The Yugoslav Experiment 1948–1974*. London: C. Hurst & Company.
Saračević, Narcis. 2004. *Rječnik sarajevskog žargona: prilog leksikografiji bosanskoga jezika*. Zenica: Vrijeme.
Sardar, Ziauddin, and Merryl Wyn Davies. 2003. *Zašto ljudi mrze Ameriku?* Zagreb: Naklada Jesenski i Turk.
Sayre, Nora. 1979. *Running Time: Films of the Cold War*. New York: Dial Press.
Sharp, Joanne P. 2000. *Condensing the Cold War: Reader's Digest and American Identity*. Minneapolis: University of Minnesota Press.
Shaw, Tony. 2006. *British Cinema and the Cold War: The State, Propaganda and Consensus*. London and New York: I.B. Tauris.
Slater, David. 1993. 'The Geopolitical Imagination and the Enframing of Development Theory'. *Transactions of the Institute of British Geographers* 18 (4):419–437.
Tribuson, Goran. 1986. *Made in U.S.A.* Zagreb: Znanje.
Vučetić, Radina. 2012. *Koka-kola socijalizam: amerikanizacija jugoslovenske popularne kulture šezdesetih godina XX veka*. Beograd: Službeni glasnik.
Williams, Raymond. 1997. 'Culture is Ordinary'. In *Studying Culture: An Introductory Reader*, edited by Ann Gray and Jim McGuigan, 5–14. London: Arnold.
Williams, Raymond. 2006. 'Analiza kulture'. In *Politika teorije, zbornik rasprava iz kulturalnih studija*, edited by Duda Dean, 35–63. Zagreb: Disput.
Yugoslav Experiment National Intelligence Estimate. 1967. U.S. Central Intelligence Agency.
Yurchak, Alexei. 2006. *Everything Was Forever, Until It Was No More: The Last Soviet Generation*,. Princeton, NJ: Princeton University Press.

7 Beyond brand Bollywood

Alternative articulations of geopolitical discourse in new Indian films

Ashvin Devasundaram

Introduction

Bollywood, India's biggest contemporary cultural brand is also the nation's most popular global stereotype. Its purchasing power notwithstanding, this commercially successful industry has often been accused of self-exoticism—deemed 'guilty of re-Orientalising India' (Banerjee 2011, 133). Brand Bollywood is often willing to sanction, serialise, and self-identify with essentialist and reductionist Western constructs of India's cultural exoticism by reproducing cultural archetypes. These include seductive images in oversaturated colour, ancient religious mythologies and mysticism, celebrations of grandiose opulence, ostentation and economic excess, fantastical weddings, song and dance sequences—in essence, representation that is largely stripped bare of verisimilitude, political engagement, serious self-appraisal, or a self-critical apparatus.

Bollywood's volitional self-Orientalising seems designed to buttress a global branding strategy that showcases and sells the idea of India as a neoliberal nation open for business. This is accomplished by popularising a simplified and silo-style image of India that entails a familiar, easily palatable, predictably exotic and mass producible representational template. This is part of a contemporary phenomenon where globally dominant mass media forms, such as Bollywood, 'are continually offering up images of nations for consumption', and consequently adopting the role of dominant information providers and exclusive perception shapers about 'foreign countries in the (post)modern neoliberal era' (Saunders 2017, 71). It is important to note that in Bollywood's dominant popular cultural narrative, there are 'two kinds of Orientalisms at work: the confirmation for a Western observer, that "India" is indeed a feast of colour; and the triumph of Hindu India, in all its colourful exuberance over a Muslim minority' (Banerjee 2011, 124). In other words, Bollywood largely satiates the status quo of religious and nationalist 'traditions, morals and values' within India, positioning itself as the binding common currency of popular self-identification in an inestimably heterogeneous nation. On the other hand, Bollywood's global-level branding is assiduously aligned with and woven into the Indian state's neoliberal global economic grand design.

Bollywood therefore enjoys the official endorsement of the Indian state, which brandishes the film industry as its sceptre of national cultural soft power. It seems paradoxical then, that Bollywood, the nation's most powerful popular cultural brand and an emerging competitor to Hollywood, reinscribes and reflects a mono-cultural model of India as an exotic and mystical zone, when 'it is obviously contradictory to orientalise oneself while in a position of power' (Hase 2007, 39). This prompts further exploration of the basis for the dual dimensions of Bollywood's internal and external popular geopolitics; the double dimensions of its self-Orientalising disposition.

This bipolarity reflected in the approaches of both Bollywood and the Indian state is indicative of the extent to which popular culture and geopolitics are intertwined in the terrain of popular geopolitics (Dittmer 2010, 27). The internal and external dimensions of India's popular geopolitics could be viewed through the lens of Bollywood's official 'brand ambassador' role, where the industry packages cinematic versions of consumer-friendly national ideology to create a 'coherent national image in the domestic realm', simultaneously striving to communicate a 'positive country image' on the global stage (Saunders 2012, 51). At the same time, under the partisan rule of the right-wing nationalist Bharatiya Janata Party (BJP), 'subversive spiritualities' (Nandy 2013, 70), and extreme forms of religion-based politics have gained ground, and arguably, have been permitted to flourish within the inner geopolitical precincts of the Indian nation-state (Wildman 2011).

Bollywood films have mostly been a blind spot when it comes to representing traumatic and controversial politico-religious national events. Significant amongst these is the demolition in 1992 of the ancient Babri mosque in the Hindu holy city of Ayodhya, by a multitude of far-right Hindu extremists backed by political groups.[1] Apropos Bollywood's dominant popular cultural discourse, Kanishka Chowdhury argues that 'the heroic subject of this narrative is a Hindu upper-caste male who is able to preserve India's cultural heritage while at the same time increase the nation's standing in the world' (2011, 27).

This chapter aims to interpret the internal and external dynamics of Bollywood's popular geopolitics—its contrarian double-Orientalist logic of cinematically asserting a majoritarian Hindu-based version of Indianness, whilst assiduously selling a self-exoticised product in the global market. I will reveal how the popular geopolitics of Bollywood's outward-facing neoliberal narrative is dissonant with the turbulent inner divisions occurring within the Indian geopolitical space—a disconcerting reality that Bollywood consistently glosses over in the greater interests of fulfilling its commercial interests. I argue that this dual strategy deployed by Bollywood is a mimesis of the Indian state's similar scheme. I will inspect the emergence of alternative articulators—new wave Indian 'Indie' films and Bollywood/Indie 'hybrids' that are increasingly interrogating the normative precepts of Bollywood's popular culture and the ruling establishment's geopolitics.

This analysis of Bollywood's double-tasking brand of global and national popular geopolitics will be conducted through an analysis of the discourses in which two contemporary films are enmeshed, alongside textual readings of these films. Sections in this chapter will first analyse contradictions between Bollywood's outward-facing popular geopolitics and the turbulent internal religio-political dimensions, drawing from discourse analysis and comparative intercultural perspectives on popular geopolitics, citing key discursive events that exemplify the instability and incommensurability of India's internal geopolitical narrative with Bollywood's sanitised and apolitical external articulation of popular geopolitics. Subsequent sections will conduct textual readings of two selected films to indicate how Bollywood's double narrative is prone to unravelling from within, when its dominant mode of representation is disrupted by hybridity and critical self-reflexivity.

Director Chaitanya Tamhane's political independent film *Court* (2015) and Rajkumar Hirani's heterodox Bollywood blockbuster *PK* (2014) are good examples to demonstrate how even brand Bollywood is open to deconstruction that is effectively instigated by the antinomies and contradictions that inhere within its own superstructure. Whilst Bollywood could be viewed as a self-Orientalising 'Other' in relation to Hollywood, Bollywood is increasingly being interrogated by its own internal cultural other, in the form of an increasingly popular new wave of alternative Indian Indie films such as *Peepli Live* (2010), *Ship of Theseus* (2013), and *Court* (2015) (Devasundaram 2016). The influence of this challenge to Bollywood mounted by the Indies is significant enough for the Indie elements of verisimilitude and socio-political critique to infiltrate into the status quo-serving Bollywood superstructure. This permeation of the marginal outside into the mainstream inside (Indies into Bollywood) can result in hybrid, self-reflexive mainstream cinematic contestations of the overarching hegemonic socio-religious, cultural and political systems and structures, in which mainstream hybrid Bollywood films like *PK* find provenance.

As will be explored in detail, *PK* is an atypical Bollywood blockbuster with a trenchant satirical sci-fi premise that blends the Indian commercial cinema genre's own exotic aesthetic codes and tropes with unconventional and polemical Indian Indie-style thematic content. An analogy could be drawn with the popular geopolitics of American TV franchise *Battlestar Galactica* (2004), with its 'unique fusion of realism and science fiction' acting as a 'hybrid space for critically analysing and representing genocide' (Bohland 2013, 106). *PK* occupies a similar strategic position from where it can destabilise some of the Bollywood genre's self-Orientalising stereotypes as well as challenge the ideology of the dominant socio-political order.

Starring longstanding Bollywood actor/producer Aamir Khan, *PK* parodies the widespread prevalence of unquestioning religious faith, uncritical belief in mythological and supernatural forces, and the veneration of spurious self-anointed 'holy' men in modern Indian society. Apart from his Bollywood credentials, Aamir Khan is known for his financial and artistic patronage of off-beat alternative Indian Indie cinema through the aegis of his production company AKP (Aamir Khan Productions). This has manifested in the release of several seminal Indie films, including *Peepli Live* (2010) and *Dhobi Ghat* (2010), which grapple with topical Indian socio-political themes such as the spate of farmer suicides and the nation's widening socio-economic disparity.

An analytical counterpoint to the Bollywood industry's largely one-dimensional branding of national image is the Indie Marathi (the regional language of Maharashtra state) film *Court* (2015). The film is an indictment of apathy and corruption in an Indian juridical system that effectively silences minority voices and denies them justice. *Court* exposes the discriminating field of local-level governing policies that are insuperably stacked against downtrodden sections of the Indian polity. *Court*'s candid representation of an unflattering internal constellation serves to magnify the ruling order's contradictory national image portrayal to a wider global externality. The next section delves into the dyadic dimensions of Bollywood's popular geopolitics by inspecting the national geopolitical discourses in contemporary India, which constitute the crucible in which films such as *Court* and *PK* are conceived.

Bollywood's outer narrative and India's inner reality

Through its popular geopolitics, Bollywood's hegemonic 'branded imagination seeks to infiltrate and subsume the symbolic order of nationhood' (Kaneva 2012, 11). Tracing the 'pocketbook' philosophy of nation branding, Saunders describes the phenomenon as being assiduously motivated by a monetary stimulus and using 'affectual content and other sorts of non-tangibles' that favour 'the sublime over the rational, affect over effect, and feeling over knowing' (Saunders 2017, 50). This interpretation of nation branding fits squarely with the popular representation blueprint of neoliberal-era Bollywood popcorn melodramas.

Bollywood's outer narrative brings into focus the inner contours of India's contemporaneous geopolitical narrative. In 2015, India's global image was corroded by a raft of incidents involving religious antagonisms and the suppression of free expression. Prominent among these events, was the Hindu mob lynching of a Muslim man in the northern state of Uttar Pradesh, over his alleged consumption of beef, which is forbidden in the Hindu religion. Under the BJP-led government, the nation is also countenancing an assault on free speech and expression of alternative views, with several prominent exponents of rationalism and science murdered by religious radicals in different regions of the country (Widmalm 2016, 95).

There is also a growing mobilisation of the Indian intelligentsia and the arts and culture community, manifested in counter-establishment Indie films and the returning of national awards by prominent Indian writers, artists, filmmakers, historians and scientists. In light of these events, the nation's outward-facing neoliberal narrative appears antithetical to its tumultuous internal geopolitical configuration. This scenario is reminiscent of how 'Britain under Thatcher became a nation of national brand names' that forged national identity 'under the reign of the "The Image"' (Friedman 2006, 49). Under this firmament, jingoistic popular cinematic celebrations of British nationalism such as *Chariots of Fire* (dir. Hugh Hudson, 1981) were contested by alternative 1980s 'state of the nation' films such as *Made in Britain* (dir. Alan Clarke, 1982), *Letter to Brezhnev* (dir. Chris Bernard, 1985), *My Beautiful Laundrette* (dir. Stephen Frears, 1985), *Defence of the Realm* (dir. David Drury, 1986), and *The Last of England* (dir. Derek Jarman, 1988).

India's economic foreign policy is punctuated by Prime Minister Modi's relentless international mission to propagate his 'Make in India' campaign, which is largely shorthand for an exhortation to multinational conglomerations to benefit from tax exemptions, whilst investing and manufacturing in the Indian market, and effectively prescribes the privatisation of the public sector and nationalised industries. In line with a national thrust towards capitalism and a consumerist model, India has shrugged off the cloak of its pre-liberalisation socialism and with it any vestiges of the tacit *entente cordiale* with the then USSR, gravitating towards more strategic and lucrative geopolitical alignments by courting enhanced commercial and cultural ties with the U.S. and Britain, in addition to increased free trade and defence deals with Israel.

Cinema is a particularly useful tool in the building of brand India, and Bollywood as a showpiece of popular culture, mass entertainment and soft power, constitutes an able ally in this wooing of foreign investment on the global stage. News reports consistently conflate Modi's frenetic international tours with the ubiquitous global selling points of neoliberal India—Bollywood and an economically ascendant contingent of entrepreneurial non-resident Indians (NRIs). This is illustrated in an article in *The Guardian*, which pre-empts the Indian Prime Minister's reception by the Indian diaspora during his UK tour in 2015:

> He will also address a rally in Wembley stadium on Friday attended by supporters from among Britain's 1.5 million-strong Indian-origin community. If it follows previous precedent on both coasts of the US, the rally will be a stunning celebration of Indian identity and influence. There will be a Bollywood-influenced mix of dance, displays and speeches, rounded off with fireworks. Modi's speech will be a familiar masala of nationalism, his own brand of economics and appeals to his audience's cultural roots.
>
> (Burke 2015)

The public reaction mentioned above is symptomatic of the recurring long-distance nationalism (Anderson 1998) of diasporic Indians in the U.S., UK, and Australia, who compensate for the absent sense of bona fide belonging in their new nations of domicile, with an 'exaggerated sense of pride' (Anderson 2007) in their country of origin. As Anderson astutely asserts, these overzealous nationalists' investment in long-distance politics is decoupled from any real responsibility for actual developments on the ground in their antecedent nation. This is particularly relevant to U.S. and UK-based NRI communities, who from their remote locations campaign actively and contribute financially to the BJP and its ultra-right affiliates, including the Rashtriya Swayam Sevak Sangh (RSS) and Vishwa Hindu Parishad (VHP).[2] Anderson (1998, 73) asserts that the 1992 demolition of the Babri mosque in Ayodhya 'which plunged India into her biggest crisis since Partition … was officially sponsored by the Vishwa Hindu Parishad (World Hindu Council)'. He notes how immense sums of money were contributed by supporters in the U.S. and UK, a preponderance of whom were 'Indians living permanently overseas' (Anderson 1998, 73). Congruent with their long-distance nationalism, these diasporic donors who are detached from the geopolitical lived experience of domicile in India are largely oblivious to the implications of their actions on socio-religious cohesion and the ramifications on notions of identity, space, and place.

In this cartography of expatriate apartness and diasporic distantiation from ground-zero of the Indian geopolitical space, the global networking of organisations such as the RSS and VHP facilitates an overlapping of 'geopolitical and organizational practice' leading to a 'cultivation of geopower through shared national identities, that is "popular geopolitics"' (Sage 2014, 156). Sage explores how America's geopower is boosted by space agency NASA's cosmological colonising—its role in making the popular notion of Space synonymous with 'a sense of what America was, is, and could yet become' (2014, 10). He observes that 'very similar transcendental narratives appear in other ex-colonized and ex-colonizing cultures' (Sage 2014, 10). It would not constitute one giant leap to state that Bollywood plays a similar popular geopolitical role in the building of Indian geopower. It fulfils this role by acting as a universalising brand; a binding agent or common touchstone for diasporic Indian communities to connect with increasingly powerful transnational organisational networks that promote a transcendent longing for a historically homogenous Hindu national identity.

Pillai states that 'contemporary universalising tendencies of capital' have compelled Bollywood to cater to 'deterritorialized and portable identities of "'Indian"' audiences both outside and inside the boundaries of the nation' (2012, 42). This can be imputed to the fact that in the neoliberal age where nation-states 'sell a desired national image to both domestic and global audiences', national brands are largely conceived 'with international audiences in mind and privilege their needs and interests' (Kulcsar and Yum 2012, 197). In this respect, Bollywood, sanctioned by the Indian state as cultural soft power and commercially dominant, both in India and amongst the Indian diaspora, largely functions as an exported authorised heritage discourse (Smith 2006). For example, the marketing manager of Rajshri Productions, a company synonymous with films targeting the diaspora, states that 'Indians in America exist in two worlds. They have spent many years here and they know what it means to live in America. But they have an Indian side, and Bollywood connects them to India' (Punathambekar 2013, 30).

In this formulation, it would be specious to perpetuate the largely popular conception of Bollywood as a political no-man's-land; a fantasia whose sole function is to entertain and enthral. It is pertinent to note that ideology is triumphantly present where there is vociferous denial that there is no ideology (Zizek 2009, 39). Agnew and Corbridge contend that in hegemony, be it by coercion or consensus, resides an 'emphasis on the routinized and incorporated nature of the practices and ideological representations that give an order its "normality" and "commonsensical" acceptability to the actors involved in it' (Agnew and Corbridge 1995, 17).

Bollywood's hegemony is perceivable even at the linguistic level of the national homogenising language—Hindi, spoken mainly in North India, and used exclusively in Bollywood films, demonstrating that 'the lineage of Hindi was ideologically constructed as the language of "Indianness" and the Hindu nation' (Dimitrova 2014, 86). This chimes with the crescendo of calls from religious nationalist power structures for a 'return' to ethnic homogeneity of Hindu *rashtra* (Hindu nation-state). Dimitrova argues that 'going back to Aryan roots had shaped the idea of a modern Hindu-Indian nation from which Muslims, other religious minorities, women and untouchables were excluded' (2014, 86). Contemporary ructions in Bollywood are punctuated by the drawing of differentiating lines around its Muslim actors, Aamir Khan and Shah Rukh Khan, as a reaction to their public expressions of uneasiness about rising intolerance in India. This may prove a litmus test of the veneer of neutrality and masquerade of apolitical benignity that the commercial industry often seems compelled to adopt in the face of larger global capital interests.

Whilst alternative films such as *Court* provide narrations from multiple viewpoints, particularly from minority perspectives, the popular geopolitics of Bollywood has largely sculpted a static and one-size-fits all narrative of the nation. Bollywood's popular geopolitics perpetuates an imagined geography (Said 1978) of India by recycling a 'collection of facts and stereotypes' that have become concretised as 'the idea of the nation' (Dittmer 2010, 18).

Indeed, Bollywood has cast an amaranthine cultural blanket across the temporality of India's post-liberalisation timeline, starting in the early 1990s. Since its ascendency as the plenipotentiary of Indian arts and culture, Bollywood's status as commercial cinematic Leviathan and national cultural superstructure seems an immutable component of an otherwise labile often volatile geopolitical landscape that is cutting a distinct path towards free market capitalism. Salutary accounts of India's economic ascendency are typified by Karan Singh, who declares that in-between the '"romantic India" of maharajas and snake charmers', and 'the negative image of endemic poverty and corruption' is a picture of 'a modern vibrant nation growing rapidly with tremendous talent and manpower reserves' (Singh 2013, ix). It is incontrovertible that the national image of India has tended to 'suffer from Orientalist distortion/Occidental self-adulation' (Saunders 2017, 53). I argue that the imagination of India as an antiquarian land of turbaned kings, arcane rituals, and snake charmers has received a postmodern reboot through re-Orientalism engineered by the popular geopolitics of Bollywood, reflecting a more contemporary configuration of the earlier romanticised, exoticised, and idyllic tropes. In Bollywood's 'Version 2.0' for the new millennium, an affluent elite capitalist class revels in the nouveau riche economic powerhouse of neoliberal India, whilst unwaveringly flying the flag of a Hindu nationalist brand of geopolitical ideology.

Despite Bollywood's hubris-soaked celebrations of India's economic ascendency, it is difficult not to take cognisance of the rise of religious extremism that takes the shape of a Hindu 'saffronising' of India's conceptually secular constitutional credentials. Couched in this tumultuous crucible of geopolitics; culture, cinema and the arts in India are inevitably menaced by the 'BJP's long term vision' of a uniform logo-centric Hindu *rashtra* (Basu 2005, 281). Amrita Basu finds this political paradigm paradoxical, because in her view the BJP 'never concede[s] that the nation-state itself might be of Western inspiration, though they often level this charge at secularism and democracy' (Basu 2005, 281). Brand Bollywood manages its balancing act as sanctioning agent of a majority nationalist ideology and global cinematic commodity branded as benign entertainment-based cultural form. The double-Orientalism in Bollywood's internal/external strategies provides clues to this conundrum. For example, some sections of scholarship in India continue the trend of validating the hegemonic cultural narrative. For instance, Kishwar declares:

> Even though the bulk of Bombay films are melodramas, they play a vital role in shaping both individual and collective aspirations regarding India's 'uniqueness' in the fast globalizing world. They have consciously worked to build a social consensus on traditions and values that need to be upheld and preserved.
>
> (2013, 94)

Positioning Bollywood as a sanctifier of ancient Hindu mythologies and scriptures, Kishwar exults that 'Bollywood cinema has assumed the mantle of upholding a distinct moral code, just as the pauranic kathas [i.e. ancient religious myths] once did' (2013, 95). By valorising Bollywood as the custodian of ancient Hindu scriptures, this logo-centric line of argumentation seems to conveniently disregard the mainstream industry's minority communities who comprise a significant part of its workforce. The wider reverberations of this discourse were manifested in 2015, when Aamir Khan and another Muslim Bollywood star Shah Rukh Khan, were pilloried for publicly expressing alarm about rising intolerance in India (NDTV 2015). Significantly, several established and senior Bollywood stars with BJP affiliations, including Anupam Kher, Ram Gopal Varma and Raveena Tandon criticised Aamir Khan for expressing 'unpatriotic' views, while members of the ruling BJP Government branded the comments a 'conspiracy to tarnish the image of India' (Ghosh 2015).

As will be seen in the next sections, the Hindu 'saffronisation' and normative popular geopolitics promoted by Bollywood and the state are contested by *Court* and *PK*—the latter film standing apart from its mainstream Bollywood film counterparts with its scathing contestation of politically incited intolerance, superstitious belief and religious dogma. We will now examine the ideas and points raised thus far through the lenses of the two films, commencing with the Indie *Court*, whose critical approach is considerably different from *PK*.

Court: an independent cinematic investigation of local level legal and social space

Chaitanya Tamhane's *Court* charts the protracted trial of an elderly left-wing revolutionary folk singer Narayan Kamble (Vira Sathidar), who belongs to the 'lower-caste' Dalit community. Kamble is arrested for sedition, on the accusation that his politically critical song lyrics incited a man to commit suicide. Set in multicultural metropolis, Mumbai - the capital of Maharashtra state, the film performs a class dissection, revealing the after-court life of the two lawyers fighting the case and the sessions court Judge Sadavarte (Pradeep Joshi), who presides over proceedings. The prosecution lawyer, Nutan (Geetanjali Kulkarni), subscribes to a religious nationalist credo. Her desire for an ethnically pure 'higher-caste' Hindu Maharashtra state is rendered explicit in the film's vignettes portraying her everyday life. When divorced from her court duties, the modern career woman Nutan morphs into a 'traditional' homemaker, dutifully fulfilling her wifely duties of supplying meals to her husband and escorting the children to and from school. At a micro-level, Nutan's double role largely mirrors the external and internal contradictions displayed by Bollywood, which is saddled with its own dual function of peddling both tradition and modernity.

The geopolitical significance of India's conflictual religious politics is starkly presented in a sequence that follows Nutan and her family on a recreational outing to a vernacular drama performance, whose audience is largely comprised of right-wing Hindu Maharashtrian segregationists. The prosecution lawyer's personal political proclivities are mirrored in the play's theme, which explicitly advocates the forceful ejection from Maharashtra state, of all non-Maharashtrian migrant workers who hail from other Indian states and who largely compose Mumbai's blue-collar labour force.

The drama's message is a microcosmic rendition of a larger imagined 'narrative of the new nation focussed on the assertion of the "imagined" Hindu glory of India' (Dimitrova 2014, 85–86). This metanarrative of Hindu triumphalism has been the bedrock of the ruling BJP's power structure in Indian politics and is particularly strong in states, such as Gujarat and Maharashtra, dominated by the BJP and/or its political allies. This outlook also resonates with Bollywood's own Orientalising of minority cultures, through a track record of films where 'Muslims were excluded and women's rights and the rights of Dalits (untouchables) were erased from the sphere of the national imagination and societal concerns' (Dimitrova 2014, 85–86). Compared to Bollywood's blinkered popular cultural representation of space and place, where dominant power is privileged, and majority ideology upheld, the narratives emerging from the alternative space of increasingly popular new Indian Indie cinema are significantly more diverse. The diegetic frames of several Indie films represent the composite ethnic, religious, linguistic, gender and class layers in contemporary Indian society and culture.

In *Court*'s broad palette of representation, the defence lawyer, Vinay Vora (Vivek Gomber) is a bourgeois liberal whose middle-class tastes extend to fine wines and modern jazz. Representing the opposite pole of the socio-political spectrum from his right-wing adversary, Nutan, the young lawyer goes out of his way to demonstrate his willingness to engage in social interaction with members of the Dalit community, in line with his work on the case. The background of the Dalit defendant, Kamble, however, is not elaborated in the film, apart from fragments that emerge before and during the court trial. In this respect, the subaltern Dalit largely remains storyless and historyless—a man without a past, whose life is trapped within the confines of the court. In terms of both representation (in the film) and reality (of the often-persecuted Dalit community)—Kamble seems suspended in the purgatory of an erased past and an uncertain future.

The absence of information about Kamble's antecedents, whilst the backgrounds of the two other central characters receive greater attention, development and representation, is a narrative ploy that has geopolitical connotations. It signifies India's historiographical omission of narratives 'from below'

(Chaturvedi 2012, ix–xi), eliding subaltern peasants and labourers in the writing of the postcolonial Indian narrative. The most striking representation of this silencing of the subaltern is witnessed at the start of the film, when Kamble is seen performing a revolutionary song to a live audience from the local Dalit community. His song constitutes a stirring call to overthrow the established social and political order, which he castigates for propagating 'religious jungles, racist jungles' and 'fancy malls'. Ironically, neoliberal lifestyles, material luxury, mall culture, and religious iconography are some of the elements pervasively visualised in the popular geopolitical aesthetic frame of Bollywood films. This example of self-reflexive critique in *Court* therefore denotes a very different geopolitical aesthetic in Indie films, where connections are made between 'localised experience of the individual and the globalised totality that is late capitalism' (Dixon and Zonn 2007, 101). The strategy of representation in *Court* also emphasises how 'lower caste' Kamble is granted 'space' in this Indie film to articulate his dissenting views, whilst Bollywood largely sidelines or silences marginal characters and communities, particularly Dalits, whose inclusion may prove inimical to Bollywood's neoliberal/nationalist brand of popular geopolitics.

Unlike the xenophobic drama watched by Nutan's family, Kamble's musical performance is disrupted by the intervention of the police who proceed to arrest the singer. This sequence cinematically underscores this chapter's proposition that Bollywood's popular cultural artefacts often elide or suppress representation of alternative cultural expressions and contestations of authoritarian political power at the local level, in the interests of brandishing an investment-friendly brand image of the nation. Geopolitics, seen as the impact of transnational cultural politics on the local terrain, illustrates how the local can become a battlefield for the state's ambivalent yet coterminous nationalising and globalising drives. The geopolitical aesthetic imagined at the local level in *Court*, is a 'codification of the real' (Dixon and Zonn 2007, 106) enactments, machinations and configurations in India's agonistic neoliberal and nationalist trajectory. This contrasts with the mythologised popular geopolitical aesthetic fashioned by Bollywood in which capitalism is so serialised and normalised, it 'renders itself invisible and thereby unassailable' (Dixon and Zonn 2007, 106).

The exclusion of the marginalised and oppressed from mainstream political and cultural spaces is a real and present phenomenon in India. Independent filmmakers attempting to provide representational space for these peripheralised voices often face repressive measures. This is exemplified by the police action dispersing a private screening of *Papilio Buddha* (2013), Jayan Cherian's controversial independent film about the political resistance movement mounted by the Dalit community in Kerala in response to their routine brutalisation and oppression at the hands of the local police and the state (New Indian Express 2012).[3] There are distinct parallels between Kamble's experience in the above scene in *Court* and the actual invasion of private consensual exhibition space in relation to *Papilio Buddha*, a film featuring and produced by Dalits. In fact, Vira Sathidar, who plays Kamble in *Court*, is a real life leftist radical Dalit activist, who has repeatedly been victimised by the police and booked on unfounded charges. Sathidar states about his experience 'I had to deal with the police and court for about four years ... Finally, I was acquitted. There are many like me who have suffered due [to] the delays and denial of justice in court. The film highlight this' (Haygunde 2015).

Court's filmic performance of Kamble's arrest by the repressive apparatus of the police and his experience of non-justice in a court space that belies its own *raison d'être*, contains a convergence of both the mimetic and aesthetic modes of analysing representations of geopolitics. On prima facie evidence, the film seems to convey a 'firm vision' of what local politics and law in India 'is', by aiming to represent politics as 'realistically and authentically as possible' (Bleiker 2001, 510). However, it is not the fixed realism of international relations that *Court* strives for. The aspect that distinguishes *Court* is its multi-layered, poly-discursive aesthetic and parallel narrative strategies that alert the viewer to the fact that 'texts (including images) always have another possible meaning, another story to tell and another dialogue which we can enter' (Moore and Farrands 2013, 231). In this volume, Kyle Grayson, taking cognisance of the aesthetic turn in international relations (see Bleiker 2001), argues the case for cultural artefacts to be regarded 'as theories and practices of geopolitics in their own right' rather than just mimetic reflectors of the geopolitics of an external 'real' world. In *Court*, there is interplay between the dimensions of aesthetic and mimetic, where the film's apparent mimesis of the 'real' world serves to 'highlight the [multiple] allegorical, explanatory, and/or ethico-political properties' aesthetically communicated by the film (Grayson 2013, ix).

Alternative cultural underwritings in the Indian national narrative are largely overlooked, overwritten and eventually subsumed by the brand hegemony of Bollywood. In his analysis of Hollywood's popular Western genre of films, Shapiro illustrates how 'state-centric macro-level politics hides the continuing struggles of Native American peoples' (2007, 32). Similarly, Bollywood has been enlisted in the construction of 'cinematic nationhood' (Shapiro 2007, 32)—as a cultural articulator of the state's neoliberal/nationalist project. By its lack of proportional representation, Bollywood is implicated in the cinematic erasure of India's marginalised micro narratives. However, in a world where popular geopolitics is ascendant across technologically interconnected information societies, dominant state-endorsed national brands such as Bollywood 'must anticipate, negate, and even accommodate its unwanted "branding" by alternative actors' (Saunders 2012, 49). As will be demonstrated, *PK*, a 'mutant' Bollywood film—owing to its hybrid 'Indie' sensibilities—has embraced divergent discourses, incorporating Bollywood's own popular geopolitics as a platform to critique the dominant national discourse. *Court*, on the other hand, exemplifies an alternative Indie agent that adopts a more direct, multi-perspective approach, to destabilise the mainstream narrative of nationalism and neoliberal aspiration.

With its grassroots-level study of class, caste and political fissures in the modern Indian urban and legal space, *Court* emphasises Bollywood's prolific acts of preterition—of leaving out elements that would be detrimental or inimical to its brand building process. India's turbulent internal struggles are impacted by the nation's relentless external push towards neoliberalism, which has widened the wealth gap, deepened class and caste divisions, and amplified globalisation-fuelled anxieties about the erosion of 'authentic' Indian moral values and traditions. Once again, this gestures towards a dual dimension, where the BJP Government either maintains tight-lipped silence, or feigns ignorance of internal communal atrocities and violent ethnic purges by deflecting attention through the flag-waving of neoliberal progress. This simultaneous disavowal of endogenous socio-religious difference and showcasing of consumer capitalist affluence typifies Bollywood's popular culture and the state's geopolitics, and is part of a branding strategy deemed suitable for the global geopolitical sphere. In this dominant milieu, alternative articulators—new Indies such as *Court*, use their autonomy to contest the official state-sanctioned narrative and its erasure of minority discourses, and in doing so, stand apart from the hegemony of Bollywood's popular geopolitics.

In one of the most arresting scenes in *Court*, the camera is stationed at the rear of the courtroom, providing a fly-on-the-wall perspective of the Kafkaesque trial of the Dalit defendant. As yet another session concludes with the apathetic judge proroguing the trial to yet another date, the framed shot of the courtroom remains static, as the cast of characters incrementally exit the 'stage'. We remain locked in on the scene as the courtroom disgorges the last of its occupants; the clerks and stenographers. The court orderly performs the final act of switching off the courtroom lights, walking out and locking the door, leaving us, the audience, in the dark confines of an inconsequential juridical space, where the legislative process has failed its weakest and marginalised citizens. This symbolic indictment of India's legislative and judicial systems as being entrenched in colonial-era bureaucracy and stasis suggests that the hope for grassroots-level justice in the nation is shrouded in darkness. This scene lends credence to the notion that there is a stark disjuncture between the notions of space as it is *conceived*, *perceived* and *lived* (Lefebvre 1991). The court space, as depicted in the film, could be read as a metaphor for the larger inconsistencies between the idealised spatial concept of the secular Indian nation-state and the contradictions and conflicts that inhere in its contemporary geopolitics.

There is another more tangible dimension to the discourse of space, place, and politics evoked by the film. *Court* exemplifies the Indian political manoeuvre of proffering independent arthouse films to the foreign language category of the Oscar Awards, as a symbolic statement of the nation's willingness to tolerate self-critical free expression and to affirm India's credentials as a progressive, democratic, secular space.[4] *Court* (official entry in 2016) follows a line of independent arthouse Indian films with strong critical themes, such as *Peepli Live*, that have been sent to the Oscars. However, within the spatial contours of India, several socio-politically censorious films, such as *Papilio Buddha* (2013), *Unfreedom* (2015), *Gandu* (2010), *Harud* (2010), are routinely censored by the state-controlled Central Board of Film Certification (CBFC), commonly regarded as the Indian Censor Board, operating under the custodianship of the Ministry of Information and Broadcasting. An example of state power over cinematic content is the mass resignation in 2015, of CBFC members, who attributed their action to protracted political interference in the certification process. This was followed by the Board's strategic repopulation with sympathisers and affiliates of the ruling BJP Government.

In addition, the alternative views of real life Dalit activists including *Court*'s actor Vira Sathidar are regularly monitored and censored. This provides a peek into the Indian state's schismatic approach to global geopolitics from a cinema industry perspective, where 'worthy' arthouse Indies are put forward to represent the progressive face of the nation, whilst at the same time, the larger corpus of non-mainstream cinema is subjected to rigorous censorship within the boundaries of India.

The filmic coda in *Court* again draws on the motif of the interiority of discourse within the local-level legal space and its links to the exteriority of the civil space. The court Judge Sadavarte is seen in civilian clothes with a busload of friends and family, out on a group trip to a holiday resort. The judge's conversations with his companions are ubiquitous of the ambitions of India's rising bourgeoisie class. The judge extols the virtues of the I.T sector in India and declares how securing an MBA degree from the elite Indian Institute of Management (IIM) would guarantee a graduate a handsome salary. Sadavarte cites an article in *The Times of India* to inject veracity into his claims. The dominant narrative of nation forms a through-line in the scene. The judge, who is a proponent of state power, validates the hegemonic neoliberal national narrative of capital accumulation, whilst endorsing the state's dominant educational institution for finance studies—IIM, citing this configuration as a winning formula to achieve capital gains. Sadavarte completes the ideological cycle by turning to the above-mentioned prominent national newspaper to bolster his truth claim.

Sadavarte's personal *Weltanschauung* is revealed in another scene, when he engages in private conversation with a man, and enquires about the progress of the latter's young son, who suffers from a speech disability. When the man answers that there has not been any improvement in his child's condition and that he is consulting a speech therapist, the judge proceeds to offer a non-medical panacea. He directs the man to change his son's name, asks the boy's father to bypass the therapist and consult a numerologist instead, and prescribes astrological gemstones, as a set of measures to remediate the boy's condition. This implicates the court's presiding authority figure, revealing his predisposition to myth and superstition, which is coeval with his simultaneous desire for and denial of materialistic neoliberal values.

The judge's contradictory worldview mirrors the double-Orientalism promoted by Bollywood and the dual model of geopolitics adopted by the Indian state. Sadavarte's sensibilities also highlight the hegemony of unscientific and irrational religiosity, which, as the film *PK* aptly demonstrates, wraps itself around multiple dimensions of Indian geopolitics, cultural space and place. The above scenario draws attention to Partha Chatterjee's contention that the '"material" and "outer" was linked to the West and economic progress … By contrast the "spiritual" and "inner" was linked to the East, to the issues of religious and cultural identity and values' (Dimitrova 2014, 97). As Said has demonstrated eruditely, this is precisely the template that has been used by the West to undertake its Orientalising of the East as an exotic and mystical other.

It is interesting to note that the judge's personal beliefs mirror the material/spiritual bipolarity which informs Bollywood's and the state's geopolitical dual narrative of negotiating between the external pressures of globalisation and the internal demands of religious traditionalism. We shall now turn to *PK,* whose discourse deals with the rising tide of religious nationalism in India.

Breaking the Bollywood barrier: *PK*'s dismantling of religious dogma

PK melds elements of sci-fi with satirical comedy in its depiction of an eponymous humanoid extra-terrestrial on a reconnaissance mission, who is left stranded on Earth when his remote control constituting the sole means of his repatriation to his home planet is stolen during his quest to recover the lost device, PK, played by Aamir Khan, harnesses his extra-sensory powers to learn the local language, explore and familiarise himself with his new space, and in the process, ethnographically evaluate the intense fervour and passion with which the native populace seem to pursue their various brands of religion—Hinduism, Islam, Christianity, Sikhism and Jainism, amongst a motley host of others. PK's search leads him to parlous situations, where he encroaches on the sacred space of religious buildings and encounters religious custodians who preside over their respective religions with unforgiving fervour. PK is mystified by the irrationality and exiguity of scientific basis in the religious rituals he witnesses. The alien then meets Jaggu (Anushka Sharma), a young TV journalist, whose family is indoctrinated by the religious dogma of a self-professed 'godman' named Tapasvi Maharaj (Saurabh Shukla). Jaggu's parents are part of a multitude of devotees, who have all but deified Tapasvi for his alleged ability to communicate directly with God.

With their affiliations firmly rooted in this modern-day cult, Jaggu's parents are vehemently opposed to their daughter's romance with a Pakistani Muslim man, Sarfaraz (Sushant Singh Rajput), whom she meets during a sojourn in Bruges. Defying Jaggu's parents' injunction on their inter-religious and cross-national relationship, the couple arrange to get married immediately. However, when a letter is wrongly delivered to Jaggu, she is left standing at the altar, as Sarfaraz apparently fails to turn up. *PK*'s portrayal of an India-Pakistan love relationship is a significant intervention in India's geopolitical narrative of ascendant Hindu nationalism under BJP rule. In this context, the film marks a distinct divergence from the enduring popular geopolitics of Bollywood representation that has traditionally favoured an ultra-patriotic brand of gung-ho nationalism and one-upmanship over India's neighbouring arch-rival. This has largely been the Bollywood norm manifested in a raft of mainstream 'anti-Pakistan, anti-Muslim slanted films' including *Main Hoon Na* ('I'm Here Now', dir. Farah Khan, 2004), *Gadar—Ek Prem Katha* ('Revolution—A Love Story', dir. Anil Sharma, 2001) and *The Hero* (dir. Anil Sharma, 2003) (Dudrah 2012, 18).

The cartography of antagonistic international relations between India and Pakistan collapses into what has been inscribed historically as a 'conflict of interest' (Agnew and Corbridge 1995, 95) between the two nations, where the misfortunes of one, are perceived as a victory for the other. *PK*'s portrayal of reconciliatory miscegenation between an Indian woman and a Pakistani man exemplifies how the above-mentioned antagonistic formulation can be destabilised by 'more liberal and idealistic accounts', through which the opposing nations can 'negotiate a temporary regime of cooperation' (Agnew and Corbridge 1995, 95).

Whilst the BJP-led government in India, commensurate with its hard-line nationalist policies (Schwecke 2011, 93), has been criticised for not reinitiating a dialogue with Pakistan, *PK* serves as a cinematic surrogate, creating a momentary detente for the duration of the film's running time. So, where liberal or constructivist political diplomacy has been side-lined by the state, *PK*'s portrayal of a cross-national romance between a Hindu Indian woman and a Muslim Pakistani man is all the more noteworthy in the Indian cultural space. This assertion is predicated on the thesis that India as a nation is represented symbolically as feminine—'Mother India' (Hirji 2010, 30). Therefore, by positing an Indian woman-Pakistani man configuration, rather than the other way around, *PK* contests the mythologised gendering of the Indian geopolitical space by the Bollywood genre and by agents of religious nationalism. In addition, *PK*'s conceit of presenting the lead character as an extra-terrestrial alien could be read as a strategy to harness the popularity of the sci-fi genre, using its allegorical licence to imagine 'alternatives to hegemonic power', and akin to American popular culture such as *Battlestar Galactica* (2004), *Mars Attacks!* (1996), and *Starship Troopers* (1998), investigating 'sensitive political topics with more freedom and creativity' (Bohland 2013, 99).

At a later stage in the film, PK unveils Tapasvi's elaborate deception, exposing him as a charlatan and communicating this information to Jaggu. PK also discovers that Tapasvi has acquired his remote control and is using it as a celestial talisman or 'divine' relic to attract more acolytes. Inspired by PK's revelation, Jaggu and her TV producer boss, Cherry Bajwa (Boman Irani) launch a nationwide initiative to expose the illogicality of blind belief and religious rituals and uncloak the villainy of religious conmen across all faiths. Their all-India mission eventually leads them to a direct showdown with Tapasvi.

In a prime-time national television debate, featuring a verbal duel between PK and Tapasvi, the godman manipulates existing prejudices against Muslims and Pakistan embedded in the Indian social consciousness. Tapasvi's inflammatory statements are symptomatic of the 'old Hindu nationalist allegations of Muslim disloyalty to the Indian nation and the image of Muslims as a violent 'Other' in significant parts of the Indian public even outside the Hindu Nationalist movement' (Schwecke 2011, 93). The mountebank holy man also undertakes an elaborate scheme to publicly discredit Jaggu's ex-lover Sarfaraz on the premise that he is Muslim and Pakistani and hence 'doubly dubious'. Tapasvi mockingly challenges PK to address Sarfaraz's always-already 'untrustworthiness' as a Muslim, which Tapasvi claims is demonstrated by Sarfaraz's 'abandoning' of Jaggu at the altar, thereby betraying his promise of fidelity and marriage. This imaginary of Muslims invokes comparisons with stereotypes cemented by a 'continuation of pre-existing Orientalism' (Upstone 2010, 39) in popular geopolitical film representations post-9/11. Indeed, pre- 9/11 portrayals of Muslims in global Hollywood blockbusters, such as *True Lies* (1994), exemplify this pre-existing Oriental typecasting of an '"imaginary identity" of Islam' (Upstone 2010, 39). Relating this to Bollywood's popular geopolitics re-invokes Banerjee's (2011) assertion about the genre's dominant re-Orientalising representation that privileges the majority Hindu self over minority Muslim other.

The invidious wedge of discord implanted by Tapasvi in the minds of the people is the film's metonymic reference to en masse political manipulation and the insistent thrust by Hindutva political groups campaigning for a Hindu *rashtra*. In one scene, Tapasvi exhorts his audience to mobilise funds in order to rebuild a new Hindu temple. This is a direct reference to the earlier mentioned demolition of the Babri Mosque by Hindu extremists in 1992, and the continued demand to construct a Ram temple over the razed mosque site at Ayodhya, which remains etched in the future agenda of the *Sangh Parivar*.

In his publicity strategy, Tapasvi brings a stylised branding to his own persona as well as the fallacious ideological creed he espouses. He is portrayed as charismatic, tech-savvy, and adroit in the use of smart phones and social media. This trope in the film recalls the ruling BJP's election strategy of assiduous engagement with the electorate through Facebook and Twitter. In addition, the party used futuristic technologies, such as the projection of holographic images of Narendra Modi to remote locations, as an auratic simulacrum of the leader to compensate for his physical absence, when he was unable to attend a mass gathering in person. Indeed, a suffusion of Modi merchandise is available in the Indian market, and runs the gamut of personalised products, from android smart phones bearing the brand name 'NAMO—Narendra Modi and Next Generation Android Mobile Odyssey', to the 'Modi condom' (India Today 2015)—prophylactics carrying an image of the Prime Minister. In its depiction, *PK* extends an irreverent, culturally specific barb at Indian politicians and religious figureheads, who have co-opted globalisation's instruments of technology and communication, and wielded them to advance a mythologised religious nationalist geopolitical state ideology. This twinning of state-of-the-art media technologies with religious dogma as represented in *PK*, largely reflects the neoliberal brand of religiosity that seems increasingly indispensable in India's external/internal narrative. As Nandy states:

> We are probably entering a period when the decisive battle will not be between fundamentalism and secularism or between identity politics and normal, interest-based politics. The battle may well be between religion in its new, packaged, consumer-friendly version as a political ideological platform and the subversive spiritualities.
>
> (2013, 70)

Following on from Nandy's cautionary assertion, it is also worth noting in *PK*, how time and space are compacted and renegotiated in an intermedial zone of the public sphere, where religion-based political ideology can be interrogated and contested. This is evoked through Jaggu and Cherry's 'Wrong Number' campaign, inspired by PK's empirical observations about the meretricious and misleading nature of organised religion. Drawing on citizen journalism, the campaign exhorts all members of Indian civil society from across the length and breadth of the nation's geopolitical coordinates to publicly film and share their own experiences of uncovering scams perpetrated by so-called religious leaders. This is analogous to envisioning a mass movement by the 'multitude' (Hardt and Negri 2004, xiv) that organically co-opts popular media as a conduit to express 'counter-elite' narratives (Nandi and Devasundaram 2017). In this televised exposé, public agents across the nation are democratic and undiscriminating in their dismantling of charlatan conceits and sacred subterfuges in Islam, Hinduism, Christianity, and several other Indian religions. This libertarian mass investigation results in a nationwide unveiling of doctrinaire religious customs, rituals and superstitious practices that are out of step with modern life. It is notable that this transgressive and indeed revolutionary portrayal of popular mobilisation in the Indian public sphere is another divergence from Bollywood's universalised template. The global box-office success of the film also indicates that the external reception of Bollywood's monolithic brand image of India is not immutable, and that there is space for outliers and alternative narrations.

Ultimately, *PK* is a cogent example of a Bollywood film performing a subversion of its own encompassing genre's predictable metanarrative of validating hegemonic systems and practices. To reiterate, the trace of the radical and politically self-reflexive new Indie film genre inhabits mainstream *PK*. This 'adulteration' destabilises the purported purity and linearity of Bollywood's popular geopolitical branding as a cinematic mouthpiece for a majoritarian Indian narrative. In this regard, the dialectically delineated interiority/exteriority dynamic of India's image construction is perturbed and fragmented by *PK*'s displacement of the accepted order.

Connecting the dots

Bollywood's paradigm of operations has a dual configuration—a double-Orientalising schema, aimed at dominating the internal domain of the domestic Indian space and appeasing the external arena of popular global acceptance. In effect, Bollywood mythologisations and colourful fantasies are designed to encapsulate an inviolably majority Hindu version of traditional Indian culture, social mores, customs and values. To a significant extent, this representation is achieved by side-stepping critical engagement with urgent themes and issues confronting the nation. Bollywood's mono-cultural model that collapses an inestimably multi-dimensional and polytheistic nation, permits the productification, packaging, and proffering of a cohesive saleable national narrative to the outside world for popular consumption.

Bollywood is likely to continue to court the state's official neoliberal agenda and encourage an undifferentiated and essentialist vision of modern India, thereby re-Orientalising representation of India's national image. However, the 'contamination' of the commercial industry's apparently invincible narrative, through hybrid amalgamations with alternative and marginalised cultural forms, could spawn new counter-narratives and agents of cultural resistance. This thesis is affirmed by the rise of new Indian Indies such as *Court*, and by *PK*—an atypical Bollywood film that has been intentionally tinctured by Indie influence—to a point where any archetypal notion of infrangibility and identifiability of bona fide 'Bollywoodness' has been irrevocably subverted. The dialectical split—external and internal dimensions of both Bollywood and the Indian state, is therefore disrupted by the incursive intervention of third space cultural agents—the Indies. As new provocateurs providing alternatives to Bollywood's popular geopolitics, the Indie intermediaries not only distort the dyadic (dis)equilibrium of the official national image artificed by Bollywood and the state, but they also shatter the illusion of unity, self-Orientalising sameness, and rupture the purportedly timeless linearity of traditional values and religious practices.

It remains to be seen how Bollywood will navigate the internal fissures developing along religious and sectarian lines that jeopardise further, the integrity of its purported unity of representation, and therefore places in peril the continued veracity and representativeness of its popular geopolitics. It is important to ask what future mechanisms the Bollywood brand will deploy to continue to obfuscate the internal tremors rocking the 'Incredible India' narrative. Notably, newspapers have announced in 2016 that Pahlaj Nihalani, Bollywood producer and controversial chairman of censoring authority—the CBFC, is to produce a mainstream film about the Babri mosque demolition, starring Bollywood's most iconic actor, Amitabh Bachchan and popular action star Ajay Devgan (Kaur 2016). It is noteworthy that all three above-mentioned individuals have BJP leanings. Also of note is the galvanising of liberal sections of the nation's arts and culture community through popular gestures of protest that seem determined to address the transformation from incredible to intolerant India, in an increasingly parochial geopolitical space and place.

The arguments raised in this chapter indicate that Bollywood's invocation of a 'geopolitical aesthetic' is through the 'portrayal of "archaic" categories such as nationality and myth' (Dixon and Zonn 2007, 101). By contrast, the new Indies, of which *Court* is a pertinent example, extend a self-evaluation of the 'connection between the localised experience of the individual and the globalised totality that is late capitalism' (Dixon and Zonn 2007, 101). These films represent how 'states are historically constructed (and reconstructed) in the nexus between global and domestic/local social relations' (Agnew and Corbridge 1995, 16–17). *PK* falls into the fluid space of independent representation, where it positions itself as a Bollywood film, but nevertheless performs a narratival debunking of normative religious practices, nationalism and myth. Therefore, *PK* is a cogent cinematic example of how the accepted goalposts of popular culture can be pushed, or even uprooted.

This chapter demonstrates that films 'exist in an intertextual relationship with other geopolitical knowledges' and possess the ability to 'undermine and challenge hegemonic geopolitical discourse' (Power and Crampton 2007, 3). The discourses of *Court* and *PK* interrogate Bollywood's double narrative in their own ways. *PK* subverts the popular imagination of Bollywood's popular geopolitics through a discursive, deconstructive approach—one that uses the Bollywood's genre's own aesthetic of colour, myth, religion, mysticism and exoticism to destabilise the dominant Indian religious and geopolitical discourse. *PK* accomplishes this by revealing the shadowlines in Bollywood's double-Orientalising strategy, discursively interpellating both the internal Indian and external global representational strategies of conventional Bollywood.

New non-aligned independent films, such as *Court,* adopt a more direct, interrogative approach whilst performing the task of cinematically contesting dominant cultural narratives—of both Bollywood and the nation. These new films have in some part arisen from within the Bollywood morphology and territorial space. As the outside fringe of Bollywood's inside hegemony, these marginal Indie forms bear the potential of slowly diffusing into and influencing the mainstream, as much as the mainstream tries to subordinate and subsume the Indies. This increasing inter-osculation between India and Bollywood is signposted by *PK*, positioning the film as a bellwether of possible future realignments in Bollywood.

Notes

1 The extremist Hindu groups argued that the Babri mosque stood on the birthplace of Hindu god, Lord Ram. They claimed that sixteenth-century Muslim Mughal Emperor Babar had destroyed a Hindu temple that previously marked the spot and erected the mosque in its place.
2 The RSS, VHP amongst several other groups are collectively known as the Sangh Parivar—a family of Hindu nationalist organisations, with the BJP as the overarching political spearhead of the movement's ideology.
3 This film had earlier borne the brunt of an injunction from the Indian censoring authority—the CBFC—and was excluded from the International Film Festival of Kerala (IFFK) in 2012.
4 Arguably, this is similar to Russia's selection of *Leviathan* (dir: Andrey Zvyagintsev, 2014), a film largely antithetical to the grand nationalist design autocratically inscribed into the contemporary socio-political Russian narrative, but nevertheless dispatched to the Oscars; see relevant articles by Vlad Strukov in *The Calvert Journal* (2014) and in *The Conversation* (2015).

References

Agnew, John, and Stuart Corbridge. 1995. *Mastering Space*. London: Routledge.

Anderson, Benedict. 1998. *The Spectre of Comparisons*. London: Verso.

Anderson, Benedict. 2007. 'Benedict Anderson on Diaspora Nationalism'. *Euro Topics*, avilable at http://www.eurotopics.net/en/home/medienindex/media_articles/archiv_article/ARTICLE19423-Benedict-Anderson-on-diaspora-nationalism [last accessed 20 December 2016].

Banerjee, Mita. 2011. 'More than Meets the Eye: Two Kinds of Re-Orientalism in Naseeruddin Shah's *What If?*' In *Re-Orientalism and South Asian Identity Politics*, edited by Lisa Lau and Ana Cristina Mendes, 124–143. London: Routledge.

Basu, Amrita. 2005. 'Mass Movement or Elite Conspiracy? The Puzzle of Hindu Nationalism'. In *Comparative Politics Vol III: The Politics of Asia*, edited by Howard Wiarda and Lena Tan, 277–300. London: Routledge.

Bleiker, Roland. 2001. 'The Aesthetic Turn in International Political Theory'. *Millennium* 30 (3):509–533.

Bohland, Jon D. 2013. ''And They Have a Plan': Critical Reflections on *Battlestar Galactica* and the Hyperreal Genocide'. In *Battlestar Galactica and International Relations*, edited by Nicholas J. Kiersey and Iver B. Neumann, 98–118. London and New York: Routledge.

Burke, Jason. 2015. 'Narendra Modi Retains Core Support at Home as World Tour Reaches UK'. *The Guardian*, available at http://www.theguardian.com/world/2015/nov/11/narendra-modi-support-india-world-tour-uk [accessed 12 November 2015].

Chaturvedi, Vinay. 2012. 'Introduction'. In *Mapping Subaltern Studies and the Postcolonial, 2nd. ed.*, edited by Vinay Chaturvedi. London: Verso.

Chowdhury, Kanishka. 2011. *The New India*. New York: Palgrave Macmillan.

Devasundaram, Ashvin. 2016. *India's New Independent Cinema: Rise of the Hybrid*. London: Routledge Advances in Film Studies.

Dimitrova, Diana. 2014. 'The Politics of 'Otherness': The Hindi Plays of Urdu-Hindi Author Upendranath Ashk (1910–1996)'. In *The Other in South Asian Religion, Literature and Film*, edited by Diana Dimitrova, 84–99. London: Routledge.

Dittmer, Jason. 2010. *Popular Culture, Geopolitics, and Identity*. Lanham, MD: Rowman & Littlefield.

Dixon, Deborah, and Leo Zonn. 2007. 'Confronting the Geopolitical Aesthetic: Fredric Jameson, *The Perfumed Nightmare* and the Perilous Place of Third Cinema'. In *Cinema and Popular Geo-Politics*, edited by Marcus Power and Andrew Crampton, 95–120. London: Routledge.

Dudrah, Rajinder. 2012. *Bollywood Travels: Culture, Diaspora and Border Crossings in Popular Hindi Cinema*. London: Routledge.

Friedman, Lester. 2006. *Fires Were Started, 2nd ed*. London: Wallflower.

Ghosh, Deepshikha. 2015. 'Aamir Khan Stands By Comments, Says "Wife And I Don't Intend To Leave India"'. *NDTV*, available at http://www.ndtv.com/india-news/proud-of-being-indian-stand-by-everything-i-said-aamir-khan-1247513?pfrom=home-latest stories [accessed 20 December 2016].

Grayson, Kyle. 2013. 'Series Editor's Preface'. In *Battlestar Galactica and International Relations*, edited by Nicholas J. Kiersey and Iver B. Neumann, vii–xi. London: Routledge.

Hardt, Michael, and Antonio Negri. 2004. *Multitude*. New York: The Penguin Press.

Hase, Birgit. 2007. *Culture and Economic Development*. Frankfurt: IKO Verlag für Interkulturelle Kommunikation.

Haygunde, Chandan. 2015. 'Court Lead Actor Says Film Gave His Activism a Boost, Will Continue Fight Against Justice Delivery System'. *Indian Express*, available at http://indianexpress.com/article/cities/pune/court-lead-actor-says-film-gave-his-activism-a-boost-will-continue-fight-against-justice-delivery-system/ [accessed 20 December 2016].

Hirji, Faiza. 2010. *Dreaming In Canadian*. Vancouver: UBC Press.

India Today. 2015. 'From Condoms to Rakhis: Narendra Modi, The Brand'. *India Today*, available at http://indiatoday.intoday.in/story/from-condoms-to-rakhis-narendra-modi-the-brand/1/460490.html [last accessed 20 December 2016].

Kaneva, Nadia. 2012. 'Nation Branding in Post-Communist Europe: Identities, Market, and Democracy'. In *Branding Post-Communist Nations: Marketizing National Identities in the 'New' Europe*, edited by Nadia Kaneva, 3–22. New York and London: Routledge.

Kaur, Kiran. 2016. 'Amitabh Bachchan and Ajay Devgan to Team Up for a Film Based on Babri Masjid Demolition'. *Times of India*, available at http://timesofindia.indiatimes.com/

entertainment/hindi/bollywood/news/Amitabh-Bachchan-and-Ajay-Devgn-to-team-up-for-a-film-based-on-Babri-Masjid-demolition/articleshow/51922636.cms [accessed 17 June 2016].

Kishwar, Madhu. 2013. 'Bridging Divide: The Triumph of Bollywood'. In *On India: Self-Image and Counter-Image*, edited by Anindita N. Balslev, 72–89. New Delhi: Sage.

Kulcsar, Laszlo, and Young-Ok Yum. 2012. 'One Nation, One Brand? Nation Branding and Identity Reconstruction in Post-Communist Hungary'. In *Branding Post-Communist Nations: Marketizing National Identities in the 'New' Europe*, edited by Nadia Kaneva, 193–212. New York and London: Routledge.

Lefebvre, Henri. 1991. *The Production of Space*. Cambridge: Blackwell.

Moore, Cerwyn, and Chris Farrands. 2013. 'Visual Analysis'. In *Critical Approaches to Security*, edited by Laura J. Shepherd, 223–235. London: Routledge.

Nandi, Anik, and Ashvin Devasundaram. 2017. 'Contesting the Conventionalising of Castilian: The Role of Galician Parents as Counter-Elites'. In *Bilingualism and Minority Languages in Europe: Current Trends and Developments*, edited by Maria Couto and Fraser Lauchlan. Newcastle upon Tyne: Cambridge Scholars Publishing.

Nandy, Ashis. 2013. 'Politics of Democracy and the Politics of Religion in a Post-secular Age'. In *On India: Self-Image and Counter-Image*, edited by Anindita N. Balslev, 59–71. New Delhi: Sage.

NDTV. 2015. 'Aamir Khan Joins 'Intolerance' Debate, Says Wife Even Suggested Leaving India'. available at http://www.ndtv.com/india-news/aamir-khan-joins-voices-against-intolerance-says-wife-even-suggested-leaving-india-1246725 [accessed 20 December 2016].

New Indian Express, The. 2012. 'Cancellation Of Parallel Screening Of 'Papilio Buddha' Sparks Protest'. available at http://www.newindianexpress.com/cities/thiruvananthapuram/article1378964.ece accessed 20 December 2016].

Pillai, Meena. 2012. 'Post-national B(H)ollywood and the National Imaginary'. In *The Magic of Bollywood at Home and Abroad*, edited by Anjali Gera Roy, 42–56. New Delhi: Sage.

Power, Marcus, and Andrew Crampton. 2007. 'Reel Geopolitics: Cinemato-graphing Political Space'. In *Geopolitics*, edited by Marcus Power and Andrew Crampton, 1–11. London: Routledge.

Punathambekar, Aswin. 2013. *From Bombay to Bollywood*. New York: New York University Press.

Sage, Daniel. 2014. *How Outer Space Made America*. Farnham, Surrey, UK: Ashgate.

Said, Edward. 1978. *Orientalism*. New York: Vintage Books.

Saunders, Robert A. 2012. 'Brand Interrupted: The Impact of Alternative Narrators on Nation Branding in the Former Second World'. In *Branding Post-Communist Nations: Marketizing National Identities in the 'New' Europe*, edited by Nadia Kaneva, 49–78. New York and London: Routledge.

Saunders, Robert A. 2017. *Popular Geopolitics and Nation Branding in the Post-Soviet Realm*. London and New York: Routledge.

Schwecke, Sebastian. 2011. *New Cultural Identitarian Political Movements in Developing Societies*. London: Routledge.

Shapiro, Michael J. 2007. 'The Demise of 'International Relations': America's Western Palimpsest'. In *Cinema and Popular Geo-Politics*, edited by Marcus Power and Andrew Crampton, 29–50. London: Routledge.

Singh, Karan. 2013. 'Foreword'. In *On India: Self-Image and Counter-Image*, edited by Anindita N. Balslev, ix–x. New Delhi: Sage.

Smith, Laurajane. 2006. *Uses of Heritage*. London: Routledge.
Upstone, Sara. 2010. '9/11, British Muslims, and Popular Literary Fiction'. In *Reframing 9/11: Film, Popular Culture and the 'War on Terror'*, edited by Jeff Birkenstein, Anna Froula and Karen Randell, 35–44. New York: Continuum.
Widmalm, Sten. 2016. *Political Tolerance in the Global South*. London: Routledge.
Wildman, Wesley J. 2011. *Religious and Spiritual Experiences*. Cambridge: Cambridge University Press.
Zizek, Slavoj. 2009. *First as Tragedy, Then as Farce*. London: Verso.

8 'Directed by Hollywood, edited by China'?

Chinese soft power, geo-imaginaries, and neo-Orientalism(s) in recent U.S. blockbusters

Chris Homewood

Introduction

This chapter analyses the impact that shifting sources of box office growth are exerting on the messages Hollywood communicates about China to the Western(ised) world. It aims to provide conceptualisations for Hollywood's ostensibly affirmative remaking of space and place regarding China, which is being driven by the increasing economic heft of the Chinese marketplace in film distribution. On the one hand, the chapter argues that, with access to this lucrative market at stake, recent large-scale blockbusters have apparently dispensed with hoary stereotypes of China and Chinese characters, instead offering a more ecumenical understanding of global political space that transcends the dominant paradigm of patriotism/enemy othering and, crucially, even allows for the accommodation and endorsement of Chinese strategic narratives. On the other hand, the chapter interrogates the widespread notion that Hollywood's decision to alter long-standing norms when it comes to popular geopolitical representation has allowed China to straightforwardly co-opt the dream factory as an adjunct tool of Chinese statecraft, especially soft power. While Hollywood's most popular movies appear to be communicating many of the messages China wants the rest of the world to hear, my analysis will show how they nonetheless provide an updated Orientalist account of geopolitical culture that is intended to reinforce, rather than weaken, U.S./Western hegemony in global affairs.

Hollywood has long played a significant role in framing and shaping perceptions of China, its culture, and Western diasporas. Indeed, few readers will need reminding of Hollywood's persistent tendency across the twentieth century to imagine China and 'Chineseness' through the narrow cultural lens of Orientalism, Edward Said's (1978) term to describe a fearful and contemptuous regime of Western knowledge production that proposes, preserves and perpetuates assumptions about Western superiority by reifying an ontological distinction between itself and the East. Consistent with the purpose and organising principles of this volume, I share Joanne Sharp's view that Orientalism, 'although not labelled as such, was of course a powerful account of geopolitics' (qtd. in Van Efferink 2015). From MGM's *The Mask of Fu Manchu* (1932) to the 'moody orientalism' the geographies of dangers and threats confronted by the U.S./West. In short, the entity Hollywood calls China is no longer understood exclusively as a place of repression and totalitarianism. Instead of dramatising contentious issues and highlighting value conflicts, representations of Chinese places and people in U.S. movies appear to contest dominant political understandings of the PRC as a threatening other. However, politically this development has been perceived as a double-edged sword. Whilst few would lament the decline of singularly negative and often nakedly racist assumptions about Chinese (American) identities and cultural mores, some believe that Hollywood's present understanding of the PRC is saturated with negative ideological consequences for the U.S. With dollars at the stake, the major film studios are certainly engaging in self-censorship, softening or eliminating any and all politically sensitive material that could be perceived by Chinese state media regulators—who remain candid in their appeal for 'positive Chinese images' (Pulver 2013)—as painting the Chinese government in a disapproving light. In other words, rather than submitting a fair but critical representation, Hollywood is instead offering a financially motivated 'positive' appraisal that is wilfully complacent regarding human rights abuses and other problems in the PRC. As a result of this, voices in the media, branches of U.S. government, and even in academia have expressed concern that the Majors' growing dependence on Chinese ticket sales is handing Beijing a 'tacit but diffuse influence over Hollywood' that is 'changing the complexion of American movies' (Zara 2015), with analysts such as Robert Daly (2016), Director of the Kissinger Institute on China, suggesting that a remiss Hollywood is no longer promoting foremost the global interests of the U.S. but rather those of its 'most formidable strategic, economic and ideological competitor'.

Through a discussion located at the intersection of popular culture and geopolitics, this chapter identifies and interrogates a significant reorientation of the politics of space, place, and identity regarding the entity Hollywood calls China. Drawing on a variety of textual examples, but focusing on the Sino-American 'joint ventures' *Looper* (Rian Johnson, 2012) and *Transformers: Age of Extinction* (Michael Bay, 2014), I argue that recent U.S. blockbusters present China as a genuinely novel example of what we might deem the 'new popular geopolitics', wherein the dominant framework is remade to achieve discreet outcomes beyond patriotism/enemy othering. Unless otherwise motivated, Hollywood tends to do what is easy, relying on stereotypes about foreign peoples and polities that implicitly nourish a U.S./Western perspective, but market forces are urging the dream factory to adopt a more ecumenical understanding of global political space. The films under question reframe China in an affirmative or approving light, creating a framework in which popular ideas about China as belonging to the 'Orient' can be contested rather than reinforced and aspects of Chinese foreign policy discourse—especially Chinese strategic narratives, which is to say the story, or stories, a nation must tell the world, and itself, to maintain its competitive edge in the international system—can be accommodated and endorsed.

Through this move away from inimical to more cordial patterns of recognition, recent representations of China therefore appear to be aimed at aiding the PRC in the task of building confidence, legitimacy and influence on mass, even elite, levels in key countries and regions around the world, especially North America and Western Europe. As such, I am therefore also interested in testing claims that the astonishing rise of the Chinese film market has allowed China to co-opt Hollywood as an adjunct Chinese nation branding and soft power asset (see, for example, Richburg 2010; Barboza 2015). As Saunders notes, these two concepts are frequently, yet mistakenly, conflated: 'Whereas soft power is a preferred tool of those seeking to motivate other states (usually in a clearly defined "sphere of influence") *to do something*', nation branding aims more to motivate others to '*buy, feel, or visit something*' (2017, 71, my emphasis). Although I maintain this distinction, my argument reflects the way that, as tools of motivation, the two concepts can often establish a reciprocal relationship. With this in mind, I therefore theorise how, on the one hand, recent U.S. movies, understood as a form of mediated cultural diplomacy, encourage us to think and feel differently (i.e. better!) about China and attempt to buoy the development of a sphere of Chinese influence in the Western(ised) world. On the other hand, the chapter problematises straightforward assumptions that Hollywood's apparent change of heart regarding China is handing Beijing the upper hand in global affairs.

In response to Klaus Dodds's call for greater 'attentive[ness] to the nature and practices associated with film' in popular geopolitical analyses of the medium (2005, 268), the discussion will consider not only content (what the films say) but also the formal devices (how they say it) which underpin Hollywood's attempt to rebrand all things Chinese in an attractive light. To this end, through close textual analysis the chapter will also argue that Hollywood's shift to practices of 'positive' recognition regarding China are redolent of aspects of a 'new' form of China-specific Orientalism, identified by Daniel Vukovich (2012), the operational logic of which permits movies to simultaneously promote and attenuate the impact of the messages the PRC wants the world to hear. Although recent blockbusters upend dominant paradigms, ultimately they do so without threatening the superstructure of U.S./Western hegemony.

Accommodating and endorsing Chinese strategic narratives

A concept developed by political scientist Joseph S. Nye (2004), soft power is the ability of a country to persuade others to do what it wants through the maintenance and projection of a positive national image. Unlike hard power, which rests upon the use of coercion and force, soft power is the ability to entice and attract others to what you want. For Nye, 'culture, values and foreign policies' are the primary sites of soft power, with culture standing as the most important (2004, 11). In a recent article, Laura Roselle et al. reflect growing opinion when they suggest that soft power, in its current, widely understood form, 'has become a catch-all term that has lost explanatory power, just as hard power once did' (2014, 70). This is because '[a]nalyses of soft power overwhelmingly focus on soft power "assets" or capabilities and how to wield them, not how influence does or does not take place' (2014, 70). In a bid to square this conceptual circle, the authors advocate that more attention should be paid to strategic narratives because they can provide 'intellectual purchase on the complexities of international politics today, especially in regard to how influence works in a new media environment' (2014, 70). For this reason, they suggest that discussions of soft power should focus on 'the analysis of the formation, projection and – critically – the reception of strategic narratives and the interactions that follow' (2014, 71). Although it is too early to be able to measure the reception (which is also to say the benefits) of China's use of Hollywood as an 'asset' for the transmission of Chinese strategic narratives, it is nonetheless possible to investigate the formation and projection of what China hopes will prove to be winning stories in the context of American movies. As Hartig observes, the first the first of these is actually:

> [A] narrative used by other states: a narrative about China, exemplifying how other states such as the US think China may behave in the future, how they wish it would behave and how other states aim to deal with such a rising China. In this narrative, China is described and perceived 'alternatively as an aspiring normal great power to balance others or as a rising hegemon'.
> (2015, 49)

However, as Hartig also notes, this 'international system narrative about China' has at the same time been largely internalised by China, becoming closely related to its self-understanding and thus the principal narrative communicated by the PRC, which 'is based on the idea of Peaceful Rise/Development' (2015, 49). Herein, China projects a point-of-view that 'is primarily about reassuring the rest of the world that the rise of China will not pose a threat to peace and stability and that other nations will actually benefit from China's growing power and influence' (2015, 249). Overwhelmingly, recent U.S. blockbusters foster this merged point-of-view by endorsing the (self-)perception of China as a rising hegemon in depictions that are simultaneously intended to allay fears in other states about a China threat.

This shift to a more inclusive geopolitics that exceeds a patriotism/enemy othering binary is in large part predicated on the genre of science fiction, which shares a long, complex and sometimes contradictory relationship to world politics (see Weldes 2003; Dittmer and Gray 2010; Dittmer and Dodds 2013). The focus of research in not only film and cultural studies but also popular geopolitics, science fiction, as Saunders reminds us, 'is a genre of *space* (terrain, topography, 'zones', etc), as well as *outer space*', which since the Cold War has 'steadily morphed into a medium for global ideological contestation and identity negotiation' (2015). To date, the vast majority of recent U.S. movies to feature Chinese story elements are structured around the imaginative concepts (such as future science and technology, space exploration, extra-terrestrials, and time travel) and spaces (alternate or (near) future versions of earth that are similar but different, and/or outer space), which describe science fiction, and on the one hand, the movement of China away from Earth and into other spaces is providing fertile ground for the transmission of messages the Chinese polity wants the rest of the world to hear.

Space capability is considered to be an attribute of great power and indication of global leadership status, and after years of investment and strategy China has its eyes set firmly on becoming a space superpower. As Dillow et al. (2016) suggest, 'To gain an edge here on earth, China is pushing ahead in space'. Running contrary to alarmist readings of China's amped-up space programme in the U.S. news media, which perceive a belligerent challenge to U.S. dominance (Baculiano 2016), recent Hollywood movies draw on what Strukov and Goscilo term the 'hierarchical iconography and discourse of ascent/descent' to reproduce the beliefs of Chinese polity and grant China legitimacy in the area of space flight (2017, 1). China's 'heavenly rise' (Dillow, Lin, and Singer 2016) is perceived not as a threat to U.S. hegemony, or global peace and stability, but as a benefit. In *Gravity* (dir. Alfonso Cuarón, 2013), Sandra Bullock's stranded astronaut saves herself by utilising a planned but as-yet fictional Chinese space station, *Tiangong* ('heavenly palace'), and lands on Earth in a Chinese space capsule, *Shenzhou* ('Divine Craft'). Along similar lines, the aid offered to the U.S. by the Chinese space agency proves pivotal in the rescue of another stranded American astronaut in *The Martian* (dir. Ridley Scott, 2015). Meanwhile in the more fanciful *Independence Day: Resurgence* (dir. Roland Emmerich, 2016), the fictional Earth Space Defence (ESD) Moon Base (the globe's first line of defence against alien threats) is commanded by China's Jiang Lao (depicted by Singapore-born actor Chin Han), and in *Arrival* (dir. Denis Villeneuve, 2016) non-Earthbound safety concerns on global security policy discourse are used to ratify the centrality of Chinese geopolitical decision-making. These and other imaginative attempts to project China's point-of-view and shape an approving perception of the nation's role in global affairs are contingent on Hollywood's sidestepping of contentious issues and measured refrainment from a great many of the perceptibly negative stereotypes associated with Orientalist distortion. This is especially the case in two movies made under the auspices of China Film Group Corporation's (CFGC) Sino-foreign 'joint venture' agreement:[3] *Looper* and *Transformers: Age of Extinction* (hereafter *Transformers 4*), both of which pay particular attention to China's place in global political space.

In *Looper*, the U.S. of 2044 is depicted as a failed superstate beset by economic and social collapse, and it appears to have ceded its position as world centre power to China, which is depicted less as a rising than a risen hegemon. However, the notion of China's ascendency provokes neither fear nor contempt from within the diegesis of the film. Instead, it 'just is', a fact that becomes all the more significant if one considers the heavy debt of aesthetic inspiration that *Looper* owes to Ridley Scott's seminal *Blade Runner* (1982). The latter secured its place in film history both as a cornerstone in the pantheon of American science fiction cinema and a 'techno-Orientalist' urtext. Coined by Morley and Robins, techno-Orientalism describes a prevalent form of anti-Japanese and anti-East Asian racism that appeared in the 1980s, when Japan came to assume a hegemonic position in the spheres of technology, manufacturing and finance that disturbed Western notions of modernity by 'beat[ing] the West at its own game' (Park 2010, 8). According to Morley and Robins, this development promoted an array of stereotypes and deformations about Japan in Western (popular) culture, so that the country came to epitomise a hyper-technologised, dehumanised, and threatening society (1995, 147–173). The postmodern *mise-en-scène* of *Blade Runner* is cluttered with the 'enigmatic signifiers' that constitute King's notion of 'ornamental Orientalism' (2010, 3), and as Cornea has observed, the 'Oriental' within this futuristic city 'forms the underbelly of a commercial and technological culture [...], which suggests that the threat to human authenticity that the Replicants represent is associated with an underlying Eastern menace' (2007, 194). Shot almost exclusively at night, the downtown area of Kansas City in *Looper* borrows liberally from the claustrophobic, neo-noirish labyrinth of imposing skyscrapers, mob dens, seedy clubs, and the kind of 'accumulated progress' that describe the dystopian setting of *Blade Runner*. But crucially, and in stark contrast to Scott's film, this fetid warren of social decline is at the same time devoid of Asiatic signifiers of any kind. The bleak prediction of a dire American socio-economic future in *Looper* is constructed as a downfall created from *within*, rather than a result of East Asian/Chinese 'contamination' from without.[4] Although China appears to have taken possession of the future, it does not arouse fears and anxieties in the Western psyche, but instead becomes a welcoming site of escape for America's citizens. In at least one of the timelines conjured into existence by the film's complicated premise, China is coded as a redemptive space for the film's protagonist—the dissolute American hitman, Joe.[5]

Due to its focus on malevolent artificial intelligence in a (partial) East Asian setting, *Transformers 4* contains the ingredients for a techno-Orientalist recipe with Chinese characteristics. However, echoing *Looper*, the story chooses not to associate a technological threat to humanity with an underlying Eastern menace, relying on the Chinese less for their innovator intelligence around technological development than as a, cheap, efficient, and benign source of manufacturing expertise. Instead, it is rogue elements within U.S. government and industry who endanger the world. A self-important and arrogant Wunderkind, Joshua Joyce (Stanley Tucci), CEO of the fictional American tech-giant Kinetic Solutions Incorporated (KSI), is crafting man-made Transformers for the U.S. military, but as one might anticipate, his creations quickly turn on their master, whereupon they attempt to detonate 'the seed' (an extinction-level, bomb-like MacGuffin capable of transmuting organic and other materials into 'transformium') in China, home to KSI's main manufacturing plant. The conception and fatal implementation of the evil transformers is born of Joyce's corrupting egotism and attendant loss of moral and ethical standards. In order to gain access to the remnants of the alien technology that litter Chicago at the end of *Transformers: Dark of the Moon* (2011), Joyce sells his soul to immoral representatives of American Government: the unscrupulous and uncompromising CIA agent Harold Attinger (Kelsey Grammar) and his callous right-hand man, field agent and former Special Forces soldier James Savoy (Titus Welliver), both of whom pose as great a menace to the world as the mechanical monsters they wilfully unleash upon it. By figuring this threatening surfeit of oppressive state power (traditionally a chief signifier of China's dangerous Otherness to the West) not as a Chinese but an

American problem, the film implicitly endorses a view of China as benign. It even weakens the positional superiority of the U.S./West *vis-à-vis* China, which more than passive victim of venal U.S. folly assumes an active role in helping to right American wrongs.

However, aside from an isolated mention of China's central government (which amounts to a naked moment of pro-Beijing propaganda),[6] *Transformers 4* circumvents the need for a direct depiction of Chinese state power by focusing its attention on Su Yeuming, a non-state actor who, as the only prominent Chinese character in the film, comes to symbolise China. Portrayed by Chinese star Li Bingbing (recognisable to Western fans of Hollywood action cinema as Ada Wong from the *Resident Evil* franchise), Su is the head of security at KSI and also appears to act as a personal assistant to Joyce. However, despite working closely with Joyce, Su is positioned firmly outside the American web of corruption that surrounds and consumes her boss. Camera and editing take care to highlight her censorious facial expressions as she becomes aware of Joyce's nefarious dealings and the 'illegal, icky shit', as Joyce himself later terms it, occurs in a proximate space to her. Physically powerful and morally upright, Su epitomises strategic narratives about China's peaceful and world-beneficial rise/development, exerting a positive influence over her errant American boss and playing an important role in restoring world security. Leading by the force of her principled example, she provides co-stimulus—alongside the film's hero, Yeager (Mark Wahlberg) – for Joyce's halting moral and ethical rehabilitation, dragging him (literally at times) to his redemption at the film's close. In this way, Su is even granted dominance over her delinquent boss (see Figure 8.1).

As the film progresses, Su and Joyce find themselves thrust into a fractious biracial buddy relationship, whereupon it becomes her task to get Joyce and the seed (which, having begun to see the error of his ways, he has stolen back and reluctantly carries) across Hong Kong to a rooftop extraction point. Speaking of biracial buddy narratives, Ed Guerrero explains how unequal power dynamics between white and non-white characters typically serve to contain the latter, placing the non-white character 'in the protective custody [...] of a white lead or co-star and therefore in conformity with white sensibilities and expectations of what [non-white] characters should be' (1993, 128). In the case of this pairing, however, *Transformers 4* erodes racialised/Orientalised expectations of gendered behaviour and disturbs the uneven balance of power they typically engender by deploying a reversal that places Joyce, a male, white American co-star, in the protective custody of Su, a female Asian co-star. Punished in the film for his loss of moral and ethical equilibrium, Joyce is subjected to a strategy of feminisation and emasculation. Prone to emotional instability, irrationality and moments of hysteria, he exhibits behaviour more commonly ascribed to women in an action context—shrieking his way through the streets of Hong Kong and cowering from the violent action that swirls around him. In contrast, Su exhibits a calm control, and is constructed as what Purse would term 'an extremely potent action body': displaying 'agency, physical strength and skill', she protects Joyce from a squad of Attinger's American assassins (2011, 14). The unconventional buddy

Figure 8.1 Filmstill from Michael Bay's 2014 *Transformers: Age of Extinction*. Su takes charge of Joyce.

relationship between Su and Joyce emphasises and strengthens the perspective of China within the film. It creates an equally unconventional instance of intersectionality, in which gender and ability interconnect and overlap to create an interdependent system of dominance, but one that actually disadvantages the ordinarily hegemonic party. The pairing of a physically capable, virtuous Asian female—implicitly symbolising China, and embodying strategic narratives that stress China's globally beneficial role—with an ethically and physically weak white American male (the conventional trifecta of Western supremacy *vis-à-vis* 'the Orient') destabilises conventionally racialised hierarchies of Western dominance. Authority is divested from a delinquent representative of U.S./Western power and invested in a Chinese woman, thereby denying the U.S. the moral and positional upper hand in its relationship with China. Within this relationship at least, China is framed as the dominant positional entity.

Sinological-Orientalism: discreetly safeguarding U.S. positional superiority

Although recent U.S. blockbusters accommodate and endorse Chinese strategic narratives and even create moments of Chinese (moral) supremacy, they nonetheless operate ultimately to hold China in a dependent role *vis-à-vis* the U.S./West. The main difference to their twentieth-century predecessors, however, is that now they do so discreetly, which is to say without backsliding to old Orientalist practices of negative recognition regrading China and East Asia, or perhaps more accurately, without deploying those practices which are most likely to cause perceptible offence and therefore run a greater risk of harming Hollywood's market ambitions. This section therefore investigates Hollywood's strategies for the simultaneous promotion and containment of China and its message.

The first and perhaps the most straightforward of these is the use of outwardly racially and politically benign narrative and genre devices. Although the concepts and spaces associated with science fiction do on the one hand aid the promotion of China's point-of-view, on the other hand they also provide the means to simultaneously constrain its impact. For instance, *Looper* uses temporal paradoxes to furtively cast doubt on its erstwhile prediction of China's rise to global hegemony, containing it in a future that was but may never be. The second strategy is linked to tokenism, an enduring issue in Hollywood (see Yuen 2016). Despite Chinese characters and story elements becoming a mainstay within the breadth of Hollywood's blockbuster output, the depth of the individual representations is often perfunctory, reflecting a tick box approach to courting Chinese audiences and, especially in the case of intended joint ventures, Chinese regulatory requirements. American films that contain Chinese elements are precisely that, and Chinese characters are typically relegated to small supporting roles. Revealing of the way Hollywood's current geopolitical culture regarding China is still harmonious with that of the U.S., movies deploy the hierarchy governing acting roles to safely contain powerful Chinese characters in parts subordinate to those of their leading white American counterparts. Thus, whilst U.S. movies endorse and project a message that China wishes them to disseminate, they nonetheless also quietly renege on the idea of a major reorganisation of global political space. In what becomes a form of U.S. hegemonic wish-fulfilment, they reinforce prevailing power structures by having China model the preferred geopolitical order. Specifically, these films present an idealised version of the U.S.'s immediate post-millennial desire to develop a strong and stable alliance with China (as a foundation for the protection and promotion of its own national interests), which accepts China as a responsible stakeholder in global affairs, but *under* U.S. leadership. Reflecting the sentiment of a 2009 U.S. Centre for Strategic and International Studies (CSIS) report that 'Virtually no major global challenge can be met without U.S.-China cooperation', movies that imagine an American-led response to global threats and challenges (*The Arrival*; *X-Men: Days of Future Past*; *Independence Day: Resurgence*, and so on) invariably position China as the U.S.'s primary ally, even if the screen time is brief. Consider Fan Bingbing's brief turn as the heroic mutant Blink in *X-Men Days of Future Past.* That Blink is strong and virtuous is not brought into question, but (while she survives longer than the other minority and ethnic mutants) she ultimately sacrifices her life to buy the principal white American and Western European characters the time they need to prevent a dystopian future from ever coming to pass. Blink's death may be motivated rather than gratuitous, but it remains inevitable. Another cipher for China, she is accepted as a powerful yet non-threatening ally, and in line with the U.S. geopolitical imaginary, America leads the way whilst China provides a vital but supporting role.

At the same time, this positively framed but safely contained understanding of China is reliant on a 'new' and China-specific form of Orientalist knowledge, which not only serves as a form of containment in and of itself but also permits Hollywood access to new, at least where China is concerned, strategies of containment that, moreover, function to mask the racialised political anxieties underpinning them.

In *China and Orientalism: Western Knowledge Production and the PRC*, Vukovich reworks the post-colonial concept of mimicry—which describes the way a dominant group (the coloniser) actively encourages marginal groups (the colonised) to adopt its cultural habits and values—to argue that since the programme of Chinese economic reform and 'opening up' initiated in 1978, Western knowledge about China has been dominated and defined by what he terms 'Sinological-Orientalism'. Previously the irreducible Other and thus the location of absolute difference to the West, China is instead construed as the place of 'becoming-sameness' to the West (Vukovich 2012, 2). By this Vukovich means that China remains the Other—it is still not 'normal'—but is now placed within a scale of hierarchical difference in which 'China is seen as in a process of haltingly but inevitably becoming-the-same as "us": open, liberal, modern, free. Put another way, China is understood as becoming generally equivalent to the West' (2012, 1). However, China therefore also finds itself always stuck in the realm of the 'not yet': it is perceived by Western observers as wanting to become the same as the West but always lagging behind or lacking (2012, 3). As such, a paradox emerges in which discourses suggesting China is 'becoming democratic, normal, civil, creative artistic (avantgarde), liberal, and so on' simultaneously aver that China 'still lacks something (often the same items)' (2012, 9).

The knowledge produced in Hollywood's recent textual encounters with China certainly conforms to aspects of the operational logic of Sinological-Orientalism. I provide a caveat here because, stemming from the regulatory requirements for positive Chinese images, the ordinarily attendant notion of ontological Chinese lack is itself lacking, and in this way the hierarchical scale of racialised difference that underpins the idea of China's march to sameness is apparently dissolved. Thus, rather than being stuck in a perpetual, stumbling state of *becoming* generally equivalent to the U.S./West, Hollywood movies tend to interpret China as already having *become* 'normal'. Thus, despite China's economic leverage of Hollywood to transmit its voice to the world, the voice we hear is not China's own (even if it does reflect aspects of China's self-understanding) but rather the continued product of an authoritative American lens whose willingness to re-brand everything 'Chinese' in an approving light is contingent on the innately 'denigrating and condescending faith that they are, after all, becoming-the-same as us (or should be made to do so)' (Vukovich 2012, 3). Despite Beijing's alleged influence, today Hollywood is no more inclined to cede legislative rights to China than it was during the last century.

Enacting the 'should be made to do so' of Vukovich's statement, movies tend to decouple China from the perceived sources of its lack (in terms of both imaginative, old Orientalist markers of Chinese difference and its actually existing differences to the West) and impose in their place Western ethical and cultural systems. China is reframed in terms of its (becoming-)sameness to the West, and thus ascribed a politically, culturally, and even racially neutered identity. In effect, we are presented with a Chinese version of a Western cultural model. Regarding political values, doctrines and paradigms this is often achieved tacitly, by act of omission. In the absence of traces of China's autocratic system of governance (evidence of the nation's actually existing difference to 'us'), 'universal' U.S./Western doctrines and values fill the void—whether Su Yeuming or Blink,

Chinese characters intuitively adhere to U.S./Western political and ethical values and norms. Where social constructions of Chinese identity are concerned, sport becomes a prime ground for the overwriting of Chinese with Western cultural characteristics. Although a wide variety of sporting activities, both Western and traditional Chinese, have become popular in China, films tend to stress the former: in *Looper*, young Joe plays soccer with a group of children in Shanghai, and the sport of American football features prominently in *Transformers 4*. Here Yeager uses a Chinese teenager's American football to triumph in his concluding fight against the depraved Savoy. In sum, the victory of the film's principal American hero is underwritten by Chinese 'provision', which is itself made possible by China's growing cultural equivalence to, and resulting compatibility with, the U.S./West (see Figure 8.2: China's 'sameness' underwrites Yeager's U.S. victory).

Transformers 4 continues the Western sporting theme with a cameo. During a brutal encounter with Attinger's assassins in Hong Kong, Su is assisted by a local good Samaritan, who is played by Chinese professional boxer and double Olympic champion Zou Shiming. Although the use of a practitioner of a Western martial art dodges the potential of stereotyping (the 'all Chinese people know Kung-Fu' trope), which the film's setting (Hong Kong) and certain of its genre affiliations (action) might otherwise easily invite, the pervasiveness of Westernised conceptions of China and Chineseness nonetheless underscores the production of implicitly, if less perceptibly, disparaging Orientalist knowledge.

Even when largely divested of the notion of ontological Chinese lack, Hollywood's recent pursuit of knowledge which insists that China is the same as the U.S./West appears to afford Hollywood a variety of means to simultaneously promote and (regarding the preservation of U.S. global hegemony) disadvantage China's point-of-view without compromising its own market ambitions; perhaps evidence of creeping Orientalist epistemes, the representations discussed here are, after all, considered roundly acceptable by Chinese authorities. The idea of China's equivalence to 'us' also holds implications for the issue of soft power.

Figure 8.2 Filmstill from Michael Bay's 2014 *Transformers: Age of Extinction*. China's 'sameness' underwrites Yeager's U.S. victory.

Representations that remake China in the image of the West may—somewhat perversely—help foster soft power outcomes. If, as Nye contends, 'popular culture is more likely to attract people and produce soft power outcomes in situations where cultures are somewhat similar rather than widely dissimilar' (2004, 15), then Hollywood's shift from negative to 'positive' practices of recognition that, even when predicated on Orientalist knowledge production, insist on China's thoroughgoing cultural compatibility with the West may yet aid the PRC in its desire to cultivate a friendlier international environment and improve its image in the Western(ised) world. But of course, the ability of this approach to generate attraction and influence could at the same time be cancelled out by the numerous contradictory positions proposed elsewhere in Western popular culture. In more complete accord with the operational logic of Sinological-Orientalism, although Western news media reports do not deny China its achievements, they tend to focus on the notion of the nation state's lack *vis-à-vis* the West. And then there is also the question of credibility, upon which soft power relies (Nye 2011, 83).

If information is suspected to be propaganda, its credibility, along with its ability to generate influence, is lost, and it remains to be seen whether Hollywood's thorough-going erasure of not only the actually existing traces but also the consequences to U.S.-Sino relations of China's political differences to the West will serve the inverse of its intended function. The tensions between Chinese autocracy and Western democracy, communist and capitalist ideologies and the uncertainty these create around the future trajectory between leaders in Washington and Beijing could become conspicuous by their absence, generating scepticism rather than confidence. Lastly, the overwriting of Chinese culture with a Western model also poses questions for China's nation brand. Even if the pursued notion of Chinese compatibility with the West were to prove capable of generating soft power outcomes for China, that this attempt at persuasion relies on placing China in conformity with Western expectations of what it should become and envelops not only matters of political values and foreign policy but also the arts, customs, and so on, stands at odds with the unique cultural heritage China also wants to promote to the world in support of its nation brand. The notion of cultural exchange—a fundamental prop of soft power, nation branding, and cultural diplomacy alike—is to some extent undercut by a largely one-way current of flow that favours the dominant global entity.

Speaking of dominance and the safeguarding of prevailing power structures, there is one important area where Hollywood's variation on Sinological-Orientalism is indebted to a lingering 'and especially powerful discursive trope of old Orientalism'—namely the 'feminaization of East Asian nations and their cultures' (Leong 2005, 2). Park explains this trope thusly:

> The normalizing gaze of the West reduces the [Orient] to a racialized object and women to sexualized objects. In other words, Orientalism is gendered: the East is figured as the eroticized, feminized other that exists to be known, penetrated, and subordinated by the masculinized West.
>
> (2010, 3)

Apparently unsure about casting Chinese men, the sheer predominance of Chinese women in supporting roles points to the perpetuation of this trope. There are, however, some noticeable differences. Consistent with the operational logic of China's (becoming-)sameness to the West, race remains an unnamed signifier. Whilst in many cases they remain sexualised, they are at the same time de-exoticised and permitted to confound racialised expectations of gendered behaviour. For example, although Su's sway over Joyce is strengthen by his romantic infatuation (he is fascinated by her physical strength and beauty), she does not subsequently morph into the so-called 'Asian Pocahontas'. A victim of 'racist love', the Asian Pocahontas is a 'selfless caregiver of colour', who 'lives for and through an extraordinary white man' (Park 2010, 89). Resisting the libidinal contingencies of Joyce's desiring white male gaze, Su remains romantically unavailable throughout the film, and instead of becoming a 'diverted' Asian mother whose actions are driven by an emotive, 'blind protective instinct', the attention Su pays Joyce is born of grudging professional necessity and maintained by her vigilant sense of corporate social/global responsibility (Wilkie 2003, 57). Furthermore, while she 'help[s] birth and rear the new and improved [...] white male', she does so not by teaching him the 'spiritual, erotic, and kinetic secrets of the Far East' but rather by helping to (re-)educate him in the Western ethical system he should represent but fails to uphold (Park 2010, 88). The fleeting representation of old Joe's Chinese wife in *Looper* runs along similar lines. Although she is linked to nature, her salvation of Joe is predicated on a universal, rather than racially distinct, understanding of the love she gives him.

Reluctant to apply recognisably racialised stereotypes of gendered behaviour to regulate the power bestowed on physically capable, even superhuman, female Chinese characters, U.S. movies are instead making use of the conservative, yet ostensibly non-racialised, gender politics that govern Hollywood action cinema as pretence to quietly disadvantage China and recuperate the 'sameness structured by hierarchical difference' that safeguards the positional hegemony of the U.S. In *Contemporary Action Cinema*, Purse (2011), draws on the work of Yvonne Tasker to expose and critique the uneven distribution of power in recent examples of the action genre. Purse argues that Hollywood action cinema, motivated by an underlying conviction that women should not be muscling in on traditionally male territory, frequently deploys tropes designed to contain or compensate for the threat posed by the presence of physically powerful women to the genre's structuring fantasy of male hegemony. As Purse illustrates, these strategies serve to disadvantage the action woman by diminishing the plausibility of her feats in relation to her male counterpart, on whom the gender-biased 'privilege of ultimate empowerment' is ordinarily conferred (Purse 2011, 14).

In *Transformers 4*, this recipient of this privilege is Yeager, a struggling mechanic and aspiring inventor who hails from small town Texas. Yeager does not develop during the film, which he begins and ends as a paragon not only of masculinity but also American virtues. Thus, while the film may grant Su dominance over Joyce, she is not permitted to outdo Yeager. Consistent with the action genre's structuring fantasy of male hegemony, Su is not only positioned as subordinate to Yeager by dint of her supporting status but also on what Purse terms the 'credibility continuum', along which 'the relationship of the hero's feats to real-world laws of physics and physiology run' (2011, 79). Although the physics governing both Su's and Yeager's action exploits occupy a position close to the naturalistic end of the credibility range, the film weakens the authority of Su's physical power by depicting her exploits as having less plausible physiological consequences. Whereas Yeager is 'permitted to sweat, strain and be bloodied as he engages in combat', Su is subjected to action cinema's 'persistent tendency to downplay significantly the physical consequences of [female] action, such as pain and injury' (Purse 2011, 81). (She walks away from what should be a bone-crunching brawl with Attinger's men without displaying physical consequences.) As Purse notes, 'the investment seems to be in preserving the woman's status as sexual object in spite of the action, so that she can retain her erotic appeal.' The Teflon-like durability of the action women becomes a means to re-inscribe her in 'the traditionally binary conceptions of gendered behaviour' that her potent agency disturbs. Thus, when exhibited by a woman, the traditionally masculine, 'heroic qualities of toughness and resolve are re-framed in relation to gender-biased notions of decorum: grace and dignity; perfect hair and make-up; and unconvincing physical action' (Purse 2011, 82).

Transformers 4 does little to disabuse us of Su's correspondence to this 'sexualised object/active subject dualism' (Purse 2011, 85), but although she is posited a 'fetishistic figure of fantasy' for the spectator (Tasker 1998, 69), her erotic appeal is not constructed along the exotic, racialised lines that typically inform representations of Asiatic women. Instead, Su, consistent with homologising logic of sameness is treated equivalently, as an equal among her already unequal white action women counterparts. The intersectionality of gender and race is in operation (to the extent that the insistence on holding Su in a position of less authority *vis-à-vis* Yeager reflects implicitly racialised U.S. geopolitical anxiety about the threat a rising China may pose to its global dominance), but is masked by the film's observance of non-culturally specific tropes, which diminish the threatening presence of powerful women *in the main*. Although *Transformers 4* uses gender status to quietly disadvantage Su in relation to Yeager, it notably spares her from those strategies that might threaten to undercut significantly her erstwhile empowerment. Su's subjugation to Yeager is not made explicit; in fact, the two characters hardly come into direct contact, and the film resists the genre-conventional strategy of re-containing the action woman by awarding her to the male hero at the film's close (Purse 2011, 84–85). Neither Yeager nor Joyce (the most likely candidate given his amorous intent) gets the only available 'girl'. To have moved Su to a heterosexual union with either Joyce or Yeager would have risked the revivification of racialised tropes about subaltern Asian women which the film otherwise forecloses on. Neither does she have to make way for the primary male hero through death, which as Purse reminds us, and as we witnessed with Blink, is an equally common outcome for action heroines (Purse 2011, 84). Instead, Su is permitted to coexist alongside Yeager as a largely self-governing and powerfully capable, but never quite as credible, action body.

U.S. 'smart' narratives and China

Thus far, I have considered how Hollywood has promoted Chinese national interests whilst quietly preserving the dominance of the U.S. in the existing geopolitical order. In this final section, I develop the latter idea further, by exploring Hollywood's strategic pursuit of concrete benefits for the U.S. in the form of rebuilding confidence, legitimacy, and political influence in countries and regions around the world, and not least China, where simmering tensions between the two nations can make real-world examples of the cooperation imagined by U.S. blockbusters elusive.

The U.S. invasion of Iraq in 2003 as part of the ill-defined 'war on terrorism' stands as an example for the ineffectiveness of basing foreign policy solely on hard power strategies. According to Steinberg, 'the strategy [of military intervention] failed to understand what elements of power were needed most to defeat the emerging threat' (2008, 159). Summarising Steinberg's argument, Wagner (2014) explains that:

> This misunderstanding resulted in ignoring two key elements of soft power: the Bush administration firstly forgot about the USA's dependence on their allies' intelligence and policy forces and on global public support; and secondly, the question of the legitimacy of the invasion was not attributed any importance (Steinberg 2008, 160). In the short term, these mistakes led to the failure of the action. In the long term, they have caused the degradation of American soft power as 'the strategy undermined the U.S. global position' (Steinberg 2008, 160) and 'global public confidence in U.S. leadership.
>
> (Steinberg 2008, 157)

In 2009, the newly elected President Barack Obama sought to undo the mistakes of the past and restore public confidence in U.S. global leadership by pushing the concept of smart power, which became a core principle of his administration's foreign policy strategy. The convergence of more hawkish 'hard' and more internationalist 'soft' power tools, smart power is defined by Nye and the Centre for Strategic and International Studies as 'an approach that underscores the necessity of a strong military, but also invests heavily in alliances, partnerships, and institutions of all levels to expand American influence and establish the legitimacy of American action' (CSIS 2007). In the last stages of Obama's presidency, this approach appeared to be paying dividends. The success of the U.S. in the 2016 Soft Power 30 index (it overtook the United Kingdom to claim the top spot) suggests that the U.S. president had moved some way to repairing the damage dealt to his country's international reputation following the 2003 Iraq war (Portland Communications 2016). Although perceptions of 'ugly' U.S. foreign policy linger and remain a weakness (America still fares best on ratings around higher education, culture and technological innovation), Obama's investment in a series of diplomatic moves (Iran, Cuba, and Asia) have yielded a positive return.

Reflecting the manner in which 'ideas and images associated with world politics "leak" into popular culture' (Dodds 2005, 267), recent U.S. movies often reflect and reinforce this 'smarter' understanding of U.S. foreign policy in relation to global political space. Thus, despite claims in some quarters that Hollywood had become concerned foremost with the pursuit of China's international agenda, its largest-scale blockbusters are nonetheless promoting the national interest of the U.S. by offering representations that satisfy Obama's inaugural call for a new era of militarily strong yet responsible American leadership in the world. Although they declare support for the U.S. military power, they enshrine it in a legitimising framework of international partnership and collaboration aimed at persuading global audiences that American leadership in the world is still wanted. Whereas *Independence Day* (1995) extoled U.S. military intervention as the only force capable of standing between the world and its destruction at the hands of invading alien forces, its 2016 sequel, *Independence Day: Resurgence*, suggests that U.S. power is greatest when exercised with the support of others and embedded in internationally accepted institutions (here the aforementioned United Nations-founded Earth Space Defence). Other films render the 'necessity' of American military power/intervention more obliquely, displacing it in venerable (and oftentimes already globally popular) characters that are either non-state actors operating in partnership with representatives of the nations that global threats necessitate they work across or characters which, when they are tied to institutions of U.S. government/military, often stand at odds with dubious policy and practices. Numerous examples of such trends can be found in the relatively young superhero genre. From the aforementioned *X-Men: Days of Future Past* to the many films that comprise the interconnected, so-called Marvel Cinematic Universe, we consistently encounter powerful characters that, while they ultimately promote the national interest of the U.S., simultaneously embody a foreign policy approach which recognises that, 'in a globalized world, challenges are increasingly transnational, and so too must be their solutions' (CSIS 2009).

Consistent with the notion of a more ecumenical popular understanding of global political space this message is intended for the world at large rather than just China. As we have seen, however, consistent with Hollywood's current market priorities and (now perhaps erstwhile) U.S. foreign policy, China frequently receives special attention. More often than not in terms of screen prominence, the 'global' dimension in ostensibly multilateral, imagined mobilisations of economic and military resources in pursuit of the global good amounts to China, which is positioned towards the centre of the world-saving stage as the U.S.'s primary partner. In this way, the messages communicated about China are also directed at China and again operate as a form of geopolitical wish-fulfilment, as a projection and promotion not only of what the U.S. wants China to become but also what it wants its relationship with China to be. Movies such as *Independence Day: Resurgence* shepherd an optimistic trajectory of U.S.–China relations by performing the stable relationship the U.S. would like to have with China, but which strategic mistrust between the two powers can often impede.

Thus, erring on the side of caution, direct representations of American military power (a primary source of the simmering tensions that can act as a barrier to cooperation) tend to be avoided in movies whose primary foreign market target is the PRC, and perhaps for good reason. The introduction to this chapter considered how *Pacific Rim*, owing to its distorted understanding of Chinese social and cultural mores, was as an ill-conceived vehicle of cultural diplomacy, but from the perspective of generating political influence it also achieved little in the way of drawing China to America's side. Writing in an op-ed, an officer of the People's Liberation Army perceived in the film a brazen case of pro-U.S. propaganda regarding the latter's military determent of Beijing by the forward deployment of its forces in the South China Sea, and suggested that soldiers 'should strengthen their combat capability to safeguard [the PRC's] national security and interests' (Patten 2013). Made with the intention of securing full joint venture status, *Transformers 4* takes careful measures to attenuate the contentious issue of U.S. military might and its presence in the Asia Pacific region and persuade China (along with the rest of the world) of the prudence, justness, and benefit of U.S. global leadership. Firstly, in a decisive shift away from the gung-ho, unilateral militarism that described both the Bush administration and the first three films in the *Transformers* franchise, the crack group of internationally incursive U.S. Army Rangers that previously proved pivotal against the evil Decepticons are noticeably absent in *Transformers 4*, replaced by a rag-tag American-led international civilian militia. It therefore becomes the task of everyman Yeager—working in partnership with the stateless, if American-sounding, heroic Autobots, his daughter Tessa (Nicole Peitz), her Irish boyfriend Shane (Jack Reynor), of course China's Su, and, eventually, Joyce—to stop his corrupt compatriots, who believe their hard (economic and military) power alone protects them and entitles them to do as they please in the world. The belief that U.S. leadership generally advances global security and is needed to prevent chaos is still present in the film, but following the tactic of displacement outlined above, a direct depiction of the military power that underpins this U.S. self-understanding is avoided. Power is divested from immediate representatives of the U.S. government/military and (re-)invested a non-state actor (and thus in what is taken to be a more palatable vessel for the communication of this message) whose leadership arises from within a legitimising framework of international (or, in truth, Chinese) partnership.

Drawing on the often-overlooked role that sport can play as a vehicle of cultural diplomacy and soft power, Yeager is also associated with American football, which is deployed as a means to nurture the film's furtive establishment of a dominant U.S. hegemonic position from within Chinese space. On the one hand, the use of America's national sport might appear to be a risky strategy. With its military-like terminology (i.e. 'aerial assault', 'blitz pass', 'ground assault', 'long bomb' and such), American football is often perceived as a metaphor for the less appealing hard dimension of 'smart' American foreign policy, and both at home and abroad there exist those for whom the synonymy between football and military terminology serves as a disturbing and therefore unattractive marker of underlying American aggression and dominance. However, as Nye reminds us, 'attraction is dependent on the mind of the perceiver' (2004, 100), and on the other hand Chinese minds are, for the time being at least, receptive to the idea of American football, which in recent years has exploded in popularity in the PRC. Chinese parents appear to share (in aspects of) the sport's affirmative self-perception as a repository of positive, quintessentially American qualities and values,[7] believing it can 'help their youngsters develop character and self-reliance' (Dong 2015).

Taking advantage of this appeal, *Transformers 4* adopts the sport in a bid to foster Chinese public confidence in Yeager and, by extension, the U.S. Following a dramatic rooftop chase close to the film's end, Yeager and Savoy crash through the window of a Hong Kong high-rise, much to the shock of the apartment's occupants, a Chinese mother and her teenage son. As a highly trained field operative, Savoy clearly has the upper hand in the ensuing fight—at least, that is, until Yeager spies the teenager's football, which has been given pride of place on a display self, next a picture of the boy waving it proudly on the field. Mimicking the actions of a quarterback (also known as the field general), Yeager launches the football with projectile-like force and deadly accuracy. With Savoy momentarily stunned by the impact, Yeager seizes his opportuning and tackles the knife-wielding assassin out of the window to his justified demise. The fusing together of an American non-military wielder of power with appealing cultural stimuli appears to be aimed at manoeuvring (Chinese) opinion in the U.S's favour. An accepted vehicle of cultural influence in China, the sport serves as an attractive foundation for the projection and promotion of the interests of the U.S., and refracted through the cultural lens of football, U.S. military power and foreign policy appears especially beneficent where China is concerned. Shortly before his victory, Yeager admonishes Savoy, telling him 'You shouldn't have threatened my family'. In this moment, however, his daughter and friends are elsewhere, and so Yeager's words instead fall on the Chinese mother and her son, who stand immediately behind and are shielded by him. As much as the film transmits the message that a strong China is beneficial rather than a threat to U.S. and global security, so too does it attempt to persuade China of the same about the U.S.

Concluding remarks

In what is akin to an instance of hegemonic negotiation, shifts in the flow of global market forces have opened Hollywood's geopolitical culture regarding China and Chinese identity to change. Keen to court a seemingly ever-expanding Chinese audience base and gain the goodwill of Chinese authorities, recent U.S. movies are distancing themselves from the old Orientalist tropes and stereotypes that, consistent with the prevailing geopolitical anxieties, understood China as an inscrutable site of U.S./Western imperilment, and are instead adopting more convivial patterns of representation. Stemming from this development, there is an extent to which U.S. blockbusters have become the site for a new popular geopolitics, which transcends the dominant paradigm of patriotism/enemy othering. The movies under question in this chapter fashion an ecumenical framework of global political space, one that allows them to oblige multiple world-political viewpoints, even if they ultimately operate to preserve and perpetuate a dominant U.S. hegemonic position. Although Chinese strategic narratives are accommodated and endorsed, they are simultaneously contained by filmic narratives which seek foremost to serve U.S. national interests by providing an ideated stage for upgraded Sino-American relations and reminding audiences of the necessity of U.S. global leadership. In this regard, the spectre of Orientalism lingers, and novel stereotypes that insist on China's 'sameness' or 'equivalence' to the U.S./West provide an updated Orientalist account of geopolitical culture: although patterns and practices of recognition are approving rather than negative, the films' understanding of China still stems from and enacts the politically hegemonic determination that the PRC should aspire to be not merely a powerful force in the world but also a moral inspiration to it. From this vantage point, it is unclear whether Hollywood will succeed in its de facto commissioned function as a vehicle for the transmission of Chinese soft power in the Western(ised) world.

Notes

1 Although English-language news organisations can be an undeniably important resource in the global narrative of a rising superpower whose domestic media is strictly controlled, their reports tend to be weighted heavily towards a depiction of China as oppressed, dangerous and bizarre Other to the U.S./West.
2 China posted total box office of $2.8 billion (a rise of 35% on 2011). In 2013, the rate of growth dipped slightly (to 27%) but revenue rose again to reach $3.6 billion, and in 2014 a renewed surge in growth of 36% saw the Chinese market end the year with a total value of $4.82 billion (Frater 2015). In 2015 this figure grew to $6.78 billion (Brezski, 2015). Although China still has a way to go until it matches North America's current $10.4 billion annual spend, the pace of growth is such that Ernst and Jung predict China will become the world's biggest film market by 2020 (see Child 2012), while IMAX CEO Richard Gelfond believes this will happen as soon as 2018 (see Shao 2013).
3 For more detail, see http://www.chinahollywood.org/about-co-productions.
4 The notion of East Asian contamination of Western civilisation can be discerned in Spike Jonze's small, cerebral film *Her* (2013), which uses Shanghai to create the near-future Los Angeles, and the Netflix three-part miniseries *Residue* (dir. Alex Garcia Lopez, 2015).
5 Joe is portrayed in his younger form by Joseph Levitt Gordon and in his older form by Bruce Willis.
6 When the evil Deceptions start wreaking havoc in Hong Kong, the local authorities immediately contact Beijing for guidance.
7 Such as profound belief in family, home and patriotism, perseverance, courage, equality of opportunity, independence, initiative and self-reliance expressive of a high degree of merit-based individualism, profits, fame and glory, even when these values are readily discernible in other cultures.

References

Baculiano, Eric. 2016. 'China Amps Up Space Program in Race to Challenge U.S'. *NBC News*, available at http://china-amps-space-program-race-challenge-u-s-n666391 [last accessed 19 May 2017].

Barboza, David. 2015. 'China Escalates Hollywood Partnerships, Aiming to Compete One Day'. *New York Times*, available at https://www.nytimes.com/2015/04/06/business/media/china-escalates-hollywood-partnerships-aiming-to-compete-one-day.html [last accessed 24 May 2016].

Brezski, Patrick. 2015. 'China Box Office Grows Astonishing 48.7 Percent in 2015, Hits $6.78 Billion'. *The Hollywood Reporter*, available at http://www.hollywoodreporter.com/news/china-box-office-grows-astonishing-851629 [last accessed 24 May 2016].
Child, Ben. 2012. 'China Will Be the World's Biggest Film Market by 2020'. *The Guardian*, available at https://www.theguardian.com/film/2012/nov/29/china-biggest-film-market-2020 [last accessed 24 May 2017].
Cornea, Christine. 2007. *Science Fiction Cinema: Between Fantasy and Reality*. Edinburgh: Edinburgh University Press.
CSIS. 2007. *CSIS Commission on Smart Power: A Smarter, More Secure America* Washington, D.C.: Center for Strategic and International Studies.
CSIS. 2009. *Smart Power in U.S.-China Relations: A Report of the CSIS Commission on China*. Washington, D.C.: Center for Strategic and International Studies.
Daly, Robert. 2016. 'Hollywood's Dangerous Obsession with China'. *Los Angeles Times*, available at http://www.latimes.com/opinion/op-ed/la-oe-daly-hollywood-20161007-snap-story.html [last accessed 19 May 2017].
Dillow, Clay, Jeffrey Lin, and P.W. Singer. 2016. 'China's Race to Space Domination'. *Popular Science*, available at http://www.popsci.com/chinas-race-to-space-domination [last accessed 19 May 2017].
Dittmer, Jason, and Klaus Dodds. 2013. 'The Geopolitical Audience: Watching *Quantum of Solace* (2008) in London'. *Popular Communication* 11:76–91.
Dittmer, Jason, and Nicholas Gray. 2010. 'Popular Geopolitics 2.0: Towards Methodologies of the Everyday'. *Geography Compass* 4 (11):1664–1677.
Dodds, Klaus. 2005. 'Screening Geopolitics: James Bond and the Early Cold War Films (1962–1967)'. *Geopolitics* 10 (2):266–289.
Dong, Alvin. 2015. 'How American Football is Taking on China Challenge – and Winning'. *South China Morning Post*, available at http://www.scmp.com/lifestyle/article/1861046/how-american-football-taking-china-challenge-and-winning [last accessed 19 May 2017].
Frater, Patrick. 2015. 'China Surges 36% in Total Box Office Revenue'. *Variety* available at http://variety.com/2015/film/asia/china-confirmed-as-global-number-two-after-36-box-office-surge-in-2014-1201392453/ [last accessed 24 May 2017].
Guerrero, Ed. 1993. *Framing Blackness: The African American Image in Film*. Philadelphia: Temple University Press.
Hartig, Falk. 2015. 'Communicating China to the World: Confucius Institutes and China's Strategic Narratives'. *Politics* 35 (3–4):245–258.
King, Homay. 2010. *Lost in Translation: Orientalism, Cinema, and the Enigmatic Signifier*. Durham, NC: Duke University Press.
Leong, Karen J. 2005. *The China Mystique: Pearl S. Buck, Anna May Wong, Mayling Soong, and the Transformation of American Orientalism*. Berkeley and Los Angeles: University of California Press.
Morley, David, and Kevin Robins. 1995. *Spaces of Identity: Global Media, Electronic Landscapes and Cultural Boundaries*. London and New York: Routledge.
Naremore, James. 1998. *More than Night: Film Noir in its Contexts*. Berkeley and Los Angeles: University of California Press.
Nye, Joseph S., Jr. 2004. *Soft Power: The Means to Success in World Politics*. New York: Public Affairs.
Nye, Joseph S., Jr. 2011. *The Future of Power*. New York: Public Affairs.
Park, Jane Chi Hyun. 2010. *Yellow Future: Oriental Style in Hollywood Cinema*. Minneapolis: University of Minnesota Press.

Patten, Dominic. 2013. 'Pacific Rim is 'American Propaganda,' Chinese Military Official Says'. *Deadline Hollywood*, available at http://deadline.com/2013/08/pacific-rim-american-propaganda-chinese-military-says-572510/ [last accessed 19 May 2017].

Portland Communications. 2016. 'The Soft Power 30'. *Portland Communications*, available at http://softpower30.portland-communications.com/ranking/ [last accessed 30 May 2017].

Pulver, Andrew. 2013. 'China Demands 'Positive Images' in Return for Access to Markets'. *The Guardian*, available at https://www.theguardian.com/film/2013/nov/06/china-hollywood-positive-images-movies [last accessed 19 May 2017].

Purse, Lisa. 2011. *Contemporary Action Cinema*. Edinburgh: Edinburgh University Press.

Richburg, Keith B. 2010. 'Chinese Firms Look to Team Up with Hollywood'. *Washington Post*, available at http://www.washingtonpost.com/wp-dyn/content/article/2010/09/03/AR2010090302853.html [last accessed 24 May 2016].

Roselle, Laura, Alister Miskimmon, and Ben O'Loughlin. 2014. 'Strategic Narrative: A New Means to Understand Soft Power'. *Media War & Conflict* 7 (1):70–84.

Said, Edward. 1978. *Orientalism*. New York: Vintage Books.

Saunders, Robert A. 2015. 'Imperial Imaginaries: Employing Science Fiction to Talk about Geopolitics'. In *Popular Culture and World Politics: Theories, Methods, Pedagogies*, edited by Federica Caso and Caitlin Hamilton, 149–159. Bristol, UK: E-International Relations.

Saunders, Robert A. 2017. *Popular Geopolitics and Nation Branding in the Post-Soviet Realm*. London and New York: Routledge.

Shao, Heng. 2013. 'China's Box Office to Surpass the U.S. by 2018, Says IMAX CEO'. *Forbes* available at https://www.forbes.com/sites/hengshao/2013/10/17/chinas-box-office-to-surpass-the-u-s-by-2018-says-imax-ceo/#5d98693f5f74 [last accessed 24 May 2017].

Steinberg, James B. 2008. 'Real Leaders do Soft Power: Learning Lessons of Iraq'. *Washington Quarterly* 31 (2):155–164.

Strukov, Vlad, and Helena Goscilo. 2017. 'Introduction: The Aerial Ways of Aspiration andlinspiration, or the Russian Chronot(r)ope of Transcendence'. In *Russian Aviation, Space Flight and Visual Culture*, edited by Vlad Strukov and Helena Goscilo. London and New York: Routledge.

Tasker, Yvonne. 1998. *Working Girls: Gender and Sexuality in Popular Cinema*. London and New York: Routledge.

Van Efferink, Leonhardt. 2015. 'Jo Sharp: Feminist Geopolitics, Postcolonialism, Common Sense, Reader's Digest' *Exploring Geopolitics: The Academic Faces in the Geopolitical Debate*, available at http://www.exploringgeopolitics.org/interview_sharp_jo_joanne_feminist_geopolitics_postcolonialism_common_sense_readers_digest_security_development_gramscian_foucaultian_approaches_geographies/ [last accessed 19 May 2017].

Vukovich, Daniel. 2012. *China and Orientalism: Western Knowledge Production and the PRC*. London and New York: Routledge.

Wagner, Jan-Philipp N. E. 2014. 'The Effectiveness of Soft & Hard Power in Contemporary International Relations'. *E-International Relations*, available at http://www.e-ir.info/2014/05/14/the-effectiveness-of-soft-hard-power-in-contemporary-international-relations/ [last accessed 12 April 2016]

Weldes, Jutta. 2003. 'Popular Culture, Science Fiction, and World Politics'. In *To Seek Out New Worlds: Exploring Links between Science Fiction and World Politics*, edited by Jutta Weldes, 1–22. Houndsmills, UK: Palgrave Macmillan.

Wilkie, Laurie A. 2003. *The Archaeology of Mothering: An American Midwife's Tale*. New York: Routledge.

Yuen, Nancy Wang. 2016. *Reel Inequality: Hollywood Actors and Racism*. New Brunswick, NJ: Rutgers University Press.

Zara, Christopher. 2015. 'Hollywood Studios are Self-Censoring Movies to Appease Communist Censors in China, Says US Report'. *International Business Times*, available at http://www.ibtimes.com/hollywood-studios-are-self-censoring-movies-appease-communist-censors-china-says-us-2163232 [last accessed 19 May 2017].

9 'Warning! Zombies Ahead …'

Determinate negation, predatory capitalism, and globalised place-making

Roxanne Chaitowitz and Shannon Brincat

Introduction

Soon after the horrific attacks by Rudy Eugene, aka the Miami Zombie, a street sign warning 'Zombies Ahead' was posted on the 520 in Washington State by a humorous maintenance worker.[1] The seemingly possible potential of such an 'outbreak' reminds us of Adorno's trenchant warning that 'the world – which could be paradise here and now – can become hell itself tomorrow (1998, 14). While a zombie apocalypse was not upon us, the message captured the essence of contemporary society's intensified fascination with the zombie as an allegory of our future: how through the exploitation and destruction that we commit to ourselves, society, and nature, we come step-by-step closer to embodying this creature of horror. The zombie of today is no longer undead—it has evolved. And this evolution is inextricably linked to the social pathologies that have metastasised under the geopolitical and economic strictures of contemporary capitalism as it enters the phase of what we call predatory capitalism. This concept refers to the intensification of capitalist processes of surplus value extraction and accumulation: the ubiquitous pursuit of profit to the point that it consumes the system's very basis of reproduction. From the labouring zombie or mind-controlled slave of colonialism to the ghoul of Romero's Golden Era[2] that no longer laboured but could only mindlessly consume to the fast zombie of today and its frenzied consumption that spreads like an epidemic across the globe, each phase of zombie evolution reflects our temporal fears of economic exploitation: from slavery to consumption, and now predation. We aim to show how the zombie of popular culture not only reflects the production of these subjectivities under Western capitalism, but highlights the importance of social relations in resisting them.

Linking the zombie of popular culture to geopolitics and the construction of political narratives is not new. Serious survival guides have been written to meet the existential threat posed by zombies (Brooks 2003, 2006). Perhaps most telling is that the U.S. military has used zombies as a training tool, creating CONPLAN 8888 to combat the threat of an imminent zombie apocalypse (Lubold 2014). Popular culture has become a common sense site to challenge the elite spaces in which (geo)political imaginations are received, produced, consumed, and legitimised (Dodds 2006; Dittmer and Dodds 2008; Dittmer and Gray 2010). Given the rich allegorical content zombies provide in shaping perceptions and discourses of threat, security, and surveillance, the zombie has become a useful pedagogical tool to assist in teaching International Relations (IR) Theory (MacDonald, Hughes, and Dodds 2010; Caso and Hamilton 2015). Some have advanced an ultra-realist or hegemonic narrative (see Drezner 2011; Blanton 2013), while others have been more critical of the allegorical purchase zombies provide by using it to challenge (rather than mirror) the discursive formation, identity constructions, and circulations of power in world politics (see Morrissette 2014; Hannah and Wilkinson 2014). The undead have similarly been used to interrogate the way power, human nature, and the colonisation of space and resources map on to geopolitical representations found in popular culture (see May 2010; Saunders 2012; Pasko 2015). With the mindless zombie knowing no limits nor self-restraint in its insatiable hunger, it is unsurprising that the zombie has become a trope in popular geopolitics for critiquing the current social conditions under capitalism's reign of consumption (see Shaviro 2002; Lauro and Embry 2008; Webb and Byrnand 2008). While popular culture can authenticate or reify representations of people, spaces, and places as 'mimetic of the real world' (Power and Crampton 2005, 197), it is also a site from which to resist these dominant geopolitical narratives, in spite of being shaped by them. It is this co-construction of popular culture and geopolitics (see Grayson in this volume) that allows the zombie to incarnate both the manifestation *and* critique of predatory capitalism. We later take up this challenge by using the zombie to highlight the importance of revivifying social relations as a way to resist zombification.

Zombies are political creatures precisely because they give visual depiction and bring forth the chaos of the very real violence hidden under the ideological veil of Western capitalism as it moves into a peculiarly predatory phase. Our fascination with this macabre zombie genre is as potent as ever because, like all survivor-fantasies, it touches upon our deepest instinctual fears and social anxieties situated within existing social conditions. Through them we can process strategies of survival, fantasise acts of heroism or escape, and feel the solidarity of a close-knit community that we no longer experience, given the evisceration of social relations under predatory capitalism. The tragic endings of most zombie movies also help us develop coping mechanisms to deal with the inevitable mastery of consumption *over* us. In this way, what the zombie symbolically consumes is not simply flesh nor man-made borders (we can thank neoliberalism for the latter), but the *socius* itself. By externalising social deformation as mere horror-fantasy, the zombie paradoxically both blinds and reveals what we need to overcome: the erosion of social-life, arguably the most precious of all the world's resources. In the first two parts of this article, we posit the zombie as an allegory of the determinate negation of the human-being—of our uniqueness, will, labour, desires/needs *and* our intersubjective relations with others—under this global economic order. We then chart the changes of zombie physiology, behaviour, and narratives, in cinematic history showing how these reflect the transitions in global political economy—a movement in which the labouring zombie has been displaced by the ravenous horde of consuming ghouls in the last decade.[3] We conclude by speculating a (re)turn to 'the social', as the means to reverse this process of zombification.

Zombies and determinate negation

Zombies represent the determinate negation of human existence under the diminishing social relations of predatory capitalism,[4] an allegory of the deformation of even the most basic forms of human sociality. While the term 'predatory' has been used to describe material aspects of late capitalism—including deregulation, monopolisation, production of harmful goods, and exploitation of labour and the environment—we use it to emphasise how individuals and society writ large objectifies, hunts, and consumes, with absolute non-reflexivity. Predatory capitalism is marked by the pauperisation of the very consumers it needs to satisfy its incessant drive for profit, by how it ruptures the moral fabric of formal civil relations essential for its function, and how it destroys the natural environment necessary to sustain itself. This represents an acute stage of crisis of capitalism, where it can no longer expand into new markets, but faces a declining rate of profit that it cannot solve under the weight of the dead ideas of neoliberalism. Its only option is to devour itself; hence the increasing push towards monopolisation and corporatised state structures to facilitate market *capture* over market *creation*. Coupled with these economic contradictions is the deterioration of socio-political life in which polarisation within states is matched by greater recourse to violence and imperialism between them. This parasitic metaphor articulates what Fromm would have called a culturally patterned defect or what Honneth would call social pathology. In this new horrific vision of the future, Hobbes' 'war of all against all' is retained but recast as some rapturous End Time, in which subjective caprice—and the pursuit of its satisfaction—is mistaken as freedom itself. An apocalyptic 'end of history', the zombie horde gives representation to the zenith of predatory capitalism that, finally, consumes its own makers in some orgiastic war of all against all.

Hegel was one of the first to consider the implications of an intersubjective politics contingent upon our social relations with others. In his dialectical system, *negation* highlights the relations of opposition and difference. It is through the process of negation that ideological representations which have become habituated or static are met with their contradictions and sublated (*Aufhebon*): a movement that preserves and overcomes in ways that supersede the contradiction. Adorno's correction of the dialectic—away from an emphasis solely on its affirmative traits in which change always issues in unity and coherence—is also necessary to understand the reflective relation between predatory capitalism and contemporary zombie films. Insofar as *Negative Dialectics* reconfigured the unity of Hegel's dialectic (see, especially, Adorno 1973 [1966]), it sought to overcome the domination of object by the subject, rendering the zombie a particularly well-suited motif for describing contemporary capitalism. For Adorno, determinate negations are those that identify specific contradictions in society that are otherwise covered up, misunderstood, or obscured by identitarian thought—the type of thought that imposes unity or homogeneity on to the object. In so doing, identitarian thought supresses all difference and heterogeneity in the object, which Adorno labels 'the nonidentical'. While the zombie gives effect to human immiseration under capitalism—where consumerist drive exceeds capabilities of thought, empathy, and sensibility—the zombie is peculiarly suited to unlocking the nonidentical that has been subsumed under identitarian thinking.[5] That is, the zombie is symptomatic of the contemporary human-being and the evisceration of social relations that cannot be expressed in existent conditions dominated, as they are, by the profit motif and exchange principle. For Horkheimer and Adorno:

> determinate negation does not simply reject imperfect representation of the absolute, idols, by confronting them with the idea they are unable to match. Rather, dialectic discloses each image as script. It teaches us to read from its features the admission of falseness which cancels its power and hands it over to truth.
>
> (1994, 17–18)

For Adorno, thought can only have access to the nonidentical via conceptual, representational, or aphoristic criticisms of false identifications. These determinate negations that identify contradictions covered up by identitarian thought offer indirect expressions of those aspects of society misidentified by such thought. As argued by Zuidervaart, '[the] only way to expose these antagonisms, and thereby to point toward their possible resolution, is to think against thought – in other words, to think in contradictions' (2011, sec. 5, par. 9).

By pushing contemporary capitalism to its dialectical extreme, the zombie movie therefore performs this radical function by exposing—albeit hyperbolically—the very real limits of contemporary capitalism. Take Gunn and Treat's account as an example: following the Althusserian notion of the pre-ideological subject, they find that the undead of Romero's films illustrate the 'individual who has yet to become self-conscious or called into the service of larger social organization, community, or state' (Gunn and Treat 2005, 155). They observe in the labouring zombie the obverse of the Kantian subject who is 'gloriously autonomous and independent of the socius' (2005, 165). Yet they are only half right, for while the zombie clearly lacks the autonomy capable only of a rationally directed will, the ghoul is a radicalised example of what being 'gloriously independent' of the social is. With no familial, community, or ethical bonds; with no normative prescriptions/proscriptions governing their behaviour; with no intersubjective recognition or communication with others, they heed their internal drives for consumption without any moral restraint. Zombies then, are not the obverse of the liberal subject, but its sublime manifestation.

Understanding the zombie as the determinate negation of human social relations under predatory capitalism reveals how our frenzied consumption is based on the predation of others and how this predation severs social relations, creating a swirling mass of isolated, atomised individuals. The antagonism generated by the push for consumption and profit is shown in zombie films to lead to either the annihilation of society under an asocial mass of violent consumers (the zombie horde), or nihilistic struggles in which a handful of survivors seek out an existence at the fringes of society. Arguably, Acevedo (2016) would identify both as manifestations of the hyper-individualisation of what he calls the 'zombie factor'. The humanity of the zombie can never be reclaimed in this framing, and neither can our own, given both reflect worlds where people are no longer social creatures but atomised entities acting in isolation. But how did we fall so far from the socius for the zombie to reflect our fears in this manner? To understand this, we have to trace the evolution of the zombie in popular culture—something that reveals a startlingly close relation between our collective fears of the zombie and the changes in world capitalism.

Cinematic zombies in the history of modern capitalism

By taking an historical approach, we can observe how the changing threats of market forces have been reflected across the cinematic history of the zombie genre (see Table 9.1). As argued by Shaviro, 'zombies present the "human face" of [the] capitalist monstrosity' (2002, 288) and as capitalism changes and is experienced differently (temporally, spatially, and subjectively), so too do the nightmarish visions it gives rise to in the collective conscious translated into film (see McNally 2011).

Table 9.1 The changing physiology of the zombie at different stages of capitalist development

Timeline	*Zombie evolution*	*Changes in capitalism*	*Development in popular culture and film (exemplars)*
1920–1940	Labouring zombie	Fear of eternal 'undead'; slavery/fear of the Master (*Bokor*) and becoming the colonized subject; continued theme of slavery, fear of reverse colonialism, racism	*The Cabinet of Dr. Caligari* (1920); *White Zombie* (1932); *Revolt of the Zombies* (1936); *Things to Come* (1936)
1940–1960	Decline of the zombie	Golden age of Bretton Woods System and fear of external threats to decline of domestic economy	*King of the Zombies* (1941); *Revenge of the Zombies* (1943); I *Walked with a Zombie* (1943); *Zombies of the Stratosphere* (1952)
1960–1980	Shift from undead to 'consuming ghoul'	Onset of neoliberalism and the 'Golden Era' of zombie films	*Last Man on Earth* (1964); *Night of the Living Dead* (1968); *Curse of the Living Dead* (1973); *Dawn of the Dead* (1978)
1980–2000	Consuming ghoul	Reaganomics and fall of communism; increased consumptive patterns	*Creepshow* (1982); *Re-Animator* (1985); *Day of the Dead* (1985); *The Serpent and the Rainbow* (1988)
2000–2013	Quick zombie	Predatory capitalism	*28 Days Later* (2002); *Shaun of the Dead* (2004); *Land of the Dead* (2005); *I am Legend* (2007); *Zombieland* (2009); *World War Z* (2013); *Walking Dead* (TV series 2010-present)

We are not concerned with the fear of mortality, the horror of rotting or consuming human flesh, or the metaphysical conundrum of an undead being that exists in a shell of decaying matter. Rather, we seek to draw out what studies in literature typically overlook through their triangulation between alleged universal human fears, local cultural conditions, and the peculiarities of individual identity that are assumed to explain the appeal of the horror film (see Boyd 2009). What is missing is an account of the deformation of social relations under the different forms of capitalist development that reveals itself to be the truly horrific aspect of the zombie film throughout history. As we show, the precise attack on social relations shifts at different stages of capitalist development and this is reflected in the changing physicality of the zombie and the vulnerability of the social world to this threat.

The original Haitian zombie was no villain, but a victim of the most pitiable form of exploitation imaginable: slavery from beyond the grave. These re-animated corpses were controlled through powers known only to *Bokors* or evil *Houngans*, and were directed to fulfill specific purposes. In Vodoun belief, it is said that upon death the second part of the soul—the *ti-bon-ange* that holds the individual qualities of a person—can be captured by those with knowledge to do so. There are two forms of woeful creatures that result: the *zombi astral*, a dead soul without a body never allowed to achieve final rest, and *duppy* or *zombi*, a dead body without a soul, forced to undertake specific commands, typically slave-labour or to harm someone else (send *a mort*) (see Boon 2011). That the *duppy* can take the form of an animal (see Leach 1961, 207), and the etymology of the term 'duped' originates from the dull-witted hoopoe bird *(de huppe)* in seventeenth-century France, when Haiti was under French rule, is of great significance. Though the etymology is difficult to trace, the connection to the contemporary meaning of 'being duped' is clear, particularly of being cheated or manipulated to someone else's will. It was this second type of creature that possessed the character Cesare to commit a series of murders in the first zombie movie—and classic of German expressionism—*The Cabinet of Dr. Caligari* (1920). The *Bokor* is the willing agent of slavery, serving the interests of those who could pay for their secret knowledge—the elites of post-revolutionary Haiti, or its plantation owners and colonial masters before 1791—and use it for the purposes of social control. From its earliest origins, the legend of the zombie was therefore one that coalesced around the machinations of the horrors of economic exploitation: specifically, the fears of slaves manipulated by governing elites to cajole this potentially mutinous population into acquiescing to chattel slavery, rather than the marginally worse conditions of 'undead slavery' for eternity.

How zombies of Haitian Voodoo were depicted in early film was, of course, highly disaggregated from the colonial experience: the story had been appropriated and bastardised for Western audiences (see Sheller 2003). The first zombie movies followed narrative patterns centred on the labouring zombie, voodoo ritual, and exoticism (geographic, racial, colonial and sexual in account). In *White Zombie* (1932) the tyrannical sugar mill owner populates his Haitian factory with the slaves of his enemies' walking corpses and is contracted by the wealthy plantation owner to transform a woman into a zombie, so she can 'perform his every desire'. In its sequel, *Revolt of the Zombies* (1936), a curious re-writing of history occurs through which the origins of zombification are transferred to Cambodia even though the theme of the mastery of labour remains. Its more concealed projection, however, is the fear of the white master being turned into an uncivilised, unthinking, colonial subject and labourer. What distinguishes these early zombie films from those of contemporary cinema is that zombification and the attacks of zombies upon humans are highly personalised and directed. Moreover, the white victim of zombification is to be readily sympathised with and the process of their transformation something to be prevented; 'saving the innocent' (usually a Western, hapless female) drives the entire plot structure. This personal side of zombification, and the attempt to arrest its degenerative process on a loved one, would be replaced with the anonymity of the zombie of modern cinema, in which their human biology and individuality is completely erased. This marks a dramatic shift away from solidarity with the afflicted.

From the mid-1930s, the zombie film expressed fears of fascism and a loss of social control that followed shifts in the geo-strategic interests of the democratic West. While the location of the zombie is transferred from the colonial sphere to Germany, the thematic fear of the loss of free-will is retained, as is the purposefully directed labouring zombie controlled by a master. The adaptation of H.G. Wells' book *Things to Come* (1936) was aired in the year Germany reoccupied the Rhineland. As a thinly veiled warning of the threat of Nazi domination, part of the story involves a plague of the 'wandering sickness' that is unleashed on the survivors of a decades-long international war, a plague that can only be cured by shooting the infected. In *King of the Zombies* (1941), made before America's entry in World War II and therefore taking all precautions not to refer to Nazi Germany directly, the villain is a spy from 'a government of Europe' who employs zombification to acquire war intelligence. A far less inconspicuous condemnation of Nazism was given in *Revenge of the Zombies* (1943), in which a mad scientist is attempting to create a zombie army for the Third Reich. Although these films retain the same fear of the loss of free-will as the labouring zombie of classic cinema, it is a specific loss of cognition/consciousness, shifting from the direct enslavement by another to one in which individual free-will is subsumed under a 'common mind'. This theme of struggling against a soulless-collective reflected geopolitical fears, appealing to American audiences from 1930s to the height of the Red Menace in the late 1950s.[6] However, while these external threats led to the proliferation of alien films in the 1950s, there were, in contrast, few zombie films produced in this same time because, at least for American audiences, this period bore minimal fears about the domestic economy from *internal* threats.

The existential threat of what Gunn and Treat label the 'consuming ghoul' (2005, 155) expresses very different social pathologies to the zombie of classical cinema that defined the 'Golden Era' of zombie films (1968–1983). No longer commanded by a master, they do not make highly personalised attacks of the labouring zombie, but are now 'pure, motorised instinct' (*Dawn of the Dead*, 1978). Instead, the ghoul's slow but indefatigable advance, *en masse*, has a 'sociological purpose behind it'—an attack on 'the sins of modernity' (Koven 2008, 24). Rather than the vacant stare of the automaton obeying the commands of its master, the consuming ghoul is part of a ravenous horde that follows the dictates of unreflexive consumerism. What elicits fear is not slavery to a master, but the potential reversal of this process: of being made the subject of consumption by another, and, more subtly, the fear of not being able to compete effectively against others. Hence, the paradigm shift in the zombie physiology and its behaviour reflect the relevant position of the affluent West in the geopolitical and economic struggles maintaining the Core/Periphery divide. The zombie, here, is a mirror held to Western consumerism as it preys upon peripheral economies. Romero's ghouls are no longer possessed, but only retain a desire to consume—the subconscious recognition of the costs of the West's consumption (desire) being turned against itself, to be either ripped open and consumed or, to be turned into something 'like them'. The latter is perhaps the more dreaded, suffering all the ignominies and inequities of being placed in the Periphery—a place where you can only mill aimlessly outside shopping malls full of goods that you once produced, but can no longer access (*Dawn of the Dead*, 1978)—all the while being the subject to the violence and brutality as the 'humans' seek to regain control. The consuming ghoul reminds us of the facets of Western society complicit in exploitation and cruelty and reveals how we dehumanise others and how such processes have now infected ourselves in the Core. The fear is always there, the fear of what Chang (2005) would call having the ladder kicked away beneath you and falling back into the ranks of the zombie horde.

The fact that zombies have undergone yet another series of radical physical and behavioural changes since the ascendency of neoliberal orthodoxy is unsurprising. Beginning with the film *28 Days Later* (2002) (and its sequel), continuing throughout a host of Romero remakes in the last decade, as well as the veritable explosion of low-budget, amateur, and online films in the last few years, zombies have gained intelligence and are far more physically daunting predators. No longer decaying, grey-faced, lumbering beasts, but frenzied and swarming entities, driven by pure rage. The script has been altered: infection is typically viral, outbreaks are nearly always the responsibility of either the Military Industrial Complex or government experiments, and the expected bleak conclusion is no longer a certainty with some films even ending with positive signs of hope, such as *28 Days Later* (2002), *I Am Legend* (2007), *Made Out Alive* (2009), and *Planet Terror* (2007).[7] It would be wrong, however, to see this change as merely a question of increased speed and power. Rather, what underlies this shift is the change in consumptive patterns, both social and environmental. The predatory zombie is symptomatic of our behaviour in the face of predatory capitalism: aggressive, unthinking, consumers. Even in the comedy *Zombie Strippers* (2008), strippers are compelled to become zombies due to consumer preference—the human can no longer compete. Bereft of older forms of sociality, we now view *all* others as zombies, either victims of our cruelty or our consumption. This is reflected in the distanced and atomised responses to the detritus of modernity, whether the poor, the immigrant, or the 'Third World'—all those who struggle because of necessity—who are deemed to be individually responsible for their position, rather than acknowledging the systemic causes behind their destitution. Wood observed that zombie and some slasher horror movies represent 'the consequences of the dominant social order taken to their logical extreme' that such movies 'only carry to its logical conclusion the basic (though unstated) tenet of capitalism, that people have the right to live off other people' (1979, 21–22). Here, the zombie film continues to be an unswerving means to reveal our complicity in these horrific actions of the consumption of others.

The key difference is that the quick zombie is the acceleration of this same tenet identified by Wood. While the zombie continues as a visceral threat to personal security, in this new secularised apocalypse the source of existential fear is the evisceration of societal and civil life under predatory capitalism. It is no longer the slow march of the zombie-ghoul, but the headlong rush toward the precipice of social destruction. This should not be mistaken as a quantitative change in the nature of capitalist appropriation of labour or environment, but the *intensification* of this appropriation. The bodily actions of today's zombie are so virulent because our exploitation of others has intensified in kind and these films capture the ensuing social disintegration. Individual zombies are now threatening, rather than the swarming mass, and the non-reflexivity of the maximising consumer is replaced with a driven rage that cannot be reasoned with or redirected.

At the same time, the survivors of today's cinema show a peculiar lack of sociality and their solutions are rarely cooperative joint ventures. We see this in *Survival of the Dead* (2009), where the members of the National Guard believe they 'are better off on [their] own' and are 'looking for a place where there was no "them",' a 'no place'—a clear swipe at the concept of Eutopia. Similar critiques can be laid against *I am Legend* (2007) and *World War Z* (2013), where the protagonist, despite being on a mission to save humanity, is the lone man pitted against the world. Above all, these narratives fortify the belief in atomisation, that the individual—and the individual alone—is all that can be relied on for survival; we truly have been *duppyed*. There is no socius to turn to because solidarity has itself been consumed. This reflects broader problems associated with predatory capitalism, but also it shows the way out of the problem as we shall see in the last section.

Zombies and predatory capitalism

As we have seen in the previous section, the evolution of the zombie corresponds to the shifting perceptions of threat within a changing global, political economy. The application of this insight shows how zombies can be a very effective and popular means to communicate the intersecting influence between the practices of political institutions, society, the world economy, and geopolitics (see, especially, Hall 2011). The limits of analysis of those working in this area however, lies not in their interrogation of the problems of contemporary political economy, but in the prescriptions for how such contradictions are to be overcome. For example, Chris Harman's analysis identifies zombie capitalism as an increasingly disordered and potentially violent world, breeding economic crisis, wars, and environmental calamity. For Harman, zombie capitalism 'cannot survive without more labour to feed it, just as the vampire cannot survive without fresh supplies of blood' (Harman 2009, 349). Quiggin identifies zombie capitalism as an unstable system breeding inequality, wherein the 'rich get richer and the poor go nowhere' (2010, 152, 13ff). He asks us to break with the 'dead economic ideas' related to market autonomy that were killed by the Global Financial Crisis and yet, somehow, live on. Similarly, Giroux refers to the 'casino capitalist zombie,' in which competition becomes social combat, war becomes a legitimate extension of politics, and people become redundant under the mantra of Social Darwinism (2011, 2). What typifies this rule of the 'living dead' is the dismantling of all 'social relations that embrace the common good', whilst breathing life back into financial institutions 'even when it appears that the zombie banks and investment houses have failed one last time' (Giroux 2011, 2). Yet the solutions proffered by each theorist do not adequately deal with the contradiction their analysis has unveiled. For Quiggin, the task is to return to another dead idea, this time Keynes. Giroux's list of political demands includes limited benefits to child welfare, federal job programmes, affordable housing and national health insurance (not health*care*). Under these Obamaesque political goals, Giroux leaves intact the structural conditions that led to the zombification he so derides. Even Harman, who goes the furthest, calling for a global unified proletariat (2009, 335–336), says little regarding the social relations necessary to sustain this community of solidarity, besieged as such relations are by the very economic structures he describes.

The early work of Baudrillard seems to explain the social disintegration accompanying predatory capitalism more keenly than these more recent engagements. In *Consumer Society*, Baudrillard predicted 'violent eruptions and sudden disintegration' that '*will* come' to wreck this 'white mass' and its penchant for consumerism (1998, 196, emphasis added). He described this process as symptomatic of conditions in which alienation had become so absolute that individuals could neither perceive their own true needs nor alternate ways of life (1998, 198ff). For him, when everything is a commodity that can be bought and sold—even life itself—the totalising tendencies of capital have triumphed, and transcendence becomes impossible. For Baudrillard, this portends not sublation to something potentially 'better,' but social collapse. In this context, the zombie or consuming ghoul of contemporary cinema is a perfect symbol: the endpoint of human devolution as zombification becomes not an aberration, but the norm, as society implodes from the complete erosion of ethical life. And yet, just like Harman, Quiggin, and Giroux, Baurdrillard fails to offer any solution for, as is well-known, he never developed a theory of agency or change from which the determinacy of capitalism could be overcome.

So, while many have made good use of zombification as an analytical device to highlight the contradictions of late capitalism, they offer little to reverse this process. Unsurprisingly, this lack of viable, imaginative ideas is conjoined with the imposition of austerity to the problems of Western consumption. Such authoritarian responses refuse to reengage with the social, but hold fast to atomisation, intensifying it. Consumption is not reduced but is restricted to an elite who consume greater and greater amounts (*Land of the Dead*, 2005). Accompanying this is violence that can emanate from the state or the private-sphere to protect this privilege. As Younge (2012) writes, 'Neoliberal globalisation, and the inequities that come with it, cannot exist without force or the threat of it ... [because] the system is set up not to spread wealth but to preserve and protect it, not to relieve chaos but to contain and punish it'. Quoting Friedman, 'The hidden hand of the market will never work without a hidden fist' (Younge 2012).

These things that appear to us as isolated phenomena of the excesses of capitalism are, in fact, geopolitical representations of a global pathology. For example, the decapitations and massacres taking place throughout the Mexican bordertowns are a signpost of a future that continues to follow the trajectory of predatory capitalism: gangs battling for market supremacy in some hyper-violent form of monopolisation; ineffectiveness, corruption, and even destruction of state forces, both civil police and military; direct targeting of individuals and groups who oppose the banditry and the authoritarian methods of the state or the gangs; killing those who resist or attempt to wean themselves off the products of consumption; terrorising communities into submission or complete isolation; and increased surveillance and militarisation of domestic life, without meaningful development in social relations of security. The fact that this conflict is centred around control of the cocaine trade is almost irrelevant, as the same logic of predation applies to nearly *all* forms of commodification under Western capitalism, whether that be the brutal mining extraction processes in Ok Tedi, Ogniland, Marikana, Puno, and Bagua; TRIPS and 'Big Pharma' that stifle generic competition, as millions die needlessly waiting for patents to expire or for prices to fall; and the triumvirate of automotive, oil, and rubber industries, whose narratives actively stifle meaningful climate change policy, despite irrefutable evidence of anthropomorphic climate change. The ideal of 'security' is reduced to mere containment. Problematic 'zones' are sealed off and quarantined, with the poor corralled into ghettos and slums, only to be obtrusively surveilled and policed—not to keep them safe but to keep them *in*. The gated communities walling the rich away from the have-nots are subsequently replicated in refugee detention centres and military compounds across the globe, erected in the name of border security by the state, and all servicing the same logic: securing the predation of the many by the few. The only difference is scale. We see this quarantining mentality in the television series *Walking Dead* (2010–present), where protagonist Rick Grimes leads a group of survivors to take refuge in the farmhouse, an industrial prison complex, then a town. All are retreats relatively cut-off from the world, but the same problem always resurfaces. The unsubtle parallels to the geopolitical strategies of Israel in *World War Z* (2013) is another example of the absolute failure of any attempt to segregate the living from the undead. The fact that Trump's campaign was based around 'compelling' Mexico to fund and erect a border Wall for their own quarantining—and threats of increased tariffs and visa removals to enforce it—is where life moves beyond the mimicry of art.

*On returning to 'the social', or 'We are all pulling in different directions' – (***Day of the Dead,** *1985)*

Since the late 1960s, the zombie horde and the survivors have embodied the struggle between the consuming and toiling masses in a tale of fallen humanity. Yet as they kill each other, the source of their relational deformation is never healed: the general loss of social relations. As we know from the mostly bleak endings of Romero's films, resistance is futile if it remains trapped by the old ways of thinking (especially *Day of the Dead*, 1985). Yet, as we have been 'duped' out of our sociality under the continuing ideological assault of individualism and consumerism, zombification is the inevitable outcome of the belief that we are alone, Thatcher's twisted ideology that is 'no such thing as society'. Under this atomising ideology, geographic space, cultural relativity, and time, all provide convenient foils to burst our ethical responsibilities to one another; our indifference to the horrors that occur daily in the slums and favelas are atrocities committed only *'over there'*; our exploitation of labour in the developing world (including of minors) is something *'they'* are responsible for; our complicity with colonialism and slavery is relegated to the work of past generations *'back then'*. We deny our complicity in predation; at the same time we increase our consumption. The plot twist: in order to sublate the process of zombification through a return to the social, we must first, in Adornoesque negative dialectic fashion, turn to the nonidentical, to the *lack* of the social. The trick is to position critique in the actual sites of suffering under predatory capitalism to expose the damage these do to social relations. Our doom is inevitable only if we continue to denigrate and destroy all social relations in the name of consumption and profit. This is the redemptive promise of the zombie movie, but one that is rarely articulated: how can we reclaim the humanity of the zombie?

It is this question that we hear least of all. Often it is lost through the hyper-violence of the zombie film, not merely the gore associated with the hordes surrounding a hapless human victim, but that played out *by* humans towards the zombies. From the hit and runs, snipers, and shotgun blasts, extended most in *28 Weeks Later* (2007) to wholesale city-wide incineration—the ethicality of such violence when directed at the zombie is rarely questioned. Ethical claims to a former shared humanity are lost. Our inability to deal relationally with zombies belies our real lack of reflexivity. In the 2008 remake of *Day of the Dead*, one of the characters is told to run the zombies over. When someone appeals, 'They are still somebody!' the response is, 'Not anymore'. The effacement of the zombie's humanity, and our inability to reclaim it, is a failure of recognition, giving sign to the basic pathology in intersubjective relations. This is why the solution to the zombie outbreak or apocalypse remain so one-sided. Their humanity is lost, or can no longer be recognised, through being undead or contagious. Their radical criminalisation follows from their dehumanisation. After all, if they were human, they could—at least in theory—be returned to their human form, cured, or at least remembered, for their former human lives. Arguably, it is this theme of socialisation that is beginning to emerge in films such as *Warm Bodies* (2013)—though such films remain far from the norm. When the human survivors extinguish a zombie, they are in effect murdering the possibility for them to return to their humanity. This makes a counter-point to any number of forms of contemporary dehumanisation-criminalisation nexus: those jurisdictions that retain the death penalty and consider such criminals so inhuman that they can be exterminated with no guilt attached to social processes of justice; the treatment of terrorist suspects in the legal limbo of Guantanamo, and the acceptability of rendition, assassination, and even torture against those merely suspected of terrorism.

Yet if zombies are shades of their former human selves they can be redeemed, and it is this motif that is uniquely captured in both the novel *I am Legend* and its 2007 film adaptation.[8] Richard Matheson's 1954 novel is set in post-apocalyptic Los Angeles, where a pandemic has either killed or mutated everyone on Earth, bar the main protagonist, Robert Neville, due to an immunity.[9] Neville's mental health deteriorates as he struggles with loneliness and depression in what Clasen calls 'a speculative account of what happens when basic human needs are suppressed' (2010, 315). Years of an alienated existence are finally interrupted when Neville discovers fellow survivor Ruth. However, he soon learns that Ruth is a member of a group of infected survivors desiring to rebuild a 'new race' society, albeit a brutal one. Worse still, they had sent Ruth to supplicate him. They anticipated his all too human need for contact and recognition. This is why Neville never gives himself up to the vampires, not merely because of his immunity but, because 'without knowing what it was to love and be loved' was itself 'a tragedy more terrible than becoming a vampire' (Matheson 2006 [1954], 68). Nevertheless, there are stark similarities between this new vampire race and Neville. The vampires do not merely want to feed on Neville—they are not blind predators—rather, they subject him to trial and death for crimes committed against their people. It is here we realise the cause of their aggression was not simply the desire to feed, but the need for recognition—'Their need was their only motivation' (Matheson 2006 [1954], 11). It is telling that recognition theorists, from Hegel to Taylor, have made the same claim: that recognition is a 'vital human need' (Taylor 1992, 26). We find the same claim to humanity in the closing scene of the 2007 film where, upon finding a cure to the infection, Robert Neville pleads with the vampires to 'Let me save you!' Indeed, the alternative ending takes this concept of recognition one step further, when Neville realises it is not him they are after, but the female vampire he had been experimenting on to find a cure. After Neville surrenders her to the lead vampire, the two tenderly embrace one another. Made aware of their shared human capacity for love, Neville humbly apologises. After all, the only thing that sustained him through his years of isolation was this fundamental search for connection.[9]

Across the zombie genre, it is telling that asociality is what typically dooms the survivors. Instead of a turn toward sociality or an egalitarianism formed under social stress that would seem to offer more safety to the group, we usually witness further atomisation. As some twisted allegory of the deformation of freedom in late capitalism, survivors mirror the pursuit of one's self-interest in complete isolation from ethical life. They kill zombies without reflection; they save their own skin without relation. Yet only the strong survive through their own individual cunning which is precisely how Hegel expressed individual behaviour in the market. But in their struggle for survival they become deformed and inhuman, with few exceptions. The greatest threat is not the zombie, but the inability to reflect on the deformation of social relations that both cause and perpetuate the predation of one upon the other. As Paffenroth observes, the specific cause of the zombie outbreak is rarely important, for the movies are 'always about some small group dealing with the effects, not the causes' (2006, 3). This shows an endemic weakness in the reflexive capacities of individuals within late capitalism, who are only capable of immediate short-term pursuit of self-interest and self-preservation, lacking the political ability, social awareness, and imagination, to grapple with the causes of social decay. This is the phenomenon described by Horkheimer as the *Eclipse of Reason* (2004 [1947]), in which only the purposes of the subject remain within a subjective form of 'reason' prone to conformity, consumption, and authoritarianism. *Dawn of the Dead* (1984) plays this tension out keenly. The fallacies of the 'myth of return' to the 'old lists' are admonished and the survivors begin a new society on distant utopian shores. Yet the 'old ways' remain unchanged—the dominance of the exchange principle and all the hierarchies, inequalities, and social pathologies that come with it—the very conditions that led humanity to the disaster in the first place. The old order lost to the zombie apocalypse is not called into question, instead the narrative assumes the survivors should somehow re-establish the old world anew.

While audiences entertain such survival fantasies in which the individual can make it on their own, this too abstracts away from how social-life is essential to our very being. This is why the putative solutions implicit in many zombie movies have as many dead-ends as the movies themselves. Beginning in *Night of the Living Dead* (1968), the group suffers from communicative deficiencies where they splinter and can no longer fight off the horde sufficiently. In *28 Weeks Later* (2007), the solution is to buttress the industrial military complex and the total militarisation of civil life through containment and surveillance—all in the name of British freedom, with the assistance of Uncle Sam. In *Dawn of the Dead* (1978), we see the growing dissolution of all social bonds and the intensification—rather than abandonment—of capitalism on an island of inequality and rapacious consumerism by the survivors who are compelled to scavenge as their new form of labour. In *Shaun of the Dead* (2004), the zombies are enslaved as effable servants and cheap labour for their human masters. It seems the cures for zombification are worse than the disease. Even the hopeful ending of *28 Days Later* (2002) is based on a small group of three, much like *Dawn of Dead* (1978). There is no possibility for social regeneration, just escape as part of the estranged survivors, clinging onto an already doomed civilisation (see Paffenroth 2006).

This shows our true lack of imagination in contemporary cinema where the problem of sociality can no longer be effectively communicated. This problem was anticipated by Horkheimer and Adorno who posited that capitalism was so totalising that it would circumscribe the possibility of almost every cultural expression external to it, including revolution (Gunn and Treat 2005, 144–148).

Consequently, and particularly so in American zombie movies, we find nihilistic despair as the only outcome. As argued by Adorno, 'fettering consciousness' impedes 'the development of autonomous, independent individuals who judge and decide consciously for themselves' (1991, 106). The zombie is therefore the most extreme embodiment of this alter-archetype, a deliberately one-sided creature—but it is no less a logical outcome of this process because of its exaggeration. We find counterparts in the modern psyche as it faces the contradictory nature of its pathologised social conditions: humans repress and deny, withdraw in numb apathy, or project aggression of the ego. Helplessness, alienation, and a gnawing anxiety are felt not just for the lack of connection with others but a deep-seated fear of the competition that each other represents to the self. We are *not* in this together. All others are *enemy*. I am *alone*. At the social level such processes manifest in racist attacks, aggressive war, mass mania, nihilistic despair, and paranoid protectionism. Such social neurosis is the impulse to fascism and 'Trumpism,' but which only mirrors the presuppositions that the dominant interests have asserted all along: that we live in a world that is utterly hostile, violent, indifferent, alien, and solitary. The ideological interests of the few have become *the* ontological fact that covers all things, the social as well as material universe, under which the ethical mind is completely obliterated.

But being 'duped' is not a foregone conclusion. The most recent turn in zombie films seems to suggest a possibility for social integration. In films like *Shaun of the Dead* and *Warm Bodies*, the zombie becomes the 'post-dead' and re-integrated into society as a 'citizen-other'. However, they then serve a particular subservient role—once again highlighting the pathologies of predatory capitalism—usually as a labourer. So, despite their reintegration, the zombie's existence is defined by neoliberal subjectivity of the *lowest* order (Saunders 2012).[10] For example in *Warm Bodies* (2013), Julie's father refuses to accept that the zombie 'R' can change and subsequently shoots him after he sees the two kiss. Even the name of the leading zombie, 'R', denotes the clear distinction between the re-integrated zombie and the humans who are given complete names. Similarly, when the post-dead return to the village of Roarton in the television series *In the Flesh* (2013–2014), they are faced with prejudice from the other villagers. As such, these most recent depictions are not so much upholding a return to sociality but can be seen as consistent with predatory capitalism. That is, these normalised or reintegrated citizen-zombies reside in contradiction, suffering not only continuous labour exploitation, but also experiencing hostility and segregation from non-infected humans.

It is not, as Wood once claimed in reference to Romero's films, that 'the social order ... can't be restored', but that it *must* be restored (qtd. in Moreman and Rushton 2011, 4). This is not in the sense of a 'myth of return' to the previous social order under capitalism, but in the sense of a genuinely new, rational order that could satisfy 'the needs and powers' of all humanity (Horkheimer 1972, 246) and thereby arrest zombification. The zombie film alerts us to the need for creating these organic social relations necessary to weaving a lasting social fabric. As Brooks (2006) argues convincingly in *World War Z*, how the survivors behave and socialise is the linchpin of the ethics of the zombie genre: many turn on each other, becoming worse than the zombie horde; others commit suicide. Yet the 'few good men'—typically always men—are those that care for their survival group and overcome the odds to retain their humanity. In this sense, the genre cuts across political divides by appealing to the organicism of Burke, and to the progressivism of social freedoms and community needs of the Left. Against all common sense, the message the zombie communicates is not about reaching the shores of utopian security, but the social relations necessary for this movement to occur—and what happens when these relations are absent, pathologised, or destroyed.

Conclusion

We no longer need to wait for the next cinematic zombie-event with its garish make-up and extreme gore. All we need to do is look in the mirror of a morning: dead-tired, red-eyed, preparing ourselves for yet another day of tedious monotony, of shuffling around our workplace; vacant, undertaking the same repetitious tasks as before. Or somehow worse, like those made truly aimless in capitalist order who are compelled to join the ever-expanding number of lumpenproletariat, no longer of use to capitalist production. As so fittingly illustrated in *Dawn of the Dead* (1978 and 2004), the only remaining goal of predatory capitalism is consumption: the irrepressible drive to which justifies systemic, hyper-violence that is instrumentalised to ensure the consumption of the few at the expense of all. In this new era of zombie films the spatial boundaries of the outbreak/contagion have been shattered so that zombification is no longer some localised phenomena on the Periphery of civilisation or experienced only by the underclass. The long hands of capitalism have indeed battered down 'all Chinese walls' (Marx and Engels 1976 [1848], 487–488) so that zombification is experienced across all cultures. We are all infected now. Our socio-political life is so utterly pathological that the zombie pandemic is no longer divine retribution like that of early films that emphasised the metaphysical dimensions of the 'undead', but rather the embodiment of our very predation of each other. We are no longer zombie labourers, nor consuming ghouls, but purely parasitic. This shift is not coincidental but marks the transitional phase of capitalism and the specific type of social destruction that accompanies it—the painful ripping away of 'the socius' that we experience as a lived nightmare.

As Robin Wood claimed, it is impossible for cinema to now shout 'Revolution!' All we can hope for are those films that 'imply its necessity' (2003, 342). This is the emancipatory potential of the zombie movie for geopolitics. And whilst this is clouded under the blood spatter of the violent destruction of world order as it tears itself apart, there is a choice to *sociality*, even in the midst of its gruesome decay. We can choose the barbarism of the zombie, or we can—through painstaking processes of moral learning and mutual recognition—create a new world of cosmopolitan community that nurtures, respects, and supports the diversity of all others. Perhaps this remarkable genre is telling us that the slogan should be changed from 'socialism or barbarism' to 'socialism or zombification'.

Notes

1 See 'No zombie sightings despite warning on 520' at http://mynorthwest.com/11/698090/No-zombie-sightings-despite-warning-on-520# [last accessed 25 June 2012].
2 The Golden Era refers to the series of blockbuster zombie films produced under Romero between 1968–1983.
3 While we have attempted to limit our scope in this chapter to predominantly film (and television), we do make reference to exemplary novels that were each made the subject of film where appropriate.
4 Lauro and Embry (2008) make a similar observation but see this relation as held in tension, whereas we push to a resolution.
5 While popular geopolitics has carved a space for the Frankfurt School and its theory of the Culture Industry (see Dodds 1996; McFarlane and Hay 2003), our contribution seeks to overcome those recent criticisms that have rejected the theory of the culture industry for rendering audiences as passive consumers (see Dodds 2006; Dittmer 2010, 46–56).
6 A particular correlation can be seen in alien horror films of this time, esp. *Plan 9 from Outer Space* (1959) which was one of the first to combine zombies and aliens. It should also be noted that the Nazi-zombie has not been lost, but continued in *Shock Waves* (1977) through to *Dead Snow* (2009).
7 Some doubt the authenticity of zombies in films that have zombies arising from viral infection rather than reanimation; nevertheless, these now seem to outweigh the latter.
8 While many dispute this classification, we classify *I am Legend* as part of the zombie genre due to the behavioural and physiological adaptations (see also Saunders 2012).
9 This is most acutely captured in the repeated symbolism of the butterfly throughout the film, in both original and alternate endings, that make the connection to sociality explicit.
10 We thank the editors of this book, Robert Saunders and Vlad Strukov, for bringing this recent turn to our attention. We return to this later.

References

Acevedo, L.C.M. 2016. *Zombie Factor*. Berlin: GD Publishing.

Adorno, Theodor W. 1973 [1966]. *Negative Dialectics*. Translated by E.B. Ashton. New York: Continuum.

Adorno, Theodor W. 1988. *Critical Models: Interventions and Catchwords*. Translated by H.W. Pickford. New York: Columbia University Press.

Baudrillard, Jean. 1998. *Consumer Society: Myths and Structures*. London: SAGE Publications.

Blanton, Robert G. 2013. 'Zombies and International Relations: A Simple Guide for Bringing the Undead into Your Classroom'. *International Studies Perspectives* 14 (3):1–13.

Boon, Kevin A. 2011. 'The Zombie as Other'. In *Better off Dead*, edited by Deborah Christie and Sarah Juliet Lauro, 50–60. New York: Fordham University Press.

Boyd, Brian. 2009. *On the Origin of Stories: Evolution, Cognition, and Fiction*. Cambridge: Harvard University Press.

Brooks, Max. 2003. *The Zombie Survival Guide: Complete Protection from the Living Dead*. New York: Three Rivers Press.

Brooks, Max. 2006. *World War Z: An Oral History of the Zombie War*. New York: Three Rivers Press.

Caso, Federica, and Caitlin Hamilton. 2015. *Popular Culture and World Politics: Theories, Methods, Pedagogies*. Bristol, UK: E-International Relations.

Chang, Ha-Joon. 2005. *Kicking Away the Ladder*. London: Anthem Press.

Clasen, Mathias. 2010. 'Vampire Apocalypse: A Biocultural Critique of Richard Matheson's *I Am Legend*'. *Philosophy and Literature* 34 (2):313–328.

Dittmer, Jason. 2010. *Popular Culture, Geopolitics, and Identity*. Lanham, MD: Rowman & Littlefield.

Dittmer, Jason, and Klaus Dodds. 2008. 'Popular Geopolitics Past and Future: Fandom, Identities and Audiences'. *Geopolitics* 13 (3):437–445.

Dittmer, Jason, and Nicholas Gray. 2010. 'Popular Geopolitics 2.0: Towards Methodologies of the Everyday'. *Geography Compass* 4 (11):1664–1677.

Dodds, Klaus. 1996. 'The 1982 Falklands War and a Critical Geopolitical Eye: Steve Bell and the If… Cartoons'. *Political Geography* 15 (6):571–592.

Dodds, Klaus. 2006. 'Popular Geopolitics and Audience Dispositions: James Bond and the Internet Movie Database (IMDb)'. *Transactions of the Institute of British Geographers* 31 (2):116–130.

Drezner, Daniel W. 2011. *Theories of International Politics and Zombies*. Princeton and Oxford: Princeton University Press.

Giroux, Henry A. 2011. *Zombie Politics and Culture in the Age of Casino Capitalism*. New York: Peter Lang Publishing.

Gunn, Joshua, and Shaun Treat. 2005. 'Zombie Trouble: A Propaedeutic on Ideological Subjectification and the Unconscious'. *Quarterly Journal of Speech Volume 91, 2005 – Issue 2* 91 (2):144–174.

Hall, Derek. I. 2011. 'Varieties of Zombieism: Approaching Comparative Political Economy through *28 Days Later* and *Wild Zero*'. *International Studies Perspectives* 12 (1):1–17.

Hannah, Erin, and Rorden Wilkinson. 2014. Zombies and IR: A Critical Reading. *Politics*.

Harman, Chris. 2009. *Zombie Capitalism: Global Crisis and the Relevance of Marx*. Chicago: Haymarket Books.

Horkheimer, Max. 1972. 'Traditional and Critical Theory'. in *Critical Theory: Selected Essays*, edited by Max Horkheimer, 188–243. New York: Continuum.

Horkheimer, Max. 2004 [1947]. *Eclipse of Reason*. New York: Continuum International Publishing Group.

Horkheimer, Max, and Theodor W. Ador1994. *The Dialectic of Enlightenment*. London: Continuum.

Koven, Mikel J. 2008. 'The Folklore of the Zombie Film'. In *Zombie Culture: Autopsies of the Living Dead*, edited by Shawb McIntosh and Marc Leverette, 19–34. Lanham, MD: Scarecrow Press.

Lauro, Sarah Juliet, and Karen Embry. 2008. 'A Zombie Manifesto: The Nonhuman Condition in the Era of Advanced Capitalism'. *Boundary 2* 35 (1):85–108.

Leach, MacEdward. 1961. 'Jamaican Duppy Lore'. *The Journal of American Folklore* 74 (293):207–215.

Lubold, Gordon. 2014. 'Exclusive: The Pentagon Has a Plan to Stop the Zombie Apocalypse'. *Foreign Policy*, available at http://foreignpolicy.com/2014/05/13/exclusive-the-pentagon-has-a-plan-to-stop-the-zombie-apocalypse-seriously/ [last accessed 16 December 2016].

MacDonald, Fraser, Rachel Hughes, and Klaus Dodds. 2010. *Observant States: Geopolitics and Visual Culture*. London and New York: I. B. Tauris.

McFarlane, Thomas, and Iain Hay. 2003. 'The Battle for Seattle: Protest and Popular Geopolitics in The Australian Newspaper'. *Political Geography* 22 (2):211–232.

McNally, David. 2011. *Monsters of the Market*. London: Brill.

Marx, Karl, and Frederich Engels. 1976 [1848]. *The Communist Manifesto: Karl Marx Frederich Engels Collected Works, Vol.* 6. New York: International Publishers.

Matheson, Richard. 2006 [1954]. *I am Legend*. London: Gollancz.

May, Jeff. 2010. 'Zombie Geographies and the Undead City'. *Social & Cultural Geography* 11 (3):285–298.

Moreman, Christopher M., and Cory James Rushton. 2011. *Zombies Are Us: Essays on the Humanity of the Walking Dead*. Jefferson, NC: McFarland.

Morrissette, Jason. 2014. 'Zombies, International Relations, and the Production of Danger: Critical Security Studies versus the Living Dead'. *Studies in Popular Culture* 36 (2):1–28.

Paffenroth, Kim. 2006. *Gospel of the Living Dead: George Romero's Visions of Hell on Earth*. Waco, TX: Baylor University Press.

Pasko, Kristina. 2015. ''Warm Bodies': A ZomRomCom Discourse on Counterinsurgency'. *E-International Relations*, available at http://www.e-ir.info/2015/07/13/warm-bodies-a-zomromcom-discourse-on-counterinsurgency/ [last accessed 16 December 2016].

Quiggin, John. 2010. *Zombie Economics: How Dead Ideas Still Walk among Us*. Princeton: Princeton University Press.

Saunders, Robert A. 2012. 'Undead Spaces: Fear, Globalisation, and the Popular Geopolitics of Zombiism'. *Geopolitics* 17 (1):80–104.

Shaviro, Steve. 2002. 'Capitalist Monsters'. *Historical Materialism* 10 (281–290).

Sheller, Mimi. 2003. *Consuming the Caribbean: From Arawaks to Zombies*. London and New York: Routledge.

Taylor, Charles. 1992. *Multiculturalism: Examining the Politics of Recognition*. Princeton, NJ: Princeton University Press.

Webb, Jen, and Sam Byrnand. 2008. 'Some Kind of Virus: The Zombie as Body and as Trope'. *Body & Science* 14 (2):83–98.

Wood, Robin. 2003. *Hollywood from Vietnam to Reagan. and Beyond (Expanded and Revised Edition)*. New York: Columbia University Press.

Younge, Gary. 2012. 'Nato Talks Security and Peace, Chicago Has Neither'. *The Guardian*, avilable at https://www.theguardian.com/commentisfree/cifamerica/2012/may/20/chicago-nato-g8-summit-inequality [last accessed 16 December 2016].

Zuidervaart, Lambert. 2011. 'Theodor W. Adorno'. In *The Stanford Encyclopedia of Philosophy*, edited by Edward M. Zalta, available at http://plato.stanford.edu/archives/win2011/entries/adorno/.

10 Popular geopolitics and the landscapes of virtual war

Daniel Bos

Introduction

The field of popular geopolitics has provided rich insights into the ways our political world is mediated and made meaningful via a range of popular cultural and media outlets. However, the ways in which videogames shape popular understandings of geopolitics remain largely underexplored. This lacuna is lamentable considering the remarkable growth of the overall videogame industry, with recent figures suggesting that half of U.S. households, for example, own a dedicated game console (Entertainment Software Association 2015). In particular military-themed videogames have become hugely popular and profitable within the videogame industry. They offer players escapism where they: enter historical, contemporary, or futuristic conflicts; navigate and interact with simulated war-torn environments; and virtually enact forms of military violence. Unlike other media, videogames present landscapes which players directly explore, interact and perform in. This gives rise to a number of pertinent questions. What landscapes are players entering? What geopolitical symbols, values and ideas are embedded in these spaces? And, how are these geopolitical and militarised virtual worlds experienced, navigated and interacted with?

In this chapter, I set out to consider questions of play and geopolitics through a detailed examination of the virtual environments of conflict as portrayed in the popular videogame series *Call of Duty: Modern Warfare.*[1] In doing so, I draw on the continuing interest of landscape in the human geography discipline to consider the ways geopolitics is made meaningful in and through the virtual landscapes of war. Whereas 'virtual' renderings of the world are purported to present an immaterial, digitalised alternative to the physicality of the 'real' world, this chapter considers the co-constitutive relationship between virtual landscapes and geopolitics. Virtual worlds often blend fantasy with realism and can draw on real-world places and peoples, which communicate ideas, perspectives, and ideologies pertaining to social realities. Moreover, videogames and virtual worlds have become increasingly important for military institutions in promoting recruitment and for the purposes of training and subsequent preparation for real-world military activities. As such, rather than abstracted from political life—videogames matter—they matter in the ways they naturalise and legitimise dominant geopolitical discourses and practices.

To begin, I outline the historical interest in landscapes offered by the discipline of human geography. More specifically, I will focus on how geographers have considered the cultural representation of landscapes in constituting social realities. However, as I argue, videogame landscapes should be considered beyond the representational, due to their interactive qualities. Rather than static iterations of landscapes, I consider recent interests under the banner of ludic geopolitics (see Carter, Kirby, and Woodyer 2016), to consider the ways in which virtual landscapes are playfully explored and interacted with, and how this is communicative of geopolitics. The purpose of the chapter is therefore to explore how popular geopolitical imaginations of the world are not only evoked through the representative virtual worlds but to consider the complex interplay of gameplay mechanics, narrative structures, audio-visual content and player practices that contribute and shape geopolitical interactions with the virtual landscape.

In drawing attention to specific landscapes, I argue that players of the *Modern Warfare* series encounter both 'othered' landscapes and familiarised landscapes. While the conflict in the series is initially situated in distant landscapes, mirroring contemporary places of military action, urban landscapes in the American homeland, and key Western metropolises, are also drawn into the battlespace. I argue that the varied landscapes of the *Modern Warfare* series articulate geopolitical imaginaries, in which places of security and insecurity become increasingly ambiguous, reflecting and constituting discourses concerning the War on Terror. To conclude, I consider future research trajectories that can go further in illuminating the geopolitical significance of playing war in virtual landscapes.

The empirical data discussed in this chapter draws on a detailed case study analysis of the *Call of Duty: Modern Warfare* series.[2] The series consists of three games, which following on from Šisler (2008), were played by the author 'while taking notes and screenshots of relevant visual signifiers, recording the narrative and analysing the structure of gameplay', as to attend to the game's meaning arguably requires direct interaction from the player/scholar. This presents a challenge as the subsequent engagement and interaction with the game world is contingent on the players' skillset, choices and motives which has implications on the playing of the game and the subsequent meaning conveyed. It is important to recognise that the analysis offered is a 'selective reading' (Keough qtd. in Jennings 2015, 4), a product of my own gameplay experience and research endeavours to consider the portrayal of geopolitics. Here, the games' overall plot, characters, and places were compiled in a database using screenshots, personal notes, and transcripts of the audio.

Landscapes, videogames, and geopolitics

In the historical development of the discipline of geography, landscape has remained a central concern for scholars (Cosgrove 1985; Mitchell 2002; Wylie 2007; Woodward 2014). Traditionally geography has been attentive to the physical, environmental, and morphological attributes inherent in landscapes. Interests here are very much directed to the material appearance of landscapes and the processes and practices that contribute to its subsequent appearance. Yet, considering landscapes as 'things' overlooks the ideological function that they perform in social life (Morin 2009). The emergence of the 'cultural turn' in Anglo-American human geography in the 1980s encouraged new analytical and theoretical perspectives in which landscapes could be understood. Here, scholars moved away from a purely material, physical, and objectified appreciation of landscape, to consider the multiple cultural forms in which landscapes are represented.

Within geography specific attention has been cast on visual landscapes that are depicted and that take form in a range of cultural artefacts, such as paintings (Daniels 1993; Sage 2008), film (Lukinbeal 2005; Harper and Rayner 2010) and photography (Schwartz 1996; Tolia-Kelly 2004). In these cases, scholars have sought to unpack the power relations evoked in the representations of landscapes and how they reinforce ideologies, 'supporting a set of ideas and values, unquestioned assumptions about the way a society *is*, or should be, organised' (Duncan and Duncan 1988, 123). Scholars have emphasised that landscapes have a constitutive role in shaping social realities.

Landscapes are important to studies of geopolitics, not just as material planes in which geopolitics is enacted and unfolds, but also as cultural reproductions in which the geopolitics is made meaningful. For Dittmer (2005), the comic book *Captain America* reproduces highly symbolic landscapes that cultivate a sense of American national identity. Sage (2008) considers the role of astronomical art in drawing on the aesthetical stylings of the American landscape sublime tradition in rendering outer space as a part of a wider geopolitical sensibility of American manifest destiny. Such work emphasises the geopolitical questions of landscape and how they evoke political questions and expressions of identity and power.

Despite their popularity, however, videogames and their virtual landscapes have thus far warranted little engagement. Virtual landscapes, Longan (2008, 24) writes, are an 'integral part of many video games' and they work 'to enhance gameplay, communicate useful information, and help tell a story'. Besides their apparent relation to the gameplay, Longan is keen to stress how these virtual landscapes 'are not just games but rather representations of reality that have something to tell their players about their everyday world' (2008, 26). These landscapes are important in the ways they reinforce dominant spatial ideologies and communicate particular understandings of the world through their interaction. Strukov (2012) outlines the ways ludic environments can be conceived as ideological spaces, noting the complex, layered ways in which the virtual worlds of *Allods* commemorate Russia's past, thus constituting popular imaginings and myth-making integral to Russian identity. On the other hand, Woodward (2014) has alluded to the significance of the virtual rendering of military landscapes in military-themed videogames in communicating the geopolitics of military violence. In doing so, they can be considered as representational tools that perform a role in 'justify[ing] [military] activities in places, spaces, environments and landscapes' (Woodward 2005, 729).

The representation of landscapes and places in these particular games offers a powerful understanding of the geographies of military violence. Ash (2014) contends they can do so in two different ways. Firstly, videogames can provide fantastical landscapes for players to explore—creating anonymous places that are scrubbed of stereotypes. In the military-themed video *America's Army*, the places and landscapes are generalised, and the enemy is abstracted—devoid of a political, national, or ethnic identity—what Allen (2011) has termed an 'unreal' enemy. Secondly, through real-world reference points videogames can draw on existing stereotypes. For instance, the single-campaign mode of *Modern Warfare* often draws on real-world locations to present particular ideas of place and its inhabitants. Analysing the virtual landscapes therefore offers a means of critically exploring contemporary 'militarised landscape imaginaries' (Woodward 2014, 45) and to consider their significance in shaping popular understandings of geopolitics and the geographies of military violence.

However, a focus on videogame landscapes in terms of their symbolic and representative features overlooks the specificities of the medium. Landscapes in videogames require the active involvement of the player who has to navigate through playful interactions with the virtual worlds (Schwartz 2006). As Magnet suggests videogame landscapes are not just 'static objects "to-be-looked-at", but are dynamic and require the active involvement of the player in their construction' (2006, 143). Taking this into consideration, we can usefully turn to more recent work within critical geopolitics that has drawn attention to the ludic character of geopolitics and the importance of play. Ludic geopolitics turns attention to various material and digital mediations in which playful encounters link to wider geopolitical cultures (Carter, Kirby, and Woodyer 2016). Play remains important to geopolitics, if we consider the historical use of 'war gaming' for the purposes of military preparedness, to the mundane setting of the home in which children (as well as adults), playfully engage with toys and videogames, which speak to wider geopolitical concerns (MacDonald 2008). Moreover, while not jettisoning the representational completely; bringing play to the forefront highlights its interactive, performative, and embodied qualities. Certainly, when turning to the virtual ludic worlds offered in military-themed videogames, we see how the process of geopolitical meaning-making is made intelligible through the players' playful interaction with the virtual landscapes. In taking this notion of ludic geopolitics forward, a useful approach then is to think through how the landscapes of virtual war are not only made meaningful through their visual renderings, but also through the complex interplay between player actions, gameplay mechanics, and narrative structures, which produce particular imaginations of the political world.

Turning to ludic virtual landscapes offered in videogames requires us to consider the specificities in which players encounter the landscapes. While a degree of agency is afforded to the player, it is imperative to note that the interaction with virtual landscapes is predefined by the games' designers. While appearing to offer player agency, games such as the *Call of Duty* single-campaign mode are limited by a range of gameplay mechanics and structures, which guide the player along a linear pathway. As such videogames offer particular *ways of seeing*.

Hughes (2010), for example, argues that videogames offer a new form of geopolitical visuality—a visuality that is premised on an interactive practice of looking. Key here is a consideration of how particular modes of looking are afforded to the player. The first-person perspective seeks to fold the distance between self and landscape. Players experience and see the unfolding landscape from the perspective and movement of the on-screen avatar. The vast majority of military-themed videogames are presented with a hyper-masculine avatar, in which the subsequent encounters with the virtual landscapes are mediated through a militarised, masculinised gaze. Such gendered ways of seeing reinforce hegemonic notions of masculinity tied to geopolitical cultures of unilateral interventionism, protectionist sensibilities and morally righteous violence (Ó Tuathail 2005), which becomes playable through these games.

Hughes (2010) goes further to contend that videogames offer an anticipatory form of visuality. By this he suggests how gameplay in military-themed videogames elicits a constant state of alertness, which is met by the pre-emptive gameplay action of the player. The act of looking in the game is always tied to the relations between movement and feeling. Such assertions bring forth concerns with the 'more-than-representational' character of videogames. By this, scholars have considered the ways vision should be understood as a wider embodied practice always commingling with other bodily senses (MacDonald, Hughes, and Dodds 2010) We can take, for example, the role of sound in amplifying affective states of play or the haptic technologies, such as forced feedback, where the control-pad vibrates in relation to the on-screen activities of the player. Playing virtual war thus speaks to a wider bodily sensorium that serves to cultivate a highly embodied and affective experience for the player. It moves focus away from 'not only what images or visions show of the world, but also to what images or visions *do* in the world' (Hughes 2007, 991). This moves us beyond purely visual accounts to consider the experiential elements of 'being-in-landscape' and how ideologies and cultures of militarism emanate from gameplay. Before examining the specificities of the landscapes offered in the series, I will outline the place specific narrative-arc of the *Modern Warfare* series.

The geographies of *Modern Warfare*

In the *Call of Duty* franchise, the places and landscapes of military violence profoundly shifted with the arrival of the *Modern Warfare* series. Previous iterations of the franchise focussed on the historical conflict of World War II; however, the *Modern Warfare* series brought the franchise's military-themed First-Person Shooter (FPS) to contemporary geopolitical settings. This has provided the producers an opportunity to script the world in particular ways. No longer tied to virtual rendering of the geographical historicity of World War II, the franchise's developers were able to creatively explore and narrate a contemporary fictitious narrative.[3] Set in the near future (*Modern Warfare* while released in 2007 was set in 2011) the series can be defined as a proleptic game—'set in the present or near future, and present[ing] possible future interventions into present-day "hot-spots"' (Smicker 2010, 113).

Here, the game used a geographical scripting of the world in which a range of real-world locations become the setting of contemporary military violence. *Modern Warfare* uses expansive real-world references thus actively mapping the political world through the logics, practices and imaginations of military violence.

The specific location and landscape in which the player is placed is narrated primarily through the cutscene—a narrative device used in videogames which uses, predominately, brief non-interactive cinematic clips to contextualise the games' objectives and narratives (Wolf 2001). *Modern Warfare* uses satellite imagery which manoeuvres around the globe to focus on the location of the next mission. Such aesthetical stylings invite a particular way of seeing, knowing and relating to the world—it contextualises the spaces of play and sutures the disparate locations of geopolitical action. Textual information is also displayed at the beginning of each mission, detailing the character name, regiment, time, and the location. The geographies detailed within the series vary. The locations can be declared via vague references to wider regions, i.e. the Middle East, mysterious locations only made knowable because of their proximity to other places, and particular locations, such as urban cities. Table 10.1 includes locations which are announced at the beginning of each mission.

Considering the evolving storyline of the *Modern Warfare* series, we can note the series begins with a focus on 'distant' locations within the Middle East, the Caucasus, and Russia. The series continues to draw on discourses from the Cold War, reigniting concerns of a nuclear arms race and evoking longstanding

Table 10.1 The geographies of the *Modern Warfare* series as presented in the games

Call of Duty: Modern Warfare (2007)	*Call of Duty: Modern Warfare 2 (2009)*	*Call of Duty: Modern Warfare 3 (2011)*
Credenhill, England	Fire Base Phoenix, Afghanistan	Northern India
Bering Strait	Tian Shan Range, Kazakhstan	Manhattan, New York, USA
Middle East	Moscow, Russia	Himachal Pradesh, India
Caucasus Mountains	Rio de Janeiro, Brazil	Sierra Leone, Africa
Western Russia	North Eastern Virginia, USA	Canary Wharf, London, England
Iran	Vikhorevka 36 Oil Platform, Russia	Hamburg, Germany
Pripyat, Ukraine	40 mi. E of Petropavlovsk, Russia	Boosaaso, Somalia
Altay Mountains, Russia	Washington D.C., USA	Montmartre Hill, Paris, France
	14 mi. SSE of Petropavlovsk	Prague, Czech Republic
Northern Azerbaijan	Georgian-Russian Border	Berlin, Germany
Southern Russia	160 mi. SW of Kandahar, Afghanistan	Eastern Siberia, Russia
	Site Hotel Bravo, Afghanistan	Arabian Peninsula

stereotypes of Russia as an aggressor and a nation inclined to sponsoring and harbouring terrorism (Gagnon 2010). Likewise, regions such as Central and Southern Asia have commonly and historically been associated by discourses of danger (Megoran 2005; Heathershaw and Megoran 2011). Scholars have drawn inspiration from Edward Said's *Orientalism* (1979), which considers the historical unequal power relations in which knowledge of Oriental cultures is discursively constructed by a range of cultural actors and practices in the West. This led to reductive 'othering' of the Orient as exotic, mysterious and backward, which is presented in opposition to the 'West' as modern, progressive and civilised. The resulting simplistic imaginative geography performed an important role in maintaining and justifying imperial control and the endeavours of Western powers. Such insights remain pertinent with the continuing importance of contemporary visual cultures in perpetuating Oriental representations. Military-themed videogames have become key cultural vehicles which mirror contemporary intrigue in the Middle East, which maintains a colonial imaginary.

However, the geographical intrigue changes in *Modern Warfare 2*. While still maintaining a focus on Russia and Central and Southern Asia, conflict infiltrates the American 'homeland'. Furthermore, in *Modern Warfare 3*, the conflict spreads further into Western European cities, such as London and Paris, mirroring current cultures of fear and anxieties, as well as the actual sites of terrorist violence, as seen in the attack on the London underground in 2005. As the series develops the geography of conflict increasingly turns inwards into the key Western urban locations and the American 'heartland'. The series offers a spatial imagination where place binaries of perceived safety and danger, security and insecurity, 'battlefront' and 'homefront' become increasingly blurred and ambiguous (Carter and Dodds 2014). While in the series distant locations become landscapes of projected military violence, Western locations are the sites of terroristic events which evoke the complex geographical imaginary of the War on Terror. In the next sections, I draw specific attention to the ways in which this geographical imagination is made meaningful through interactions with the virtual landscapes.

Landscapes of the 'other'

Central to critical geopolitics has been the unpacking of ideological construction of geographical imaginaries that reinforce spatial boundaries between domestic and the dangerous 'other' (Dalby 1990; Ingram and Dodds 2012). In the series there are continual references to real-world locations that are contemporary geopolitical 'hot-spots'. In the second mission of *Modern Warfare 2* '*Team Player*',[4] the player assumes the role of U.S. Army Ranger PFC Joseph Allen who assists in a military operation to overtake a militia stronghold in an unknown town located in what is designated as the 'Red Zone', Afghanistan. The rationale and context in the overall game's narrative is ambiguous, however the objective remains simple: 'Search and destroy enemy forces in Afghanistan'. The place is defined as the Red Zone, which is synonymous with the ways space was designated in the city of Baghdad, Iraq, with small enclaves controlled by coalition forces known as the

Green Zone. This was juxtaposed with the Red Zone—spaces of the city designated as insecure and dangerous. This dichotomy of Red/Green zones became a popular mediated means of understanding the conflict in that it presented a stark spatial framework of military violence in Iraq, but also became analogous for the militarised designation of the global space. This sense of undifferentiated danger is encapsulated as the player moves through the virtual landscapes.

Opening in the midst of a firefight spanning a river between Americans and the militia group, who are placed on the opposite side of the embankment, the player is told to engage the enemy in order to provide cover and allow the group to safely cross the nearby bridge. Through the vision of the avatar, the urban landscape is revealed with war-torn high-rise buildings and a minaret occupying the skyline. Upon successfully suppressing fire, and after a successful air-strike, the team is able to move further into the town via military vehicles (see Figure 10.1).

The series affords numerous ways of seeing and interacting with the virtual landscape. While the series' predominantly first-person perspective is based on the movement of the in-game avatar, other mediated ways of seeing are presented during the gameplay. In this instance the player now operates a fixed turret gun atop a military Humvee. As such the player's way of seeing the unfolding landscape is tied specifically to the movement and operation of the gun.

As many have contended, oriental tropes pervade the military videogame genre, engendering an 'otherness' in the places and the people that inhabit them (Šisler 2008). As Höglund (2008) suggests, Said's Orientalism remains relevant in thinking about cultural reproduction of the Middle East and finds expression in military-themed videogames. References to the Middle East as a site of continuing perpetual warfare serve to legitimate wider discourses concerning the War on

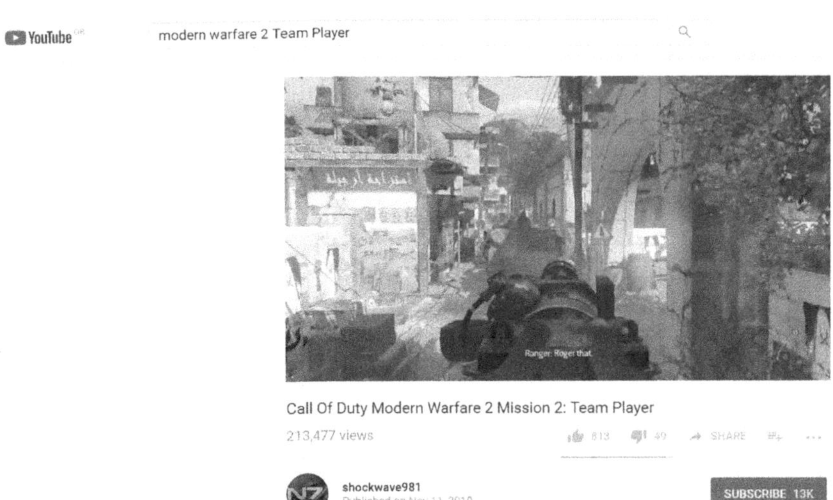

Figure 10.1 Screenshot of a YouTube video showing gameplay. The player enters the unnamed city taking control of a turret gun atop a Humvee.

Terror and American foreign policy. This is expressed in how the game landscapes are portrayed and how they are interacted with.

Upon entering the city the player is faced with numerous side streets occupied by burnt-out cars, market stalls, rubbish and walls adorned with graffiti. Dust clouds obscure the player's vision as the urban landscape unfolds as a 'dark, exotic, labyrinthine and structureless place' (Graham 2006, 256). As we move further into the urban area visual signifiers further allude to this. Two signs appear on the right, one a pedestrian sign with the black figure dressed in traditional clothing, and a second that follows a circular 'no camel' road sign.

Moreover, it is important to note who does and does not occupy the virtual landscapes. Similar to other military-themed videogames, civilians remain largely absent from gameplay, presenting a simplistic and uncomplicated imagination of the battlespace. Civilians in *Team Player* are noted running and hiding in the back allies, but as Graham (2006, 257) suggests these distant urban locations rendered in videogames become unanimously 'little more than "terrorist nest" targets to soak up U.S. military firepower'.

As the convoy moves through the town, three unarmed militia are spotted atop a balcony looking over the convoy:

Hunter 2–3:	'Three-foot-mobiles, balcony 12 o' clock. Probably militia.'
Sergeant Foley:	'Are they armed?'
Hunter 2–3:	'Negative, they're just watching us.'
Corporal Dunn:	'I bet they're scouting us.'
Ranger:	'Eh, but that don't mean we can shoot 'em.'

The enemy combatants are characterised by their suspicious behaviour; 'scouting' the area (see Figure 10.2). Despite the identity of the group being ambiguous, the characteristics are replete with recurrent orientalised motifs, such as headscarves (see Šisler 2008). The collectivisation and abstraction of the enemy is juxtaposed against the individualised protagonists. As Shaw (2010, 796) attests, this dehumanisation and abstraction services ideologically to allow players to 'feel like participants—without the attending moral dilemma that the enemy may be just like them'. Furthermore, the rules of engagement are reinforced with Sergeant Foley commanding:

> 'All Hunter two victors keep an eye out for civvies [civilians]. We're not cleared to engage unless they fire first'.

The player is unable to fire upon the military clad yet unarmed militia despite their apparent suspicious presence and behaviour. This again serves to reinforce a sense of moral righteousness opposed against an opportunist enemy. Shooting at them causes the game to immediately be stopped and a message appear stating 'you are not authorised to fire on unarmed targets'. In this sense the game structures reinforce the logics of Rules of Engagement (ROE)—albeit if disregarded the player

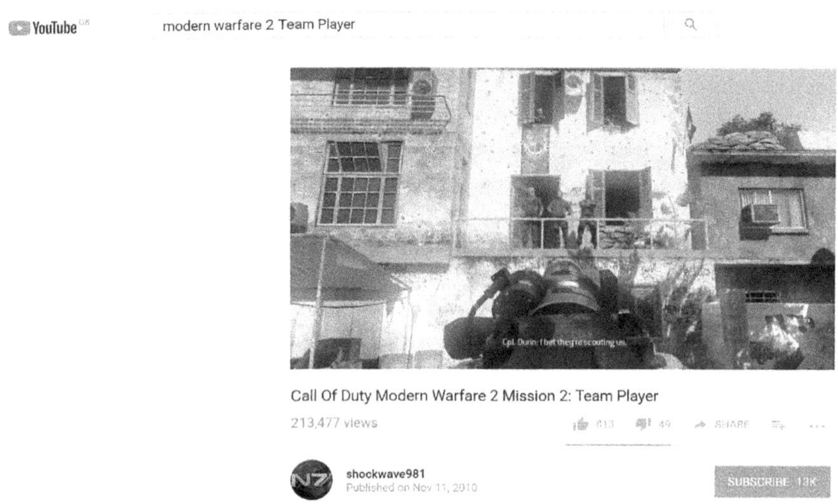

Figure 10.2 Screenshot of a YouTube video showing gameplay. Enemy combatants scout the military convoy.

Source: Activision 2009.

is able to restart from the last checkpoint. Whereas *America's Army* specifically emulates and structures gameplay around ROE, and attempts to present the social realities of the American military, the *Modern Warfare* series does not impose the same level of regulatory gameplay, nor does it permit 'to shoot anything that moves' (Salter 2011, 371). While certain aspects of the regulatory framings of the game are expressed explicitly in the narrative, the implication on the players' gameplay does not have the same regulatory ethos as the likes of *America's Army*.

While virtual landscapes can be considered in relation to their visual form, it is important to consider other aspects which evoke a sense of place. In this instance game sounds play an integral role in which the virtual landscapes are understood and experienced in 'more-than-visual' ways. Sound has a profound influence on the virtual landscape in two ways. Firstly, through diegetic means—sounds associated with the on-screen action, and secondly through the non-diegetic, sounds that appear disconnected from the on-screen action, such as 'mood' music (see Whalen 2004). Both of these are important aspects in how the player understands and relates to the landscapes that they navigate. Music in *Team Player*, for instance, works to build suspense as the player moves through the deserted streets of the Afghan cityscape. Low drumming and bass help build the tension, while the soundtrack heightens when enemy avatars are spotted and the player begins to open fire. The importance of sound should not be overstated in affectively engaging the player, augmenting tension and amplifying militaristic cultural sensibilities (Carter and McCormack 2006). Pieslak's (2009) examination of soldier's engagement with music in the Iraq War reveals the importance of music as a source of cultural awareness, combat inspiration, and psychological preparation in the battlefield. Similarly, the diegetic and non-diegetic sound in the game serves to immerse and connect players into the game world by cultivating an affective relationship with place.

Music in videogames not only generates a sense of imminent danger but can also be used to evoke particular geographical locations. As musical composer for the *Modern Warfare 2*, Hans Zimmer professed 'the locales [depicted in *Modern Warfare 2*] dictate the instruments' used in producing the soundtracks. He states 'when we are in Brazil we are more Brazilian and when we are in Russia, we are more Russian' (Snider 2009). In *Team Player*, hand drums and stringed instruments appear to be used to complement the Afghan landscape. In-game music becomes an important feature that not only works affectively generating states of 'safety' and 'danger' via players' movement through the landscapes, but is used to actively promote a sense of place and being-in-place. This opens up a 'messier, affect-orientated understanding of visuality' (MacDonald, Hughes, and Dodds 2010, 4), whereby playing virtual war connects to the wider bodily senses.

Diegetic sounds play an integral role relating to seeing the landscape. As I have already begun to indicate Non-Player Character (NPCs) dialogue is important in reinforcing in-game objectives, but also in providing descriptive accounts of the landscapes and how the player should interact with it. NPCs are avatars that are not controlled by the player and follow a predefined script set and encoded by the games' developers. In *Modern Warfare 2*, the playable character is confronted by NPCs in terms of enemy avatars, as well as team members.

In the mission *Team Player*, NPCs provide a rich militarised commentary of the landscape. The player is constantly reminded how to view, perceive and act within the landscape. As the convoy moves through the streets, NPCs remind the player who is in operation of the turret gun on the Humvee, to remain vigilant, to 'proceed with caution' and 'to watch those alleys'. Colloquial militarised phrases, such as 'stay frosty', encourage the player to stay alert. Again, further designations of this landscape are made by NPCs:

Corporal Dunn: 'I've got nothing. This place is dead.'
[...]
Corporal Dunn: 'Stay frosty, you guys. This is the Wild West.'

In the first instance the city is described as devoid of life. Despite a couple of civilians noted in the earlier part of the mission, the urban location is engendered as a place only capable of harbouring terrorists (Höglund 2008). Moreover, the 'Wild West' label presents a sense of unpredictability, lawlessness and violence, something to be tamed. In this case the distant urban location becomes reimagined in terms of popular American mythology, of territorial expansionism, the securitisation and the 'taming of "dangerous environments"', through violent means (Saunders 2012, 119). These landscapes are thus made meaningful through a combination of visual representation, gameplay mechanics, and through verbal comments made by NPCs, which shapes particular geographical imaginations through which the player navigates, interacts and responds to.

Another consideration is that it is important to recognise that the rendering of these virtual worlds and the choice of places and landscapes does not appear out of a cultural vacuum. As Wolf (2001) suggests, the videogame medium is highly intertextual, drawing upon a range of other media texts. While the *Call of Duty* series and the producers have indicated the influences of cinema, the mission detailed above draws notable similarities with the HBO series *Generation Kill* (2008).[5] The television series, a dramatisation of a book of the same name, recounts the experiences of an embedded journalist, Evan Wright, who follows a U.S. Marine Corps during the Iraq War in 2003. The parallels between the *Team Player* are evident in the second episode of the series entitled '*The Cradle of Civilisation*', which follows the Marine Corps entering an Iraqi town in military Humvees who are subsequently overcome by insurgents.

Despite being set in Iraq, the *Modern Warfare* mission presents a similar urban setting but instead is purported to be set in Afghanistan. As detailed earlier, the 'Red Zone' alludes to the labelling and mapping of space in Baghdad. Further references are made throughout, including the character dialogue and the terminology used, alongside the urban landscape, which bear strong similarities to the virtually rendered world. Ideas and representations of the military and the landscapes they operate in are translated from other texts to form and shape meanings. In this case, while the geographic locations differ, the representational portrayal remains the same: Iraq and Afghanistan are mediated as indistinguishable landscapes of violence, conflict, and danger.

Despite the game developers' proclamation that these are fictitious worlds, the geographies of military violence and geopolitical discourses of contemporary danger are reinforced in the series. As the series progresses, however, we can see how the cartographic imagination of military violence as something located in distant places and landscapes is disrupted when conflict enters Western cities and the American 'homeland'.

Landscapes of the 'homeland'

A powerful aspect of the *Modern Warfare* story is the way American and Western European urban landscapes progressively become the site of military conflict and violence. As reviews of *Modern Warfare* 2 attest, '[t]here's no place in the world where a skirmish can't go down, from airport security lines to the neighbourhood burger joint to your own backyard' (Mastrapa 2009). In particular *Modern Warfare 3* sees the conflict enter and infiltrate urban locations such as New York, London and Paris (see Table 10.1). Not only do military-themed videogames arguably manufacture consent for U.S. foreign policy, but in introducing landscapes familiar to Western audiences they ask questions around domestic politics, especially around discourses of securitisation. Popular visual forms of culture are arguably important vehicles in which geographical imaginations of threats, anxieties and danger towards national and international security find expression.

of 1940s film noir (Naremore 1998, 227) and the Chinese (American) gangsters that proliferate in the science fiction and fantasy films of the 1990s, at regular historical junctures Hollywood has employed the pejorative and denigrating representational practices of Orientalism to disseminate and also deal with greater formal and practical American anxieties about China as belonging to a geopolitical region the U.S.(qua the 'West') considers dangerously different and threatening to its global authority.

By way of contrast, I consider how, in the new millennium, the rising power of China's film market are incentivising Hollywood to expand its ordinarily limited range of possible ways of understanding the People's Republic of China (PRC) and constructing 'Chineseness'. The PRC's recent, rapid economic achievements and full-scale integration into the world economy have revitalised U.S./Western fears about the former's (and its own) place in the current world order, in turn provoking a renewed wave of anti-Chinese sentiment across large swathes of the Western mainstream (particularly news) media, but Hollywood does not appear to be following suit.[1] Keen to penetrate China's now highly lucrative, but still tightly controlled film market (which concomitant with the gradual fruition of an ideologically motivated, decade-long boom in cinema construction has grown, and continues to grow, at an exponential rate),[2] the major Hollywood studios are not only tailoring their top-tier blockbusters for Chinese audiences—by including popular Chinese locations, actors and products—but also taking greater care to avoid depictions that might offend Chinese cultural and political sensibilities.

'Progress', if we describe it as such, has not been uniform: think of *Men in Black 3* (dir. Barry Sonnenfeld, 2012), which rehearses the conventional Orientalist understanding of Chinatown as the traditional enclave of 'Chinese cruelty and malign intent' on Western soil (Sardar, 1999: 101), or *Pacific Rim* (dir. Guillermo del Toro, 2013), which depicts Hong Kong as a city of deceitful, uncivilised monster eaters. But a great many of the staple Orientalist stereotypes and tropes that encourage our regard for Chinese (and Chinese-Americans) as stubbornly uncivilised, morally dubious and even downright weird are being held in abeyance, and recent blockbusters tend to convey the impression that Hollywood—albeit motivated by the wallet rather than the soul—is mounting an attempt to make amends for its perpetuation of anti-Chinese prejudice. Consider *The Amazing Spider-Man 2* (dir. Marc Webb, 2015), which offers a more inclusive view of Chinatown. Contrasting sharply with incursive ghetto of inscrutable and baleful Chinese practices encountered by MiB agents J (Will Smith) and K (Tommy Lee Jones), *Spider-Man 2* subjects New York's Chinatown to a process of diegetic gentrification that diminishes danger and shrinks difference. A beautifully shot cavalcade of warm and welcoming colours, New York's community of expatriate Chinese is here recalibrated as a multicultural shared space that is both an integral and integrated part of the social and cultural fabric of the Big Apple.

This shift in tone extends also to China itself. As I discuss below, recent U.S. blockbusters that feature Chinese components (characters, settings, and story elements) tend towards portrayals that eschew Orientalist notions of primitivism and perilous exoticism, and reframe China as a territory now separate from

As Bialasiewicz *et al.* (2007) argue these security discourses are performative, rather than separate from the material world and produce the world they represent.

In the series, the borders and distances between the battlefronts and homelands increasingly become ambiguous and blurred. Landscapes and places, certainly familiar to Western audiences, become the site/sight of conflict as perceived iconic landscapes of domesticity and security are turned into war-torn streets, overcome by hostile, ultranationalist Russian forces. In doing so, the story develops what Sontag (1965) notes as the cultural depictions and 'imaginations of disaster'. In this case, the fears, dangers and insecurities manifest with the *Modern Warfare* series through allusions to 'World War 3', and the fear of terrorism in major Western metropolises.[6] While proclaimed by the producers as a fictive narrative, this proved contentious and uncomfortable for certain commentators. One mission set in London entitled *'Mind the Gap'* in *Modern Warfare 3* drew comparisons with the terrorist attacks that occurred on the London Underground in 2005 (Daily Mail 2011).

In both *Modern Warfare 2* and *Modern Warfare 3*, the Ultranationalists have taken control of Russia and have begun a full-blown assault on the eastern coastline of the U.S. The urban landscapes of Western Europe and North America become places where military violence is located and promotes particular understandings of national identity. Dittmer (2005, 2013) considers the ways in which the comic book series *Captain America* constructs landscapes, which are imbued with certain cultural and political values. Developing on Herb's (2004) notion of territorial bonding, landscapes 'reifies the connection of particular politics to specific territories through a variety of narrative and visual strategies' (Dittmer 2013, 102). Here, 'Americanness' is understood through either the symbolic landscapes visually depicted in the series, alongside the more quotidian landscapes of the American 'heartland'. In the mission '*Wolverines!*', *Modern Warfare 2*, the player assumes the role of U.S. Army Rangers, with the objective to repel an assault by the Russian military. The player navigates the streets of a small suburban U.S. town. Flags fly outside the homes of stereotypical white, detached properties complemented with porches and pristine, picketed lawns; symbolism of the mythic American 'heartland', which becomes reimagined as spaces and places of vulnerability and insecurity.

The essence of America is further induced as players enter iconic landscapes of the nation. In one of the later missions '*Whiskey Hotel*', players must claim back America's most iconic political landmark from complete annihilation—the White House.[7] Washington D.C. and its historic and iconic landmarks become the battleground as the player attempts to repel the Russian forces.

This cataclysmic vision of a traumatised American homeland succumbing to an invasion force is a continuing popular cultural trend.[8] In *Modern Warfare*, cityscapes imbued with national importance and power become visualised as the targets and places of violence. This use of national icons affords a distinct marker of national identity for people within that nation and a key signifier of that nation for those outside. As Edensor (2002) states, nations are not just defined in terms of their borders, but also through iconic sites and buildings, which possess symbolic power that reifies a sense of national identity. Iconic American landmarks portrayed in the series, such as the Stock Exchange, the National Monument, and the White House, operate as synecdoche. As Edensor (2002, 46) goes on to explain, sites and iconic landscapes act 'both as signifiers of [America] for outsiders and as ideological statements about [Americanness]… within'. In this case, the White House has strong connections to the imaginations of the nation, being the epicentre of American political power. Not only do these repetitious visualisations of national monuments and architecture secure a sense of place for American and Western audiences, the use of these iconic sites in the midst of destruction coincides with a geopolitical currency, which is very much dominated by a sense of danger and fear (Pain 2010). The Russian invasion and the occupation of the White House, as well as occupying the wider region, presents a geopolitical imagination of American insecurity in both the metropolitan sites of power, and the domestic setting of the suburban hinterlands (see Figure 10.3).

In the finale of the mission *Whiskey Hotel*, the player is told to advance via the left flank and, after heavy Russian resistance, enters the White House via the Oval Office. A loudspeaker advises that 'Hammerdown Protocol' has been initiated, in order to take out the Russian forces and that the destruction of the city is imminent. The player is forced to head to the roof of the White House to avert the aerial bombardment of Washington. The player must quickly use green flares to ward off the incoming military jets. Upon completing this objective and overlooking the smouldering Washington, D.C. skyline, a brief exchange occurs between NPCs:

Figure 10.3 Screenshot of a YouTube video showing gameplay. The White House in *Modern Warfare 2*.

Ranger: 'So when are we going to Moscow?'
Corporal Dunn: 'Not soon enough man. But I know we're going to burn it down when we get there.'
Sergeant Foley: 'When the time is right, Corporal, when the time is right.'

The undue trauma suffered at the hands of the Russian forces provokes an immediate call for an affectively charged retributive justice. Robinson (2015) has noted the ways that the genre of military videogames from an American standpoint, can powerfully instil sentiments of American exceptionalism. The above scene chimes with a sense of a lone victimisation of America, where equivalent violent revenge is called for. The national symbols and iconic landscapes further attach a sense of belonging and familiarity to Western audiences, while concurrently the narrative disrupts notions of security and safety, as the battlefront expands into the 'homefront'. This turn to notions of homeland ties in with the perceived Western market for the games and provides an outlet in which recognisable and familiar locations are (re)imagined from places of perceived security and national importance, to places prone to military and terrorist violence.

Conclusion

In this chapter, I have explored the geographies and virtual landscapes of the *Call of Duty: Modern Warfare* series. I have argued that the virtual landscapes that players engage with in videogames are significant as they are invested with cultural and political meaning which, when interacted with, present particular ideas of how the world works. As illustrated, in exploring the geographies of military violence *Modern Warfare* articulates popular understandings of geopolitics. While initially the narrative is suggestive of violence, danger and conflict situated in distant locations, as the series progresses the American and Western urban locations become places of insecurity.

These variegated geographies and landscapes depicted in *Modern Warfare* express a contemporary imagination, where military violence is played out in distant and familiar locations. While legitimating military violence 'over there' in distant locations, dystopic imaginations of Western landscapes under threat also work to justify widening forms of security practices and militarised solutions within Western cities (Graham 2006). As an integral part of videogames, virtual landscapes should not be considered an incidental aspect of study but need to be critically unpacked to consider the ways they are reflective, but also constitutive of the pervasive conditions in which contemporary military violence is contextualised and situated.

This chapter also raises a number of possible future research trajectories. As discussed, there is a danger that the landscapes in videogames are treated and analysed purely through representative registers. In other words, it is important to consider the specificities of the media and the ways in which geopolitics is articulated. The landscapes that are encountered in the likes of videogames, such as *Call of Duty*, are made understandable through the active involvement of the player. This requires further efforts to considering the multiple, contingent and often indeterminate nature of ludic geopolitical encounters. This might take shape through ethnographic accounts of playing virtual war (Bos 2015; Carter, Kirby, and Woodyer 2016), which would provide grounded understanding into the ways the virtual landscapes are productive of geopolitical sensibilities. Moreover, military-themed videogames attract a global audience. It would also be productive in not just considering how Western audiences engage with the dominant Western geopolitical discourses, but how these landscapes are in turn experienced, understood and interacted with by players in the Middle East, for instance (Shaw 2010). Such trajectories would reveal the actual ways in which popular understandings of geopolitics and military violence are encountered and negotiated in patterns of everyday life and in diverse cultural contexts.

A fruitful way of developing popular geopolitical enquiry is to consider the experiential and elements of playing war in virtual landscapes. There is certainly more to be said about the detailed ways landscapes unfold, their interactive nature, and the experiential and affective relationship cultivated between player and virtual world (see Shaw and Warf 2009). Game landscapes need to be taken seriously, not just in respect to the ways they articulate popular geopolitical imaginations of the world, but also the ways military institutions are increasingly utilising virtual landscapes for the purposes of military recruitment and training.

Notes

1 The series compromises of three videogames: *Call of Duty 4: Modern Warfare* (2007), *Call of Duty: Modern Warfare 2* (2009), and *Call of Duty: Modern Warfare 3* (2011).
2 To be clear at the outset, I will focus specifically on the single-campaign mode of the *Modern Warfare* series. This is played by individuals who follow a predefined storyline, completing missions in order to further the game's narrative.
3 This shift to the more contemporary setting was not without its issues. Indeed, the game developers *Infinity Ward*, cited that the publishers, *Activision* were reluctant in the transition from a historical to a contemporary setting due to the perceived risky nature (*see* Chafkin 2013).
4 *Call of Duty: Modern Warfare 2* (2009) [Xbox 360] Activision: Infinity Ward, USA.
5 *Generation Kill* (2008) [Television Series]. Produced by Andrea Calderwood. USA: HBO.
6 The marketing campaign of *Modern Warfare 3* made numerous allusions to this often using the abbreviations of 'WW3' in various advertisements.
7 *Call of Duty: Modern Warfare 2* (2009) [Xbox 360] Activision: Infinity Ward, USA.
8 A number of films, past and more recent, have centred on the White House. *Independence Day* (1996) is perhaps one of the most iconic visualisations of the obliteration of the White House. More recent films such as *White House Down* (2013) and *Olympus has Fallen* (2013) both see the White House succumb to terrorist forces.

References

Allen, Robertson. 2011. 'The Unreal Enemy of *America's Army*'. *Games and Culture* 6 (1):38–60.

Ash, James. 2014. 'Videogames'. In *Ashgate Research Companion to Media Geography*, edited by Jason Dittmer, James Craine and Paul C. Adams, 119–136. Aldershot: Ashgate.

Bialasiewicz, Luiza, David Campbell, Stuart Elden, Stephen Graham, Alex Jeffrey, and Alison J. Williams. 2007. 'Performing Security: The Imaginative Geographies of Current US Strategy'. *Political Geography* 26 (4):405–422.

Bos, Daniel. 2015. 'Military Videogames, Geopolitics and Methods'. In *Popular Culture and World Politics: Theories, Methods, Pedagogies*, edited by Federica Caso and Caitlin Hamilton, 101–109. Bristol, UK: E-International Relations.

Carter, Sean, and Klaus Dodds. 2014. *International Politics and Film: Space, Vision, Power*. New York: Wallflower Press.

Carter, Sean, Phillip Kirby, and Tara Louise Woodyer. 2016. 'Ludic-or Playful-Geopolitics'. In *Children, Young People and Critical Geopolitics*, edited by Matthew C. Benwell and Peter Hopkins, 61–73. Farnham: Ashgate.

Carter, Sean, and Derek P. McCormack. 2006. 'Film, Geopolitics and the Affective Logics of Intervention'. *Political Geography* 25:228–245.

Chafkin, Max. 2013. 'Modern Warfare'. *Vanity Fair*, available at http://www.vanityfair.com/news/2013/06/lawsuit-video-game-activision-zampella-west [last accessed 22 July 2015].

Cosgrove, Denis E. 1985. 'Prospect, Perspective and the Evolution of the Landscape Idea'. *Transactions of the Institute of British Geographers* 10:45–62.

Daily Mail, The. 2011. 'Censors Approve *Call of Duty* Computer Game Featuring 7/7 Tube Bomb-style Attacks, Despite Protests'.available at: http://www.dailymail.co.uk/news/article-2058577/Call-Of-Duty-game-featuring-7-7-Tube-bomb-style-attacks-approved-censors.html [last accessed 22 July 2015].

Dalby, Simon. 1990. *Creating the Second Cold War: The Discourse of Politics*. London: Pinter.

Daniels, Stephen. 1993. *Fields of Vision: Landscape Imagery and National Identity in England and the United States*. Cambridge: Polity Press.

Dittmer, Jason. 2005. 'Captain America's Empire: Reflections on Identity, Popular Culture, and Post-9/11 Geopolitics'. *Annals of the Association of American Geographers* 95 (3):626–643.

Dittmer, Jason. 2013. *Captain America and the Nationalist Superhero: Metaphors, Narratives, and Geopolitics*. Philadelphia: Temple University Press.

Duncan, Jim, and Nancy Duncan. 1988. '(Re)Reading the Landscape'. *Environment and Planning D: Society and Space* 6 (2):117–126.

Edensor, Tim. 2002. *National Identity, Popular Culture and Everyday Life*. Oxford: Berg.

Entertainment Software Association. 2015. 'Essential Facts about the Computer and Videogame Industry'. available at http://www.theesa.com/wp-content/uploads/2015/04/ESA-Essential-Facts-2015.pdf [last accessed 7 October 2015].

Gagnon, Frédérick. 2010. 'Invading Your Hearts and Minds: Call of Duty and the (Re)Writing of Militarism in U.S. Digital Games and Popular Culture'. *European Journal of American Studies*2 (3):available at http://ejas.revues.org/8831.

Graham, Stephen. 2006. 'Cities and the 'War on Terror''. *International Journal of Urban and Regional Research* 30 (2):255–276.

Harper, Graeme, and Jonathan Rayner. 2010. *Cinema and Landscape Film, Nation and Cultural Geography*. Bristol and Chicago: Intellect.

Heathershaw, John, and Nick Megoran. 2011. 'Contesting Danger: A New Agenda for Policy and Scholarship on Central Asia'. *International Affairs* 87 (3):589–612.

Herb, Graham H. 2004. 'Double Vision: Territorial Strategies in the Construction of National Identities in Germany, 1949–1979'. *Annals of the Association of American Geographers* 94(1):140–164.

Höglund, Johan. 2008. 'Electronic Empire: Orientalism Revisited in the Military Shooter'. *The International Journal of Computer Game Research* 8 (1): available at: http://gamestudies.org/0801/articles/hoeglund

Hughes, Rachel. 2007. 'Through the Looking Blast: Geopolitics and Visual Culture'. *Geography Compass* 1 (5):976–994.

Hughes, Rachel. 2010. 'Gameworld Geopolitics and the Genre of the Quest'. In *Observant States: Geopolitics and Visual Culture*, edited by Fraser MacDonald, Rachel Hughes and Klaus Dodds, 123–141. London and New York: I. B. Tauris.

Ingram, Alan, and Klaus Dodds. 2012. *Spaces of Security and Insecurity: Geographies of the War on Terror*. Surrey: Ashgate Publishing, Ltd.

Jennings, Stephanie C. 2015. 'Passion as Method: Subjectivity in Video Games Criticism'. *Journal of Games Criticism* 2 (1):1–18.

Longan, Michael W. 2008. 'Playing with Landscape: Social Process and Spatial Form in Video Games'. *Aether* 2:23–40.

Lukinbeal, Chris. 2005. 'Cinematic Landscapes'. *Journal of Cultural Geography* 23 (1):3–22.

MacDonald, F. 2008. 'Space and the Atom: On the Popular Geopolitics of Cold War Rocketry'. *Geopolitics* 13 (4):611–634.

MacDonald, Fraser, Rachel Hughes, and Klaus Dodds. 2010. *Observant States: Geopolitics and Visual Culture*. London and New York: I. B. Tauris.

Magnet, Shoshana. 2006. 'Playing at Colonization: Interpreting Imaginary Landscapes in the Video Game *Tropico*'. *Journal of Communication Inquiry* 30 (2):142–162.

Mastrapa, Gus. 2009. 'Powerful *Modern Warfare 2* Plot Hits Close to Home'. *Wired*, available at http://www.wired.com/gamelife/2009/11/modern-warfare-2-plot/ [last accessed 22 July 2015].

Megoran, Nick. 2005. 'The Critical Geopolitics of Danger in Uzbekistan and Kyrgyzstan'. *Environment and Planning D: Society and Space* 23 (4):555–580.

Mitchell, Don. 2002. 'Cultural Landscapes: The Dialectical Landscape-Recent Landscape Research n Human Geography'. *Progress in Human Geography* 26 (3):381–390.

Morin, Karen M. 2009. 'Landscape: Representing and Interpreting the World'. In *Key Concepts in Geography, 2nd ed.*, edited by Nicholas. J. Clifford, Holloway, Sarah L., Rice, Stephen P., and Valentine, Gill. London and Los Angeles: SAGE.

Ó Tuathail, Gearóid. 2005. 'The Frustrations of Geopolitics and the Pleasures of War: *Behind Enemy Lines* and American Geopolitical Culture'. *Geopolitics* 356–377.

Pain, Rachel. 2010. 'The New Geopolitics of Fear'. *Geography Compass* 4 (3):226–240.

Pieslak, Jonathan R. 2009. *Sound Targets: American Soldiers and Music in the Iraq War*. Indianapolis: Indiana University Press.

Robinson, Nick. 2015. 'Have You Won the War on Terror? Military Videogames and the State of American Exceptionalism'. *Millennium: Journal of International Studies* 43 (1):450–470.

Sage, Daniel. 2008. 'Framing Space: A Popular Geopolitics of American Manifest Destiny in Outer Space'. *Geopolitics* 13:27–53.

Said, Edward. 1979. *Orientalism*. New York: Vintage Books.

Salter, Mark B. 2011. 'The Geographical Imaginations of Video Games: Diplomacy, Civilization, *America's Army* and *Grand Theft Auto IV*'. *Geopolitics* 16 (2):359–388.

Saunders, Robert A. 2012. 'Hungry Lands: Conquest, Cannibalism, and the Wendigo Spirit'. In *Undead in the West: Vampires, Zombies, Mummies, and Ghosts on the Cinematic Frontier*, edited by Cynthia J. Miller and A. Bowdoin Van Riper, 182–203. Lanham, MD: Scarecrow Press.

Schwartz, Joan M. 1996. 'The Geography Lesson: Photographs and the Construction of Imaginative Geographies'. *Journal of Historical Geography* 22 (1):16–45.

Schwartz, Leigh. 2006. 'Fantasy, Realism, and the Other in Recent Video Games'. *Space and Culture* 9 (3):313–325.

Shaw, Ian G. R., and Barney Warf. 2009. 'Worlds of Affect: Virtual Geographies of Videogames'. *Environment and Planning A* 41:1332–1343.

Shaw, Ian. G. R. 2010. 'Playing War'. *Social & Cultural Geography* 11 (8):789–803.

Šisler, Vít. 2008. 'Digital Arabs: Representation in Video Games'. *European Journal of Cultural Studies* 11 (2):203–220.

Smicker, Josh. 2010. 'Future Combat, Combating Futures: Temporalities of War Video Games and the Performance of Proleptic Histories'. In *Joystick Soldiers: The Politics of Play in Military Video Games*, edited by Nina Huntemann and Matthew T. Payne, 106–121. London: Routledge.

Snider, Mike. 2009. "Interview: 'Modern Warfare 2' Composer Hans Zimmer'. *USA Today*, available at http://content.usatoday.com/communities/gamehunters/post/2009/11/qa-with-modern-warfare-2-composer-hans-zimmer/1?csp=34#.VhTeNisuuDs [last accessed 7 October 2015].

Sontag, Susan. 1965. 'The Imagination of Disaster'. *Commentary*, available at https://www.commentarymagazine.com/article/the-imagination-of-disaster/ [last accessed 5 February 2015].

Strukov, Vlad. 2012. 'Spatial Imagining and Ideology of Digital Commemoration (Russian Online Gaming)'. *Europe-Asia Studies* 64 (8):1584–1604.

Tolia-Kelly, Divya P. 2004. 'Materializing Post-Colonial Geographies: Examining the Textural Landscapes of Migration in the South Asian Home'. *Geoforum* 35 (6):675–688.

Whalen, Zach. 2004. 'Play Along - An Approach to Videogame Music' Game Studies'. *The International Journal of Computer Game Research* 4 (1):http://gamestudies.org/0401/whalen/?ref=SeksDE.Co.

Wolf, Mark J. P. 2001. 'Narrative in Video Game'. In *The Medium of the Video Game*, edited by Mark J. P. Wolf. Austin: University of Texas Press.

Woodward, Rachel. 2005. 'From Military Geography to Militarism's Geographies: Disciplinary Engagements with the Geographies of Militarism and Military Activities'. *Progress in Human Geography* 29 (6):718–740.

Woodward, Rachel. 2014. 'Military Landscapes: Agendas and Approaches for Future Research'.*Progress in Human Geography* 38 (1):40–61.

Wylie, John. 2007. *Landscape*. London and New York: Routledge.

Conclusion

Further conceptualisations of the interdiscipline

Robert A. Saunders and Vlad Strukov

We started our theorisation of popular geopolitics as an interdiscipline by employing the structuralist metaphor of language. We showed how the discernible items of popular culture have their political underpinnings and how geopolitics informs our readings of and responses to popular culture. From this essential understanding of popular geopolitics, by means of critical survey of the field's history, we shifted to view popular geopolitics as a post-language insofar as popular geopolitics exceeds the domains and functions of established disciplinary categories, including the increasingly problematic categories of authorship, production, audience, citizenship, and so forth. We highlighted the relevance of these considerations vis-à-vis the current trend in everyday politics, which functions as a type of hyper-mediated, impression-based politics (e.g., the recent rise of populism in Europe, Russia, Turkey, and the United States). The last had been boosted by the digital revolution of the 1990s–2000s, which resulted not only in the new realms of the political such as Twitter and vKontakte, but in the novel modes of impact and outreach of the popular. The phenomena of celebrity politics, the multiple cross-overs and inter-'leakages' of the political and the popular, and the use of popular culture as a space in which to stage, gauge, and engage with geopolitics have led us to suggest that there are new forms of social, political, and cultural hegemony and new areas of exclusion that require a global investigation.

By treating our field of enquiry as an interdiscipline, we focussed on areas of greatest tension. We revealed how previous theoretical frameworks, such as Edward Said's concept of Orientalism, continue to inform the study of empirical material in popular geopolitics. We were consistent in our effort to emphasise the prevalence of visual and audio-visual discourses in contemporary geopolitics (the ocularcentrism), in order to challenge not only different kinds of representations but, most importantly, the different kinds of imaging and imagining the geopolitical world in popular culture (a shift from the critique of the politics of seeing to the politics of imagining). We scrutinised the existing methodological basis of popular geopolitics to indicate that the interdiscipline should pay more attention to audience studies, while also examining how the political establishment has manipulated, for their own ends, audience-level cultural production, or what was previously labelled as DYI media, prosumption, or grassroots citizenship. To resolve these problems, we advocated a greater adaptation of research methods of other interdisciplines, such as gender studies and cultural studies, and we showed how these interdisciplinary engagements can yield new theorisations of the popular and the political and identify new directions of enquiry into popular geopolitics in the increasingly globalised world.

The volume combines a critical review of the interdiscipline's history, domains, objectives, and preoccupation with its own grand narratives such as the Cold War, with conceptual framing of current and future research such as the examination of non-Anglophone popular cultures through the prism of the interdiscipline and from within the framework of popular geopolitical feedback loops. For example, our contributors have charted the emergence and proliferation of the traditionalism and conservativism in transnational settings. By insisting that we should consider all phenomena from the perspective of global information flows and global citizenship, we invited a critique of the modern conceptualisation of the relationship between the (nation-)state and media, leading to our theoretical propositions about the relationship between the discourses about identity and post-identity, citizenship (both local and global), and self/non-self. As has been acknowledged in the Introduction, we structured our volume in such a way that enables multiple forms of engagement with the material, including linear reading, insofar as Part I provides a critical reflection on the discipline of popular geopolitics, and Part II supplies a set of new concepts emanating from the analysis of previously overlooked fields of cultural and political production. Each chapter illuminates a specific angle of our interdisciplinary enquiry, such as the issues of power, state, agency, and identity. These multi-directional, multi-purpose engagements compelled us to seek innovative ways to provide a conclusion to our argument.

In the spirit of our concept of the interdiscipline, the Conclusion was authored collaboratively. We invited each contributor to the volume to source an image, which captures and visually conceptualises the main argument of their chapter. They also sent us critical reflections on the image and its relation to the main argument of the book. In response, we re-worked and re-positioned these reflective pieces to highlight some theoretical claims and to reveal the conceptual logic of our volume. In this regard, the current Conclusion is not a summary of the book's major findings, but rather an invitation to consider them in new contexts, thus reflecting on the contemporary condition and signposting new directions of research. We wish to credit our co-contributors for the original impetus for these critical reflections. We wish to invite our readers to image, imagine and critically reflect on the theoretical complexities of popular geopolitics as an interdiscipline.

The state, global threats, and the logic of participatory popular culture (Jason Dittmer)

There has been a revolution in visual culture resulting from the spread of the (camera-enabled) smartphone and communications technology. These innovations have enabled the rapid sharing of images via social media platforms and have altered the ways in which individuals can perform their geopolitical subjectivities. Now, when (Western) countries are hit by terrorist attacks or natural

Figure 11.1 Photograph of the London Eye lit up in the colours of the Colombian flag to celebrate the state visit of President Santos in 2016.

Source: *Courtesy of Flickr.*

disasters, individuals are offered the opportunity to alter their avatars in solidarity, inserting a flag behind or within the profile picture. This performance of solidarity works through an affect-laden display of colours that purportedly aligns the subject in solidarity with victims, but does so—because of the coding of the social media platform—with minimal effort or sacrifice. Of course, this coding offers affordances to show solidarity with some geopolitical subjectivities, while ignoring others. Figure 11.1, taken by the British Foreign and Commonwealth Office's media team and publicly available on their Flickr page, speaks to this contemporary trend in popular geopolitical practice. It shows how the visual grammar of Facebook has been adopted by states. Here, the iconic shape of the London Eye is lit up in the colours (and proportions!) of the Colombian flag to celebrate the first-ever state visit of the Colombian president. This kind of social media-friendly gesture is repeated endlessly in the wake of recent terrorist attacks and disasters, as iconic buildings in allied countries have flags projected on them in purportedly poignant—but now hackneyed—displays. Therefore, the affective powers of the state are being shifted—thanks to social media—to users to install a form of participatory engagement, which is co-shared with popular culture.

Temporal ruptures, or how the popular culture of the past informs the politics of the future (Kyle Grayson)

The Imperial Federation's Map Showing the Extent of the British Empire (1886) was published in the July 24 edition of *The Graphic Magazine*. The map's designer, Walter

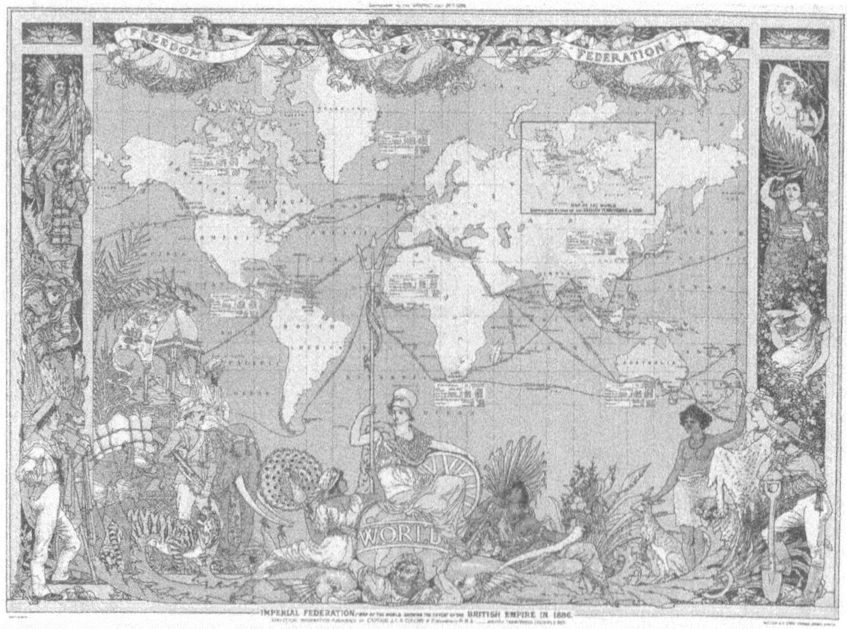

Figure 11.2 Map of the world showing the extent of the British Empire in 1886.
Source: *Courtesy of Wikimedia Commons.*

Crane (1845–1915), sought not only to capture the geographical scale and flows of British imperialism, but to also convey a sense of those people over whom Britain domineered. These subjugated populations are depicted along the frame of the map bringing tribute to Britannia. Although there are subtle visual suggestions of the exploitation upon which British colonialism was based, it is drowned out by the celebration of an Empire upon which the sun did not set. Thus, the map is considered a grand example of the British imperial geographical imagination, an imagination that continues to have political currency in a contemporary Britain, punch drunk on the hubris of Brexit and Victorian nostalgia. A return to the trading relationships forged under Empire is seen as a foregone conclusion by avid Brexiteers, who argue that bilateral agreements with the states of the Commonwealth will provide better opportunities than those offered by the European Union's single market. Moreover, scratch lightly and one quickly finds the racism of high imperialism very close to the surface of the Brexit geographical imagination. From lead Brexiteer and current Foreign Minister Boris Johnson's description of the Commonwealth as full of 'piccaninnies' with 'water melon smiles', to Tory Party whip (and Brexiteer) Ann Marie Morris' use of the n-word during a public debate on Brexit negotiations, Britain continues to look back to the future. These maps and historical iterations reveals the tensions of the globalised world, where the struggles of the imperial and post-imperial eras continue as historically inflicted popular imaginary, is utilised to articulate new geopolitical objectives.

Back to modernity (?), or re-mediating the logic of resistance in popular geopolitics (Vlad Strukov)

Figure 11.3 shows a sound installation by Rudy Decelière (Switzerland), shown in Russia in 2017. The actual work was presented at the Fourth Ural Industrial Biennial of Contemporary Art (Yekaterinburg, the Russian Federation, 2017). The work was created during Decelière's art residency at the Kyshtym copper electrolyte factory, curated by Zhenya Chaika as part of her programme of strategic partnerships between the biennial and local industrial enterprises. In Russian, the word for 'copper' [med'], especially in the name of the factory, sounds similar to the word 'media' [media], and so the artwork explores the modes of material and media production, conceived as a convergence of industrial and post-industrial discourses. This interpretation, rooted in the local cultural context, resonates with the main argument of the volume, which is the transition from and convergence of different disciplinary domains and different stages in their critical self-reflection. The metal pieces are suspended on almost invisible strings and their hovering in air symbolises the position of disciplinary discourse, always tangential and always attached to its own foundations. The repetitive clusters of the artwork mediate one of the chief concerns of popular geopolitics, namely, the objective to trace global information flows and re-appropriate subjectivity with its liberating agency beyond the paradigm of identity and centre-periphery relationships.

Figure 11.3 Photograph of a sound installation by Rudy Decelière (Switzerland), shown at the 4th Ural Industrial Biennial of Contemporary Art (Yekaterinburg, the Russian Federation, 2017). The work was developed as part of a special programme curated by Zhenya Chaika.

Source: *Courtesy of Vlad Strukov.*

Copper works as conductor for electricity thus symbolising global flows, whilst the sound is a result of changes in the electrical current, which indicates resistance and so points at the need to develop a new project of resistance. The green glass walls—the remaining structure of another, now disused factory in which the work is displayed—remind us of the previous emancipatory projects of modernity, now crystalised, and offer the possibility of a new project of resistance, pertaining to popular geopolitics, which we call for in the volume.

Looking for and seeing diversity: gendered discourses of popular geopolitics (Federica Caso)

Figure 11.4 was the front cover of the magazine *Womankind*, which is an advertisement-free newsstand women's magazine, distributed predominantly in the Anglophone West. The image is in fact a mosaic made of pieces of reggae record sleeves that together create an image of a black woman dressed in kente cloth, a fabric typical of Ghana, West Africa. The image is in the tradition of Western portraiture—it shows the head, bust and hands of the woman who is facing the

Figure 11.4 Front cover of the magazine *Womankind*.

Source: *Courtesy of Charis Tsevis.*

viewer—however, its constitutive parts make a reference to other traditions of representation and the ragged, 'shredded' notion of cultural memory and the power to see and to be seen. The strength of this image, just like the strength of popular geopolitics, comes from its multiplicity and diversity. Here the assemblage of different items, images, and the spaces in between, produces an illusion of coherence, or rather, the human ability to see, create, and impose meaning and order on the world. Popular geopolitics promises to be a venue to bring together disparate objects and subjects to grasp an increasingly complex world. In the image, the multiplicity of the record sleeve makes diversity intelligible to the eye. With its focus on the visual, popular geopolitics exalts complexities rather than the paucity of homogeneity. This is the disciplinary, political, and social hope popular geopolitics is entrusted with: to be a space where scholars interested in gender, sexuality, race, and more broadly, diversity can thrive away from parochialism of orthodox disciplinary approaches. The confident look of the woman gazing directly at the camera, proudly wearing the latest fashion derived from her cultural tradition, is the ultimate sign of empowerment and emancipation vis-à-vis white, colonial, male-dominated hegemonic culture.

From graffiti to social media: the (geo)political writing on the wall of everyday politics (Robert A. Saunders)

A form of protest dating back to Antiquity, graffiti is one of the most fundamental forms of popular geopolitics: a simple medium for simple messaging.

Figure 11.5 Photograph of a graffiti on the wall of men's toilet in a pub in central Toronto, Ontario.

Source: *Courtesy of Robert A. Saunders.*

Figure 11.5 shows a message scrawled on the wall of men's toilet in a pub in central Toronto, Ontario. It is a photograph run through a filter provided by the social media company Instagram. The photo was taken less than one week after the presidential victory of Donald Trump in November 2016. The unknown author of the graffiti (ostensibly a Canadian) sought to compel any passing U.S. citizens into action, hoping for the best outcomes for both nations. The graffiti was carved with a knife, or a similar sharp object, and while this small geopolitical intervention may physically outlast the majority of the digital graffiti (i.e. tweets) currently streaming out of the White House, in linking one of the earliest forms of popular cultural expression (graffiti) with one of the newest (social media posts), we see the timeless power of the popular to influence the political, within and across borders and for both good and ill. The interaction with and interpretation of this image was also informed by a specific intellectual context. The image was taken shortly after the ninth annual Popular Culture and World Politics conference at the Balsillie School of International Affairs in nearby Waterloo, ON, a fact which altered the perspective of the photographer/cultural producer, who is also a researcher of popular geopolitics. In an environment influenced by the recent outcome of the U.S. presidential election and the conference, this scribbling appeared to reify the notion that people, images, and messages travel across and transgress fields of knowledge, political communication, and what we may understand as popular culture.

Consumerism and popular imagination: neo-liberalism and popular geopolitics (Maša Kolanović)

T-shirts, bags, hats or pins representing a country or a region are among the most popular agents of geopolitical assumptions. In 2012, a group of Croatian designers organised a collective called Superstudio, to release tote bags branded as 'Croatia as it is' (Figure 11.6). The bags carried different slogans such as 'Between YU and EU'; 'Before Cuba, after Cote d' Ivoire', etc. These subvert the country's official touristic slogans such as 'Croatia – The Mediterranean as it once was'. Thus, on one level, the bags aim to challenge the conventions of neoliberal consumption of objects, spaces and experiences, on another, they query the official discourse about Croatia and its recent past which uses popular culture as its key domain of influence. References to Yugoslavia problematise discourses of memory, making a suggestion that the Croatian society, among many others in the post-socialist world, has not come to terms with its recent past, and that the current consumerism has obscured the pertinent social, political and cultural concerns. The bags examine the power of individuals, corporations and the state to influence popular perceptions of geopolitical entities and historical eras. They also suggest that the Yugoslavian past is some kind of baggage that must be carried around, thus intimating that perhaps the whole region is stuck in some sort of permanent post-socialist condition. The bags, which can be taken anywhere, symbolise the dynamics and directions of global citizenship, that is, the possibility of re-living the same trauma in imaginary, constructed worlds of the neoliberal order where popular culture functions as its main currency.

Figure 11.6 Photograph of a tote bag designed by Superstudio (Croatia).

Source: *Courtesy of Ranka Grgic Posavec.*

Between tradition and modernity, local, and global: popular geopolitics and its silent discontents (Ashvin Devasundaram)

Global meets local in Figure 11.7—a snapshot of everyday life in the hill station of Shimla, situated in the Himalayan foothills of northern India. The tableau crystallises contemporary India through its rendering of the casual cheek-by-jowl existence of the old and new, past and present, and traditional culture and popular geopolitics. The collage of commercial cinema posters presides over the nonchalant and unimpressed elderly local *pahari* (Himalayan) lady. The sleeping dog seems similarly oblivious to the supervening posters' promises of fantasy and escape. The vertical lines of the window grille in the centre of the frame simultaneously invite occupation and exemplify absence. They appear to beckon to the posters to territorialise them whilst sharing geometric camaraderie with the posters' linear rectangularity. The window also suggests a liminal space of undecidability—between local

Figure 11.7 Photograph a scene from showing everyday life in contemporary India.
Source: *Courtesy of Ashvin Devasundaram.*

tradition embodied by the old woman and the hegemonic narrative of Bollywood-oriented neoliberal consumer culture in contemporary India. The concomitant restriction and containment of the bars and the flights of fantasy suggested by the film posters sum up India's cinematic popular geopolitics. The tableau typifies how hegemonic popular cultural flows spurred on by globalisation and market logic, can often leave silent discontents in its wake. The question left to ponder pertains to who and what gets shut out of the cinematic precincts of Bollywood's formidable popular geopolitics fortress. Similarly, it raises the question about social, political and cultural agency and the ability of individuals to sustain and resist global challenges. Ultimately, the image queries our perception of modernity as an ongoing liberating movement of capital which seems to blur the boundaries between global and local, especially as regards the global circulation of images, and also between actual and imaginary worlds of popular geopolitics.

Comic power: popular geopolitics and the question of Western hegemony (Chris Homewood)

Recent changes to global sources and flows of capital are influencing how cultural producers can be persuaded to alter long-standing norms when it comes to popular geopolitical representation. Figure 11.8, by artist Alexander Wells, speaks to this trend in popular geopolitical practice. Drawn in 2013 in response to the Sino-Hollywood co-production agreement, the illustration captures how Hollywood, recognising the increasing economic heft of the Chinese film market in international film distribution, has moved to affirmative practices of recognition.

Figure 11.8 Alexander Wells' image of Iron Man produced for *Port Magazine*.

Source: *Courtesy of Alexander Wells.*

These practices apparently defy lingering formal, practical, and other popular U.S.–Western geopolitical anxieties about China's place in world politics, and offer an everyday view of China that is designed to constitute a sphere of influence in the West. The illustration also reflects the cognitive dissonance that characterises recent popular representations of erstwhile enemies and threatening Others: although Iron Man appears as a flag bearer for both China and the U.S., he tilts or leans towards the site of his inception. His movie, like the other blockbusters that wish to court international film markets, finds itself caught between conflicting cultural attitudes, economic barriers and international agendas. Charting a course forward, the formal expression of this double-coding of Western popular culture's most iconic and (geo)politically emblematic characters may, on the one hand, only be as durable as the flows of global capital that motivated it. Wells' picture also demonstrates the enduring place of the popular, especially illustration and animation, with regard the geopolitical. The comic book character first appeared in 1963 at the height of the Cold War. It was also used to symbolise the American power against the Soviet Union with its communist ideology. Fifty years later, almost the same iconography—the red flag and golden stars—is used to accentuate the global collaboration between the US and China.

(In)visible agency: urban spaces, human interventions, and the dominance of capital (Roxanne Chaitowitz and Shannon Brincat)

The encroachment on public space of hostile urban design—innocently badged as 'defensive architecture'—is ubiquitous. But what do they defend 'us' against? These designs mirror those within geopolitical spaces that force separation, from barbed wire to minefields. Inside state borders however, the poor face the same hostile landscape as would foreign enemies, though these struggles remain

Figure 11.9 Little Atoms' photograph of London's anti-homeless spikes.

Source: *Courtesy of Roxanne Chaitowitz and Shannon Brincat.*

vaguely hidden. From benches with solid dividers, to metal protrusions on window ledgers and floors, no other body is more insidiously rejected by the urban landscape than those experiencing homelessness. Such designs unapologetically replace human interaction with non-negotiable and non-removable force. They are a form of social cleansing to ensure the faces of homelessness remain unseen and private property secure. Figure 11.9 was taken in 2015 by a member of *Space, Not Spikes*. It captures the deliberate, 'inside/outside' quality of these processes. On the left, the bright lights of capitalism promise something new in the future; the wall that separates affirms hallowed property rights; and on the right, an example of simple, everyday resistance to such impositions of the architecture of dispossession and exclusion. The dichotomy of the mattress over the spike defies the logic of the material exclusion from private/public space, re-appropriating it under a welcoming hospitality of care. This form of 'violation' of the neoliberal order illustrates the earlier point about the project of resistance that should be developed in the field of popular geopolitics. The image also raises questions about popular geopolitics and agency, whereby the latter often remains invisible and, as in this case, excluded from discourse. Finally, the urban space captured in this image illustrates the different spaces, flows and iterations of popular geopolitics (e.g., the state reveals its presence thanks to the metal spikes), lending powers to our imagination to make sense of the (in)visible worlds of popular geopolitics.

Ludic spaces of popular geopolitics: from experiencing warfare to imaging warfare in the twenty-first century (Daniel Bos)

Current scholarly assessments of military-themed videogames have provided powerful critiques of the ways popular understandings of geopolitics are produced

Figure 11.10 Photograph of gamers playing *Kill Box* designed by Joseph Delappe and Biome Collective.

Source: *Courtesy of Malath Abbas.*

and maintained. However, there is a tendency to discount a new breed of digital games that are purposefully being developed to challenge, critique, and subvert dominate hegemonic (geo)political sensibilities. Moving beyond clichéd assertions that videogames are 'just for fun,' a host of artists, practitioners, and organisations are showing how digital games offer a powerful medium, in which to challenge geopolitical imaginations of war and conflict. Take the game *Kill Box* (Figure 11.10). Designed by Professor Joseph Delappe and Biome Collective, *Kill Box* offers a critical intervention into the politics of drone warfare. The game has two objectives; firstly, it critically examines the relationship between technology and military power through ludic engagement, and secondly, it interrogates the emotional apparatus of the gamer involved in destroying hostile subjects. Death, and specifically, killing, emerges as a form of abstraction in the virtual world of the game. However, the factual basis of the game queries the emotional undertones of the game. Based on documented drone strikes in northern Pakistan, the game is played by two players—one player occupying the view of the drone's pilot, the other player the perspective of someone on the ground. The game directly encourages the players to reflect on the broader spatial and moral implications of drone warfare. This is but one intervention coming from a broader movement that has been couched under the banner 'Games for Change'. Popular geopolitics can better attend the ways in which popular cultural artefacts, rather than seen as purely buttressing dominant geopolitical imaginaries, are actively attempting to critique the very spatialisations of military and political power that sustain them.

Contributors

Daniel Bos is a departmental lecturer in human geography at the University of Oxford. His research explores the relationship between popular culture, world politics, and militarisation. His doctoral thesis, completed at Newcastle University (2016), focussed on the role military-themed videogames have in shaping popular understandings of geopolitics and military violence. His current research interests consider how videogames are being designed and used to challenge hegemonic political discourses.

Shannon Brincat is a senior lecturer in politics and international relations at the University of the Sunshine Coast, Australia. His research focusses on international relations theory, recognition and cosmopolitanism, dialectics, tyrannicide, climate change justice, and Critical Theory. He is the editor of a number of collections, most recently *From International Relations to World Civilizations: The Contributions of Robert W. Cox, Dialectics and World Politics, Recognition, Conflict and the Problems of Ethical Community* (Routledge, 2017), and the three-volume series *Communism in the Twenty-First Century* (Praeger, 2013). He is also the co-founder and co-editor of the journal *Global Discourse*.

Federica Caso is a PhD candidate at the University of Queensland in the department of Political Science and International Studies. Her doctoral thesis looks at the role of the body in the militarisation of post-conscription societies. She also works on themes of gender and sexuality in the military and popular culture. She is the co-editor of *Popular Culture and World Politics: Theories, Methods, Pedagogies* (E-International Relations, 2015).

Roxanne Chaitowitz is a PhD candidate at the University of Queensland in the department of Political Science and International Studies. Her research looks at the value relational practices can disclose to our analyses of the experiences, struggles, and challenges of being homeless. Roxanne's research interests also include the ethics of the subject and other, the politics of autonomy and resistance, and Critical Theory and its emancipatory potential.

Ashvin Devasundaram is Lecturer in World Cinema at Queen Mary University of London. He is Programming Adviser to the London Asian Film Festival (LAFF) and Creative Director of the Edinburgh Asian Film Festival (EAFF). He is also a BBC Academy Expert Voice in Cultural Studies and Visual Arts. His monograph *India's New Independent Cinema: Rise of the Hybrid* (Routledge, 2016) is the world's first book on the topic.

Jason Dittmer is Professor of Political Geography at University College London. His current research approaches various diplomacies through the lens of assemblage theory. However, he maintains an interest in the popular geopolitics of militarism and heritage. He is the author of *Diplomatic Material: Affect, Assemblage, and Foreign Policy* (Duke University Press, forthcoming), *Captain America and the Nationalist Superhero: Metaphors, Narratives, and Geopolitics* (Temple University Press, 2014), and *Popular Culture, Geopolitics, and Identity* (2010, Rowman and Littlefield). He is also the editor or co-editor of four books, including *Geopolitics: An Introductory Reader* (Routledge, 2014) and the *Ashgate Research Companion to Media Geography* (Ashgate, 2014).

Kyle Grayson is Senior Lecturer in international politics in the School of Geography, Politics, and Sociology at Newcastle University. His research explores the intersections of security, politics, and culture and draws from the fields of international political sociology, political geography, and cultural studies. His first book, *Chasing Dragons: Security, Identity, and Illicit Drugs in Canada* was published by University of Toronto Press in 2008 and his second, *Cultural Politics of Targeted Killing*, was published by Routledge in 2016. His research articles have appeared in journals such as *Political Geography*, *British Journal of Politics and International Relations*, *Security Dialogue*, *Politics*, and the *Review of International Studies*. He is a co-editor of the Popular Culture and World Politics Book Series (Routledge).

Chris Homewood is Lecturer in German and World Cinemas at the University of Leeds, where he also manages the JH Film Studies Programme. He has published articles on the representation of the Red Army Faction (RAF) in German-language films and, more recently, the work of the so-called Berlin School of filmmakers. Recent publications include the entry 'Wind' in Roger F. Cook, Lutz Koepnick, Kristen Kopp and Brad Prager (eds.) *Berlin School Glossary: An ABC of the New Wave in German Cinema* (Intellect, 2013) and the chapter 'Politics in, and of, the Berlin School: Terrorism, Refusal, and Inertia' in Marco Abel and Jaimey Fisher (eds.) *A Transnational Art-Cinema: The Berlin School and its Global Contexts* (Wayne State University Press, 2017). His contribution to this volume forms part of a developing research interest in the portrayal of China, its culture, and Western diasporas in recent U.S. blockbusters.

Maša Kolanović is Associate Professor of Contemporary Croatian Literature at the University of Zagreb, Croatia. She is the author of *Underground Barbie* (Prospero Verlag, 2008, 2012); *Udarnik! Buntovnik? Potrošač... Popularna kultura i hrvatski roman od socijalizma do tranzicije* (Worker! Rebel? Consumer... Popular Culture and Croatian Novel from Socialism till Transition, 2011), *Jamerika: trip* (2013), and the edited volume *Komparativni postsocijalizam: slavenska iskustva* (Comparative Postsocialism: Slavic Experiences, 2013). Her field of interest is Croatian literature and popular culture in the Cold War and post-Cold War period, with a focus on Yugoslav socialism and the image of America in Eastern European cultures.

Iver B. Neumann is the outgoing Montague Burton Professor of International Relations at the London School of Economics (LSE) and a lifelong associate of the Norwegian Institute of International Affairs. He is also associate editor of *International Studies Quarterly*. His latest book is *Russia and the Idea of Europe, Second Edition* (Routledge, 2017). He is also the editor of *Battlestar Galactica and International Relations*, with Nicolas Kiersey (Routledge, 2013) and *Harry Potter and International Relations*, with Daniel C. Nexon (Rowman & Littlefield, 2006).

Robert A. Saunders is Professor in History, Politics, and Geography at Farmingdale State College, a campus of the State University of New York (SUNY). His research explores the impact of popular culture and mass media on geopolitics, nationalism, and religious identity. Dr. Saunders' scholarship has appeared in *Progress in Human Geography*, *Europe-Asia Studies*, *Slavic Review*, *Nations and Nationalism*, and *Geopolitics*, among other journals. He is the author of four books, the most recent being *Popular Geopolitics and Nation Branding in the Post-Soviet Realm* (Routledge, 2017). He is also curator of the 'Popular Culture and IR' blog at E-International Relations.

Vlad Strukov is Associate Professor in Film and Digital Culture at University of Leeds, specialising in world cinemas, visual culture, digital media, intermediality and cultural theory. He explores theories of empire and nationhood, global journalism and grassroots media, consumption and celebrity by considering the Russian Federation and the Russian-speaking world as his case study. He is the founding and principal editor of the journal 'Studies in Russian, Eurasian and Central European New Media' (www.digitalicons.org). He is the author of *Contemporary Russian Cinema: Symbols of a New Era* (Edinburgh University Press, 2016), and other publications on film.

Index

For Product Safety Concerns and Information please contact our EU representative GPSR@taylorandfrancis.com
Taylor & Francis Verlag GmbH, Kaufingerstraße 24, 80331 München, Germany

www.ingramcontent.com/pod-product-compliance
Ingram Content Group UK Ltd.
Pitfield, Milton Keynes, MK11 3LW, UK
UKHW010958180926
475701UK00004B/442

* 9 7 8 0 3 6 7 5 9 1 6 2 5 *